Studies in Fuzziness and Soft Computing

Volume 395

The series "Studies in Fuzziness and Soft Computing" contains publications on various topics in the area of soft computing, which include fuzzy sets, rough sets, neural networks, evolutionary computation, probabilistic and evidential reasoning, multi-valued logic, and related fields. The publications within "Studies in Fuzziness and Soft Computing" are primarily monographs and edited volumes. They cover significant recent developments in the field, both of a foundational and applicable character. An important feature of the series is its short publication time and world-wide distribution. This permits a rapid and broad dissemination of research results.

Indexed by ISI, DBLP and Ulrichs, SCOPUS, Zentralblatt Math, GeoRef, Current Mathematical Publications, IngentaConnect, MetaPress and Springerlink. The books of the series are submitted for indexing to Web of Science.

More information about this series at http://www.springer.com/series/2941

Said Melliani · Oscar Castillo
Editors

Recent Advances in Intuitionistic Fuzzy Logic Systems and Mathematics

Springer

Editors
Said Melliani
Department of Mathematics
Sultan Moulay Slimane University
Beni Mellal, Morocco

Oscar Castillo
Division of Graduate Studies and Research
Tijuana Institute of Technology
Tijuana, Mexico

ISSN 1434-9922 ISSN 1860-0808 (electronic)
Studies in Fuzziness and Soft Computing
ISBN 978-3-030-53928-3 ISBN 978-3-030-53929-0 (eBook)
https://doi.org/10.1007/978-3-030-53929-0

This Springer imprint is published by the registered company Springer Nature Switzerland AG
The registered company address is: Gewerbestrasse 11, 6330 Cham, Switzerland

Preface

This book contains the written versions of most of the contributions presented during the 6th Edition of the International Congress of the Moroccan Society of Applied Mathematics; it took place at Sultan Moulay Slimane University, Beni Mellal, Morocco, from 7 to 9 November 2019.

The meeting provided a setting for discussing recent developments in a wide variety of topics including fuzzy set theory, analysis, control, dynamical systems, etc. and to explore its current and future developments and applications. The congress provides a forum where researchers can share ideas, results on theory and applications.

The main goal of the event is for Moroccan mathematicians to open channels of communication with specialists from around the globe and eventually begin collaborative research projects. The audience was multidisciplinary allowing the participants to exchange diversified ideas and to show the wide attraction of different topics.

All papers had undergone a careful peer-review before being selected for publications in those volumes. This book mainly covers the topics: Fuzzy sets and their applications in partial differential equation, control of dynamical systems, and computing calculus.

We believe that this congress provided a medium for scientists and experts in the field to effectively communicate and share ideas. We would like to express our sincere thanks to all participants for their contributions and stimulating discussions. We are also grateful to keynote speakers, referees, committee members, and many others for their patience and efforts.

Beni Mellal, Morocco — Said Melliani
Tijuana, Mexico — Oscar Castillo

Contents

Fuzzy Semigroups and Fuzzy Nonlinear Evolution Equations

Said Melliani, El Hassan El Jaoui, and Lalla Saadia Chadli

Abstract In this work, we generalize the definition of a fuzzy strongly continuous semigroup and it's generator. We establish some of their proprieties and some results about the existence and uniqueness of solutions for fuzzy nonlinear evolution equation.

1 Introduction

The initial value (crisp) problem

$$x'(t) = A\,x(t) + f\Big(t, x(t)\Big); \quad x(0) = x_0$$

has a solution provided $f : I(\subset \mathbb{R}) \times X \to X$ (Banach space) is continuous and satisfies Lipschitz condition and A is the generator of a strongly continuous semigroup satisfying Hille–Yosida conditions (see [5, 12]).

Kaleva [7] proved the existence theorem for a solution of the fuzzy differential equation $x'(t) = f(t, x(t))$, and studied the Cauchy problem of fuzzy differential equations. Subrahmaniam and Sudarsanam studied existence of approximate solution of fuzzy integro-differential equations [14]. Recently, Kaleva [9] introduce an iteration semigroup of a nonlinear fuzzy-valued function, define the fuzzy exponential function and solve an autonomous fuzzy Cauchy problem.

In [7], C. G. Gal and S. G. Gal studied, with more details, fuzzy linear and semilinear (additive and positive homogeneous) operators theory, introduced semigroups

S. Melliani (✉) · E. H. El Jaoui · L. S. Chadli
LMACS, Sultan Moulay Slimane University, PO Box 523, 23000 Beni Mellal, Morocco
e-mail: saidmelliani@gmail.com

E. H. El Jaoui
e-mail: eljaouihass@gmail.com

L. S. Chadli
e-mail: sa.chadli@yahoo.fr

S. Melliani and O. Castillo (eds.), *Recent Advances in Intuitionistic Fuzzy Logic Systems and Mathematics*, Studies in Fuzziness and Soft Computing 395,
https://doi.org/10.1007/978-3-030-53929-0_1

of operators of fuzzy-number-valued functions, and gave various applications to fuzzy differential equations.

In this paper, we investigate the existence and uniqueness of fuzzy solution for the nonlinear fuzzy differential equations

$$x'(t) = A\,x(t) + f\Big(t, x(t)\Big); \quad x(0) = x_0 \tag{1}$$

provided $f : [0, a] \times E^n \to E^n$ is continuous and satisfies a Lipschitz condition and A is the generator of a strongly continuous fuzzy semigroup.

2 Preliminaries

Let $\mathscr{P}_K(\mathbb{R}^n)$ denote the family of all nonempty compact convex subsets of $\mathbb{R}^n$ and define the addition and scalar multiplication in $\mathscr{P}_K(\mathbb{R}^n)$ as usual. Let A and B be two nonempty bounded subsets of $\mathbb{R}^n$. The distance between A and B is defined by the Hausdorff metric

$$d(A, B) = \max\Big\{\sup_{a\in A}\inf_{b\in B} \|a - b\|\,,\ \sup_{b\in B}\inf_{a\in A} \|a - b\|\Big\}$$

where $\|\ \|$ denotes the usual Euclidean norm in $\mathbb{R}^n$. Then it is clear that $(\mathscr{P}_K(\mathbb{R}^n), d)$ becomes a complete and separable metric space (see [13]). Denote

$$E^n = \Big\{u : \mathbb{R}^n \to [0, 1] \mid u \text{ satisfies (i)–(iv) below}\Big\}$$

where

(i) u is normal i.e there exists an $x_0 \in \mathbb{R}^n$ such that $u(x_0) = 1$,
(ii) u is fuzzy convex,
(iii) u is upper semicontinuous,
(iv) $[u]^0 = cl\{x \in \mathbb{R}^n / u(x) > 0\}$ is compact.

For $0 < \alpha \le 1$, denote $[u]^\alpha = \{t \in \mathbb{R}^n \,/\, u(t) \ge \alpha\}$. Then from (i)–(iv), it follows that the α-level set $[u]^\alpha \in \mathscr{P}_K(\mathbb{R}^n)$ for all $0 \le \alpha \le 1$.

According to Zadeh's extension principle, we have addition and scalar multiplication in fuzzy number space E^n as follows:

$$[u + v]^\alpha = [u]^\alpha + [v]^\alpha, \quad [ku]^\alpha = k[u]^\alpha$$

where $u, v \in E^n$, $k \in \mathbb{R}^n$ and $0 \le \alpha \le 1$.

Define a mapping $D : E^n \times E^n \to \mathbb{R}^+$ as follows

$$D(u, v) = \sup_{0\le\alpha\le1} d\Big([u]^\alpha, [v]^\alpha\Big)$$

where d is the Hausdorff metric defined in $\mathscr{P}_K(\mathbb{R}^n)$. Then it is easy to see that D is a metric in E^n. Using the results in [13], we know that

(1) (E^n, D) is a complete metric space;
(2) $D(u+w, v+w) = D(u, v)$ for all $u, v, x \in E^n$;
(3) $D(k\,u, k\,v) = |k|\; D(u, v)$ for all $u, v \in E^n$ and $k \in \mathbb{R}$.

We consider $\mathscr{C}_a = \mathscr{C}\left([0, a], E^n\right)$ the space of all continuous fuzzy functions defined on $[0, a] \subset \mathbb{R}$ into E^n, where $a > 0$. For $u, v \in \mathscr{C}_a$, we define the metric

$$H(u, v) = \sup_{t \in [0,a]} D(u(t), v(t))$$

Then $(\mathscr{C}_a, H)$ is a complete metric space.

We recall some measurability, integrability properties for fuzzy set-valued mappings (see [7]).

Let $T = [c, d] \subset \mathbb{R}$ be a compact interval.

Definition 1 A mapping $F : T \to E^n$ is strongly measurable if for all $\alpha \in [0, 1]$ the set-valued function $F_\alpha : T \to \mathscr{P}_K(\mathbb{R}^n)$ defined by $F_\alpha(t) = [F(t)]^\alpha$ is Lebesgue measurable.
A mapping $F : T \to E^n$ is called integrably bounded if there exists an integrable function k such that $||x|| \leq k(t)$ for all $x \in F_0(t)$.

Definition 2 Let $F : T \to E^n$. Then the integral of F over T denoted by $\int_T F(t)dt$ or $\int_c^d F(t)dt$, is defined by the equation

$$\left[\int_T F(t)dt\right]^\alpha = \int_T F_\alpha(t)dt = \left\{\int_T f(t)dt / f : T \to \mathbb{R}^n \textbf{is a measurable selection for } F_\alpha\right\}$$

for all $\alpha \in]0, 1]$.

Also, a strongly measurable and integrably bounded mapping $F : T \to E^n$ is said to be integrable over T if

$$\int_T F(t)dt \in E^n$$

Proposition 1 (Aumann [1]) *If $F : T \to E^n$ is strongly measurable and integrably bounded, then F is integrable.*

The following definitions and theorems are given in [7].

Proposition 2 *Let $F, G : T \to E^n$ be integrable and $\lambda \in \mathbb{R}$. Then*

(i) $\int_T (F(t) + G(t))dt = \int_T F(t)dt + \int_T G(t)dt,$
(ii) $\int_T \lambda F(t)dt = \lambda \int_T F(t)dt,$

(iii) $D(F, G)$ *is integrable,*

(iv) $D(\int_T F(t)dt, \int_T G(t)dt) \leq \int_T D(F, G)(t)dt.$

For $u, v \in E^n$, if there exists $w \in E^n$ such that $u = v + w$, then w is the Hukuhara difference of u and v denoted by $u -^H v$.

Definition 3 A mapping $F : T \to E^n$ is Hukuhara differentiable at $t_0 \in T$ if there exists a $F'(t) \in E^n$ such that the following limits

$$\lim_{h \to 0^+} \frac{F(t+h) -^H F(t)}{h} \quad \text{and} \quad \lim_{h \to 0^+} \frac{F(t) -^H F(t-h)}{h}$$

exist and equal to $F'(t)$.

Here the limit is taken in the metric space (E^n, D). At the end points of T we consider only one-sided derivatives.

3 Fuzzy Strongly Continuous Semigroups

In [7] Kaleva gave a generalization of the classical Radström embedding theorem (in $\mathscr{F} = E^1$), and in [13], O'Regan et al. do the same but in E^n.

Theorem 1 (Embedding theorem) *There exists a real Banach space X such that E^n can be embedded as a convex cone C with vertex 0 in X. Furthermore the following conditions hold true:*

(i) the embedding j is isometric,
(ii) the addition in X induces the addition in E^n,
(iii) the multiplication by a nonnegative real number in X induces the corresponding operation in E^n,
(iv) $C - C = \{a - b \ / \ a, b \in E^n\}$ is dense in X,
(v) C is closed.

Remark 1 • As in [2], we can introduce another embedding by the formula $\tilde{j} : E^n \to X$, with $\tilde{j}(u) = j\Big((-1)u\Big)$, $u \in E^n$. It verify the following properties:

(i) $\|\tilde{j}(u) - \tilde{j}(v)\| = \|j((-1)u) - j((-1)v)\| = D\Big((-1)u, (-1)v\Big) = D(u, v)$

(ii) $\tilde{j}\Big(E^n\Big) = j\Big(E^n\Big) = C$, since $(-1)E^n = E^n$

(iii) For $t, s \geq 0$, $u, v \in E^n$, we have:

$$\tilde{j}(tu + sv) = j\Big((-1)(tu + sv)\Big) = j\Big[t(-1)u + s(-1)v\Big] = tj\Big((-1)u\Big) = t\tilde{j}(u) + s\tilde{j}(v)$$

- We can also generalize the property (iv) in Theorem 2.4 cited in [6]:
Let $a, b \in \mathbb{R}$ with $a, b \geq 0$ or $a, b \leq 0$ and $u \in E^n$, then $(a+b)u = au + bu$.
For general $a, b \in \mathbb{R}$, this property does not hold (at least in the case $n = 1$).
Indeed, if $a, b \geq 0$:

$$j\Big((a+b)u\Big) = (a+b)j(u) = aj(u) + bj(u)$$
$$= j(au) + j(bu) = j(au + bu)$$

Thus $(a+b)u = au + bu$. If $a, b \leq 0$:

$$j\Big((a+b)u\Big) = j\Big[(-a + (-b))\,u\Big] = (-a + (-b))\,j(u)$$
$$= -aj(u) + (-b)j(u) = j(-au) + j(-bu)$$
$$= j\Big[(-1)au + (-1)bu\Big] = j\Big[(-1)(au + bu)\Big] = \widetilde{j}(au + bu)$$

So $(a+b)u = au + bu$.

We give here a definition of a nonlinear fuzzy semigroups, which is similar to that given by Brezis and Pazy in [3].

Definition 4 By a fuzzy (one parameter strongly continuous nonlinear) semigroup on E^n, we mean a family $\{T(t), t \geq 0\}$ of operators from E^n into itself satisfying the following conditions:

(i) $T(0) = i$, the identity mapping on E^n,
(ii) $T(t+s) = T(t)T(s)$ for all $t, s \geq 0$,
(iii) the function $g : [0, \infty[\to E^n$, defined by $g(t) = T(t)(x)$ is continuous at $t = 0$ for all $x \in E^n$ i.e

$$\lim_{t \to 0^+} T(t)(x) = x$$

(iv) There exist two constants $M > 0$ and ω such that

$$D\Big(T(t)x, T(t)y\Big) \leq M\, e^{\omega t}\, D(x, y), \quad for\, t \geq 0, x, y \in E^n$$

In particular if $M = 1$ and $\omega = 0$, we say that $\{T(t), t \geq 0\}$ is a contraction fuzzy semigroup.

Remark 2 The condition (iii) implies that the function $g(t) = T(t)(x)$ is continuous on $[0, \infty[$ for all $x \in E^n$. Indeed:
for $t \geq 0$ and h very small

$$
\begin{aligned}
D\Big(g(t+h), g(t)\Big) &= D\Big(T(t)T(h)x, T(t)x\Big) \\
&\leq M\, e^{\omega t}\, D\Big(T(h)x, x\Big) \\
&= M\, e^{\omega t}\, D\Big(g(h), g(0)\Big) \to 0 \quad \text{when } h \to 0
\end{aligned}
$$

Remark 3 • Taking $t = 0$ in (iv), we can easily see that $M \geq 1$.
• The quantity $\omega_0 = \inf\Big\{\omega \in \mathbb{R} \cup \{-\infty\}/\omega \text{ satisfied (iv)}\Big\}$ is called the type of the fuzzy semigroup.

In the sequel we choose $\omega > 0$.

Definition 5 Let $\{T(t), t \geq 0\}$ be a fuzzy $\mathscr{C}^0$-semigroup on E^n and $x \in E^n$. If for $h > 0$ sufficiently small, the Hukuhara difference $T(h)x -^H x$ exits, we define

$$
Ax = \lim_{h \to 0^+} \frac{T(h)x -^H x}{h},
$$

whenever this limit exists in the metric space (E^n, D). Then the operator $A : x \mapsto Ax$ defined on

$$
D(A) = \Big\{x \in E^n \ : \ \lim_{h \to 0^+} \frac{T(h)x -^H x}{h} \text{ exists}\Big\} \subset E^n,
$$

is called the infinitesimal generator of the fuzzy semigroup $\{T(t), t \geq 0\}$.

Lemma 1 *Let $A : E^n \to E^n$ and $A_1 = jAj^{-1} : C \to C$ two (nonlinear) operators. A is the infinitesimal generator of a fuzzy semigroup $\{T(t), t \geq 0\}$ on E^n if and only if A_1 is the infinitesimal generator of the semigroup $\{T_1(t), t \geq 0\}$ defined on the convex closed set C by $T_1(t) = jT(t)j^{-1}$ for $t \geq 0$.*

Proof It is easy to verify that $\{T(t), t \geq 0\}$ is a fuzzy semigroup on E^n if and only if $\{T_1(t), t \geq 0\}$ is a nonlinear (Brezis and Pazy) semigroup on C (see [2]).
We assume that A is the generator of a fuzzy semigroup $\{T(t), t \geq 0\}$ on E^n. By the properties of j we have, for all $x \in j^{-1}(D(A))$

$$
\begin{aligned}
\lim_{h \to 0^+} \frac{T_1(h)x - x}{h} &= \lim_{h \to 0^+} \frac{jT(h)j^{-1}x - jj^{-1}x}{h} \\
&= j\left(\lim_{h \to 0^+} \frac{T(h)j^{-1}x -^H j^{-1}x}{h}\right) \\
&= j\,A\,j^{-1}\,x = A_1 x.
\end{aligned}
$$

Conversely, if A_1 is the generator of a fuzzy semigroup $\{T_1(t), t \geq 0\}$ on C, then for all $x \in D(A)$

$$\lim_{h\to 0^+}\frac{T(h)x -^H x}{h} = \lim_{h\to 0^+}\frac{j^{-1}T_1(h)jx -^H j^{-1}jx}{h}$$
$$= j^{-1}\left(\lim_{h\to 0^+}\frac{T_1(h)x - jx}{h}\right)$$
$$= j^{-1}A_1\, j\, x = A\, x$$

Remark 4 Since the infinitesimal generator A_1 of $\{T_1(t), t \geq 0\}$ is unique, we deduce that the infinitesimal generator A of a fuzzy semigroup $\{T(t), t \geq 0\}$ is also unique.

Lemma 2 *Let A be the generator of a fuzzy semigroup $\{T(t), t \geq 0\}$ on E^n, then for all $x \in E^n$ such that $T(t)x \in D(A)$ for all $t \geq 0$, the mapping $t \to g(t) = T(t)x$ is differentiable and*

$$g'(t) = A\,T(t)\,x \quad i.e. \quad \frac{d}{dt}\Big(T(t)\,x\Big) = A\,T(t)\,x, \quad \forall t \geq 0$$

Proof Let $x \in E^n$, for $t, h \geq 0$ we have

$$T(t+h)x = T(h)T(t)x$$

Since $T(t)x \in D(A)$ then

$$\lim_{h\to 0^+}\frac{g(t+h) -^H g(t)}{h} = \lim_{h\to 0^+}\frac{T(t+h)x -^H T(t)x}{h}$$
$$= \lim_{h\to 0^+}\frac{T(h)T(t)x -^H T(t)x}{h}$$
$$= AT(t)x$$

Denote $A_h = \frac{T(h)-i}{h}$, for $h > 0$. Using the continuity of g and the definition of A, we have

$$\lim_{h\to 0^+}\frac{g(t-h) -^H g(t)}{-h} = \lim_{h\to 0^+}\frac{T(t-h)x -^H T(t)x}{-h}$$
$$= \lim_{h\to 0^+}\frac{T(h)T(t-h)x -^H T(t-h)x}{h}$$
$$= \lim_{h\to 0^+} A_h T(t-h)x$$
$$= AT(t)x$$

Hence, g is differentiable and $g'(t) = AT(t)x$, for all $t \geq 0$.

Remark 5 In the linear case, we have

$$\frac{d}{dt}\Big(T(t)x\Big) = AT(t)x = T(t)Ax, \quad \forall t \geq 0$$

but in the general (fuzzy case), $AT(t) \neq T(t)A$.

Example 1 Let $k \in \mathbb{R}$ and define on E^n the linear operator $T(t) : x \mapsto e^{kt}x$, for all $t \geq 0$. Obviously, we have

(i) $T(0)x = x$, for all $x \in E^n$ i.e. $T(0) = i$.
(ii) For $t, s \geq 0$, $x \in E^n$:

$$T(t+s)x = e^{(kt+ks)}x = \left(e^{kt}e^{ks}\right)x = e^{kt}\left(e^{ks}x\right) = T(t)\left(e^{ks}x\right) = T(t)T(s)x.$$

(iii) For $t \geq 0$, $x \in E^n$, $D\left(T(t)x, x\right) = D\left(e^{kt}x, x\right)$. Suppose that $k \geq 0$, so $(e^{kt} - 1) \geq 0$, then using the remark 3.1, we deduce that: $\left(e^{kt} - 1\right)x + x = e^{kt}x$. Therefore the Hukuhara difference $e^{kt}x -^H x$ (i.e. $T(t)x -^H x$) exists and we have $T(t)x -^H x = e^{kt}x -^H x = (e^{kt} - 1)x$. Then

$$D\left(T(t)x, x\right) = D\left(e^{kt}x -^H x, \tilde{0}\right) = D\left((e^{kt} - 1)x, \tilde{0}\right) = (e^{kt} - 1)D(x, \tilde{0})$$

since $\lim\limits_{t \to 0^+} e^{kt} - 1 = 0$, then $\lim\limits_{t \to 0^+} T(t)(x) = x$.
(iv) For $t \geq 0$, $x, y \in E^n$, $D\left(T(t)x, T(t)y\right) = D\left(e^{kt}x, e^{kt}y\right) = e^{kt}D(x, y)$.

Consequently, $\{T(t), t \geq 0\}$ is a fuzzy $\mathscr{C}^0$-semigroup on E^n.

We define the linear operator $A : x \mapsto Ax = kx$ on E^n using the following identity

$$\frac{T(t)x -^H x}{t} = \left(\frac{e^{kt} - 1}{t}\right)x = \left(\sum_{p=1}^{+\infty} \frac{t^{p-1}k^p}{p!}\right)x$$

$$= kx + \left(\sum_{p=2}^{+\infty} \frac{t^{p-1}k^p}{p!}\right)x = Ax + \left(\sum_{p=2}^{+\infty} \frac{t^{p-1}k^p}{p!}\right)x.$$

Therefore the Hukuhara difference $\dfrac{T(t)x -^H x}{t} -^H Ax$ exists and we have

$$\frac{T(t)x -^H x}{t} -^H Ax = \left(\sum_{p=2}^{+\infty} \frac{t^{p-1}k^p}{p!}\right)x$$

Then

$$D\left(\frac{T(t)x -^H x}{t}, Ax\right) = D\left(\left(\sum_{p=2}^{+\infty} \frac{t^{p-1}k^p}{p!}\right)x, \tilde{0}\right)$$

$$= \left(\sum_{p=2}^{+\infty} \frac{t^{p-1} k^p}{p!} \right) D(x, \tilde{0})$$

$$= \left(\frac{e^{kt} - 1 - kt}{t} \right) D(x, \tilde{0})$$

Since $\frac{e^{kt}-1-kt}{t} = \frac{1}{2}k^2 t + o(t) \to 0$ as $t \to 0$, then $\lim\limits_{t\to 0} \frac{T(t)x -^H x}{t} = Ax$, for all $x \in E^n$. Thus, A is the infinitesimal generator of the fuzzy semigroup $\{T(t), t \geq 0\}$.

Note that if $k < 0$ and $n = 1$, the above equality $\left(e^{kt} - 1\right)x + x = e^{kt}x$ does not hold (see [6], property (iv) in Theorem 2.4), so the Hukuhara difference $e^{kt}x -^H x$ (i.e. $T(t)x -^H x$) does not exist.

Remark 6 The authors in [6] gave several examples of fuzzy semigroups $\{T(t), t \geq 0\}$ in the case $n = 1$. We define

$$T(t)x = e^{tA}x = \lim_{m\to+\infty} \sum_{p=0}^{m} \frac{t^p A^p}{p!} x, \quad t \geq 0, x \in E^1,$$

where the limit is considered in (E^1, D) and the generator A is a linear operator on E^1 (In [6] this definition was different and general because an other metric was used). In particular, $A : E^1 \to E^1$ can be defined by

$$Ax = \left[x_-(1) - \int_0^1 x_-(r)dr \right].c, \quad \forall x \in E^1$$

or

$$Ax = \left[x_+(1) - \int_0^1 x_+(r)dr \right].c, \quad \forall x \in E^1,$$

where $[x]^r = [x_-(r), x_+(r)]$ and $c \in E^1$ is a constant chosen such that

$$\mu = c_-(1) - \int_0^1 c_-(r)dr > 0.$$

4 Nonlinear Evolution Fuzzy Equation

We consider the fuzzy initial value problem

$$x'(t) = Ax(t) + f(t, x(t)); \qquad x(0) = x_0 \tag{2}$$

provided A is the generator of a strongly continuous fuzzy semigroup $\{T(t), t \geq 0\}$ on E^n and $f : [0, a] \times E^n \to E^n$ is continuous and satisfies a Lipschitz condition

$$D\Big(f(t,x), f(t,y)\Big) \leq KD(x,y), \quad t \in [0,a], \quad x, y \in E^n$$

Definition 6 We say that x is a (classical) solution of (Eq. 1) if

(i) $x \in \mathscr{C}^1([0,a], E^n)$ and;
(ii) $x(t) \in D(A), \forall t \in [0,a]$ and x satisfies the Eq. (2).

Definition 7 We say that x is a mild solution of (Eq. 1) if

(i) $x \in \mathscr{C}([0,a], E^n)$, $x(t) \in D(A)$ for all $t \in [0,a]$;
(ii) and $x(t) = T(t)x_0 + \int_0^t T(t-s)f(s,x(s))ds$ for all $t \in [0,a]$.

Theorem 2 *Suppose that $f : [0,a] \times E^n \to E^n$ is continuous and Lipschitzian with respect to the second argument, then for any $x_0 \in E^n$ such that $T(t)x_0 \in D(A)$ for all $t \geq 0$, the equation (Eq. 1) has a unique mild solution.*

Proof Denote $C_a = \mathscr{C}([0,a], E^n)$ and define a mapping $P : C_a \to C_a$ by

$$Px(t) = T(t)x_0 + \int_0^t T(t-s)f(s,x(s))ds, \quad \text{for } x \in C_a, \quad t \in [0,a]$$

Step 1:
For $x \in C_a, t \in [0,a]$ and h very small

$$\begin{aligned}
D\Big(Px(t+h), Px(t)\Big) &= D\Big(T(h)T(t)x_0 + \int_0^h T(t+h-s)f(s,x(s))ds \\
&+ \int_h^{t+h} T(t+h-s)f(s,x(s))ds, T(t)x_0 + \int_0^t T(t-s)f(s,x(s))ds\Big) \\
&= D\Big(T(h)T(t)x_0 + \int_0^h T(t+h-s)f(s,x(s))ds \\
&+ \int_0^t T(t-s)f(s+h,x(s+h))ds, T(t)x_0 + \int_0^t T(t-s)f(s,x(s))ds\Big) \\
&\leq D\left(T(h)T(t)x_0, T(t)x_0\right) + D\left(\int_0^h T(t+h-s)f(s,x(s))ds, \hat{0}\right) \\
&+ D\left(\int_0^t T(t-s)f(s+h,x(s+h))ds, \int_0^t T(t-s)f(s,x(s))ds\right) \\
&\leq D\left(T(h)T(t)x_0, T(t)x_0\right) + \int_0^h D\left(T(t+h-s)f(s,x(s)), \hat{0}\right) ds \\
&+ \int_0^t D\left(T(t-s)f(s+h,x(s+h)), T(t-s)f(s,x(s))\right) ds \\
&\leq D\left(T(h)T(t)x_0, T(t)x_0\right) + \int_0^h D\left(T(t+h-s)f(s,x(s)), \hat{0}\right) ds \\
&+ M\int_0^t e^{\omega(t-s)} D\left(f(s+h,x(s+h)), f(s,x(s))\right) ds
\end{aligned}$$

It is clear that $D\Big(T(h)T(t)x_0, T(t)x_0\Big) \to 0$ and $\int_0^h D\Big(T(t+h-s)f(s,x(s)), \hat{0}\Big)$ $ds \to 0$ as $h \to 0$.

And by the dominated convergence theorem:

$$\int_0^t e^{\omega(t-s)} D\Big(f(s+h, x(s+h)), f(s, x(s))\Big)ds \to 0 \quad \text{as} \quad h \to 0$$

Hence $Px \in C_a$ i.e P maps C_a into itself.

Step 2:
Let $x, y \in C_a$ and $t \in [0, a]$, we have

$$\begin{aligned}
D(Px(t), Py(t)) &= D\left(T(t)x_0 + \int_0^t T(t-s)f(s, x(s))ds\,,\ T(t)x_0 + \int_0^t T(t-s)f(s, y(s))ds\right)\\
&= D\left(\int_0^t T(t-s)f(s, x(s))ds\,,\ \int_0^t T(t-s)f(s, y(s))ds\right)\\
&\le \int_0^t D\Big(T(t-s)f(s, x(s))\,,\ T(t-s)f(s, y(s))\Big)ds\\
&\le M \int_0^t e^{\omega(t-s)} D\Big(f(s, x(s)), f(s, y(s))\Big)ds\\
&\le M_a\, K \int_0^t D\Big(x(s)\,,\ y(s)\Big)ds\\
&\le t\, M_a\, K\, H(x, y)
\end{aligned}$$

where the constant M_a is given by

$$M_a = M \sup_{0\le t\le a} e^{\omega t} = Me^{\omega a}$$

We can deduce that

$$\begin{aligned}
D\Big(P^2x(t)\,,\ P^2y(t)\Big) &\le M_a\, K \int_0^t D\Big(Px(s)\,,\ Py(s)\Big)ds\\
&\le M_a\, K \int_0^t sM_a\, K\, H(x, y)ds\\
&= \frac{(M_a\, K\, t)^2}{2} H(x, y)
\end{aligned}$$

And by induction, we have for all $t \in [0, a]$

$$D\Big(P^n x(t)\,,\ P^n y(t)\Big) \le \frac{(M_a\, K\, t)^n}{n!} H(x, y)$$

which implies

$$H\Big(P^n x, P^n y\Big) \le \frac{(M_a K t)^n}{n!} H(x, y)$$

Since $\frac{(M_a K t)^n}{n!} \to 0$ as $n \to \infty$, then there exists $r \in \mathbb{N}$ such that $\frac{(M_a K t)^r}{r!} < 1$. It follows that P^r is a contraction and there exists a unique $x \in C_a$ such that $P^r x = x$.

Furthermore, we have

$$P^r(Px) = P\left(P^r x\right) = Px$$

Hence Px is a unique fixed point of P^r, so we conclude that x is the unique mild solution of (Eq. 1).

5 Continuous Dependence of a Mild Solution with Initial Data

Theorem 3 *Suppose f as in Theorem 3. Let $x = x(t, x_0)$ and $y = y(t, y_0)$ be mild solutions of (Eq. 1) corresponding to x_0 and y_0 respectively. Then*

$$D\Big(x(t, x_0),\, y(t, y_0)\Big) \le M_a \exp(M_a K t) D\big(x_0,\, y_0\big)$$

Proof We have

$$\begin{aligned} D\Big(x(t, x_0),\, y(t, y_0)\Big) &= D\Big(T(t)x_0 + \int_0^t T(t-s) f(s, x(s)) ds,\, T(t)y_0 + \int_0^t T(t-s) f(s, y(s)) ds\Big) \\ &\le D\Big(T(t)x_0, T(t)y_0\Big) + D\Big(\int_0^t T(t-s) f(s, x(s)) ds, \int_0^t T(t-s) f(s, y(s)) ds\Big) \\ &\le M_a D\Big(x_0,\, y_0\Big) + \int_0^t D\Big(T(t-s) f(s, x(s)),\, T(t-s) f(s, y(s))\Big) ds \\ &\le M_a D\Big(x_0,\, y_0) + M \int_0^t e^{\omega(t-s)} D\Big(f(s, x(s)),\, f(s, y(s))\Big) ds \\ &\le M_a D\Big(x_0,\, y_0\Big) + M_a K \int_0^t D\Big(x(s),\, y(s)\Big) ds \end{aligned}$$

Applying Gronwall inequality, we obtain

$$D\Big(x(t, x_0),\, y(t, y_0)\Big) \le M_a \exp(M_a K t) D\big(x_0,\, y_0\big)$$

Remark 7 By Theorem 4, we have for all $t \in [0, a]$

$$\begin{aligned} D\Big(x(t),\, y(t)\Big) &= D\Big(x(t, x_0),\, y(t, y_0)\Big) \\ &\le M_a \exp(M_a K t) D\left(x_0,\, y_0\right) \\ &\le M_a \exp(a M_a K) D\left(x_0,\, y_0\right) \end{aligned}$$

Hence

$$H(x, y) \le M_a \exp(a M_a K) D\left(x_0,\, y_0\right)$$

References

1. Aumann, R.J.: Integrals of set-valued functions. J. Math. Anal. Appl. **12**, 1–12 (1965)
2. Brezis, H., Pazy, A.: Semigroups of nonlinear contractions on convex sets. J. Funct. Anal. **6**, 237–281 (1970)
3. Brezis, H., Pazy, A.: Convergence and approximation of semigroups of nonlinear operators on Banach spaces. J. Funct. Anal. **9**, 63–74 (1972)
4. Dorroh, J.R.: Some classes of semigroups of nonlinear transformations and their generators. J. Math. Soc. Jpn. **20**(3), 437–455 (1968)
5. Engel, K., Nagel, R.: A Short Course on Operator Semigroups. Springer, Berlin (2005)
6. Gal, C.G., Gal, S.G.: Semigroups of operators on spaces of fuzzy-number-valued functions with applications to fuzzy differential equations. arXiv:1306.3928v1 (2013). Accessed 17 Jun 2013
7. Kaleva, O.: Fuzzy differentiel equations. Fuzzy Sets Syst. **24**, 301–317 (1987)
8. Kaleva, O.: Nonlinear iteration semigroups of fuzzy Cauchy problems. Fuzzy Sets Syst. **209**, 104–110 (2012)
9. Kaleva, O.: The Cauchy problem for fuzzy differentiel equations. Fuzzy Sets Syst. **35**, 366–389 (1990)
10. Kobayashi, Y.: Difference approximation of evolution equations and generation of nonlinear semigroups. Proc. Jpn. Acad. **51**, 406–410 (1975)
11. Kobayashi, Y.: On approximation of nonlinear semigroups. Proc. Jpn. Acad. **50**, 729–734 (1974)
12. Pazy, A.: Semigroups of Linear Operators and Applications to Partiel Differentiel Equations. Springer, New York (1983)
13. Puri, M.L., Ralescu, D.A.: Fuzzy random variables. J. Math. Anal. Appl. **114**, 409–422 (1986)
14. Subrahmanyam, P.V., Sudrasanam, S.K.: A note on fuzzy voltera integral equations. Fuzzy Sets Syst. **81**, 237–240 (1996)

On Fuzzy Localized Subring

Idris Bakhadach, Said Melliani, Hamid Sadiki, and Lalla Saadia Chadli

Abstract In this paper we give a new definition of localized fuzzy subring by the fuzzy points, also we introduce the fuzzy multiplicative closed set in term to define the fuzzy field of fraction.

1 Introduction

The work of Zadeh [11] followed by that of Rosenfeld [8] led to the development of fuzzy abstract algebra. Liu [3], Mukherjee, and Bhattacharya [6] examined normal fuzzy subgroups. Liu [3] also discussed fuzzy subrings and fuzzy ideals. Wang, Ruan and Kerre [9] studied fuzzy subrings and fuzzy rings. Swamy and Swamy [10] defined and proved major theorems on fuzzy prime ideals of rings, Pu and Liu [7] introduced the notion of fuzzy points, based on this notion Melliani et al. [4]introduce the notion of a ring of fuzzy points. Alkhamees and Mordeson [1] characterize local rings in terms of certain fuzzy ideals. They also characterize rings of fractions at a prime ideal in terms of fuzzy ideals. In this paper we investigate the notion of localization of rings in terms of fuzzy points and we characterize the field of fractions of fuzzy points.

I. Bakhadach · S. Melliani (✉) · H. Sadiki · L. S. Chadli
LMACS, Sultan Moulay Slimane University, PO Box 523, 23000 Beni Mellal, Morocco
e-mail: saidmelliani@gmail.com

I. Bakhadach
e-mail: i.bakhadach@usms.ma

H. Sadiki
e-mail: sadiki.info@gmail.com

L. S. Chadli
e-mail: sa.chadli@yahoo.fr

S. Melliani and O. Castillo (eds.), *Recent Advances in Intuitionistic Fuzzy Logic Systems and Mathematics*, Studies in Fuzziness and Soft Computing 395,
https://doi.org/10.1007/978-3-030-53929-0_2

2 Fuzzy Multiplicative Closed Set

Definition 1 ([11]) Let E be a non-empty set. A fuzzy subset of the E is a function μ such that

$$\mu : E \to [0, 1].$$

Definition 2 ([5]) Let R be a ring with identity. Then $\mu \subset R$ is called a fuzzy subring if and only if $\mu(x - y) \geq \mu(x) \wedge \mu(y)$ and $\mu(xy) \geq \mu(x) \wedge \mu(y)$, $\forall x, y \in R$.

Definition 3 ([7]) Let A be a non-empty set and $x_\alpha : A \longrightarrow [0, 1]$ a fuzzy subset of A with $x \in A$ and $\alpha \in (0, 1]$ defined by:

$$x_\alpha(y) = \begin{cases} \alpha & if \quad x = y \\ 0 & if \quad x \neq y \end{cases}$$

then x_α is called a fuzzy point (singleton).

Definition 4 Let μ be a fuzzy subring of R, and x_t be a fuzzy point of R. We write $x_t \in \mu$ to express that $\mu(x) \geq t$, and we have:

$$x_t + y_s = (x + y)_{t \wedge s}$$

$$x_s y_t = (xy)_{t \wedge s}$$

Definition 5 ([2] Integral and finite ring extensions)

(a) If $R \subset R'$ are rings, we call R' an extension ring of R. We will also say in this case that $R \subset R'$ is a ring extension.
(b) Let R be a ring. An element a of an extension ring R' of R is called integral over R if there is a monic polynomial $f \in R[x]$ with $f(a) = 0$. We say that R' is integral over R if every element of R' is integral over R.

Definition 6 Let $\mu \in R$ be a fuzzy subring of R, the fuzzy subset $\nu \subset \mu$ is said to be **multiplicative closed set** if

(i) $\nu(0) = 0$,
(ii) $\nu(1) = 1$,
(iii) $\nu(xy) \leq \nu(x) \wedge \nu(y), \quad \forall x, y \in R$.

Proposition 1 *A fuzzy subset $\nu \subset \mu$ is said to be a Fuzzy multiplicative closed set of μ if and only if ν_t, $\forall t \in]0, 1]$ is multiplicative closed of R.*

Proof Let ν be a fuzzy multiplicative closed of μ

$\Rightarrow$) we have $0 \notin \nu_t$, $\forall t \in]0, 1]$, and $1 \in \nu_t$, $\forall t \in [0, 1]$.
If $x \in \nu_t$ and $y \in \nu_t$, $t \in]0, 1[$, then $\nu(x, y) \geq \nu(x) \wedge \nu(y)$. Where $\nu(x.y) \geq t$.
Hence $x, y \in \nu_t$.

$\Leftarrow$) Suppose that ν_t are multiplicative parts of R, $\forall t \in]0, 1]$. Then we have:
$0 \notin \nu_t, \forall t \in]0, 1]$, so $\nu(0) \leq t, \forall t \in]0, 1]$, hence $\nu(0) = 0$
We have $1 \in \nu_t, \forall t \in]0, 1]$, then $\nu(1) \geq 1$. Therefore $\nu(1) = 1$
Now if $\nu(x) \neq 0$ and $\nu(y) \neq 0$
Then we have $\nu(x) \geq \nu(y)$ and $\nu(y) \geq \nu(y)$.
therefore $x \in \nu_{\nu(x)\wedge\nu(y)}$ and $y \in \nu_{\nu(x)\wedge\nu(y)}$
Hence the stability of the $\alpha - cuts$, we have $xy \in \nu_{\nu(x)\wedge\nu(y)}$
Finally we have $\nu(x.y) \geq \nu(x) \wedge \nu(y)$.
Else (if $\nu(x) = 0$ or $\nu(y) = 0$)), then $\nu(x.y) \geq 0 = \nu(x) \wedge \nu(y)$
Consequently ν is a fuzzy multiplicative closed set of μ.

Proposition 2 *Let μ be a fuzzy subring of R. If ν is a fuzzy multiplicative closed of μ, then $\mathcal{F}_\nu(R)$ is it also in $\mathcal{F}_\mu(R)$.*

Proof We have $0_t \notin \nu, \quad \forall t \in]0, 1]$ (because $\nu(0) = 0$), then $0_t \notin \nu(R)$ and we have $1_t \in \nu, \forall t \in]0, 1]$ hence $1_t \in \mathcal{F}_\nu(R)$.
Let now $x_t y_t \in \mathcal{F}_\nu(R)$, then $x_t \in \mathcal{F}_\nu(R)$ and $y_s \in \mathcal{F}_\nu(R)$, which implies that $x \in \nu_t$ and $y \in \nu_s$, then $xy \in \nu_{t\wedge s}$. Consequently $(xy)_{t\wedge s} \in \mathcal{F}_\nu(R)$

3 Localization of Rings of Fuzzy Points

Let $\mathcal{F}_\nu(R)$ be a multiplicative closed set of $\mathcal{F}_\mu(R)$. Then

$$(x_{t_1}, y_{\alpha_1}) \sim (a_{t_2}, b_{\alpha_2}) \Longleftrightarrow \exists u_m \in \mathcal{F}_\nu(R); \quad u_m(x_{t_1}.b_{\alpha_2} - a_{t_2} y_{\alpha_1}) = 0.$$

is an equivalence relation on $\mathcal{F}_\mu(R) \times \mathcal{F}_\nu(R)$ We denote the equivalence class of a pair $(x_t, y_\alpha) \in \mathcal{F}_\mu(R) \times \mathcal{F}_\nu(R)$ by $\dfrac{x_t}{y_\alpha}$.

The set of all equivalence classes

$$\mathcal{F}_\mu(R) \times \mathcal{F}_\nu^{-1}(R) := \left\{ \frac{x_t}{y_s} := \left(\frac{x}{y}\right)_{t\wedge s} ; x_t \in \mathcal{F}_\mu(R) \ et \ \ y_s \in \mathcal{F}_\nu(R) \right\}$$

is called the localization of $\mathcal{F}_\mu(R)$ at the multiplicative closed set $\mathcal{F}_\nu(R)$. It is a ring together with the addition and multiplication

$$\frac{x_t}{y_s}.\frac{m_\alpha}{n_\beta} = \frac{(x.m)_{t\wedge\alpha}}{(y.n)_{s\wedge\beta}} \quad \text{and} \quad \frac{x_t}{y_s} + \frac{m_\alpha}{n_\beta} = \frac{x_t.n_\beta + m_\alpha.y_s}{y_s.n_\beta}$$

Proof The relation $\sim$ is clearly reflexive and symmetric. It is also transitive:
If $(x_{t_1}, y_{s_1}) \sim (z_{t_2}, r_{s_2})$ and $(z_{t_2}, r_{s_2}) \sim (m_{t_3}, n_{s_3})$ Then, $\exists u_{\alpha_1}, v_{\alpha_2} \in \mathcal{F}_\nu(R)$ such that

$$u_{\alpha_1}.(x_{t_1} r_{s_2} - z_{t_2}.y_{s_1}) = 0 \quad \text{and} \quad v_{\alpha_2}.(z_{t_2}.n_{s_3} - m_{t_3}.r_{s_2}) = 0$$

Therefore,

$$u_{\alpha_1}.v_{\alpha_2}.r_{s_2}(x_{t_1}.n_{s_3} - m_{t_3}.y_{s_1}) = u_{\alpha_1}.v_{\alpha_2}.n_{s_3}(x_{t_1}.r_{s_2} - z_{t_2}.y_{s_1}) + u_{\alpha_1}.v_{\alpha_2}y_{s_1}(n_{s_3}z_{t_2} - r_{s_2}.m_{t_3}) = 0$$

since $\mathcal{F}_\nu(R)$ is a multiplicative closed set then $u_{\alpha_1}.v_{\alpha_2}.n_{s_2} \in \mathcal{F}_\nu(R)$.

In a similar way we can check that the addition and multiplication in $\mathcal{F}_\mu(R) \times \mathcal{F}_\nu^{-1}(R)$ are well-defined: if $(x_{t_1}, y_{s_1}) \sim (z_{t_2}, r_{s_2})$ i.e $u_{\alpha_1}.(x_{t_1}r_{s_2} - z_{t_2}.y_{s_1}) = 0$ for some $u_{\alpha_1} \in \mathcal{F}_\nu(R)$ then

$$u_{\alpha_1}\Big((x_{t_1}n_{s_3} + m_{t_3}y_{s_1})(r_{s_2}n_{s_3}) - (z_{t_2}n_{s_3} + m_{t_3}r_{s_2})(y_{s_1}n_{s_3})\Big) = n_{s_3}^2 u_{\alpha_1}(x_{t_1}r_{s_2} - z_{t_2}y_{s_1}) = 0$$

and

$$u_{\alpha_1}(x_{t_1}m_{t_3})(r_{s_2}n_{s_3}) - (z_{t_2}m_{t_3})(y_{s_1}n_{s_3}) = m_{t_3}n_{s_3}u_{\alpha_1}(x_{t_1}r_{s_2} - z_{t_2}y_{s_1}) = 0$$

Then we have $\dfrac{x_{t_1}n_{s_3} + m_{t_3}y_{s_1}}{y_{s_1}n_{s_3}} = \dfrac{z_{t_2}n_{s_3} + m_{t_3}r_{s_2}}{r_{s_2}n_{s_3}}$ and $\dfrac{x_{t_1}m_{t_3}}{y_{s_1}n_{s_3}} = \dfrac{z_{t_2}m_{t_3}}{m_{t_3}r_{s_2}}$.

It is now verified immediately that these two operations satisfy the ring axioms.

Theorem 1 *Let $\mathcal{F}_\nu(R)$ be an integral over $\mathcal{F}_\mu(R)$, if $\mathcal{F}_\phi(R)$ is a fuzzy multiplicative closed set of $\mathcal{F}_\mu(R)$. Then, $\mathcal{F}_\nu(R) \times \mathcal{F}_\phi^{-1}(R)$ is integral over $\mathcal{F}_\mu(R) \times \mathcal{F}_\phi^{-1}(R)$.*

Proof Let $\dfrac{x_t}{y_s} \in \mathcal{F}_\nu(R) \times \mathcal{F}_\phi^{-1}(R)$, so $x_t \in \mathcal{F}_\nu(R)$, then $\exists P \in \mathcal{F}_\mu(R)[X]$ such that

$$P(x_t) = \sum_{i=0}^{i=n} a_{t_i} x_t^i = 0$$

Therefore,

$$\frac{1_1}{y_s^n} \times P(x_t) = \sum_{i=0}^{i=n} \frac{a_{t_i}}{y_s^{n-i}} \cdot \left(\frac{x}{y}\right)^i_{t\wedge s} = 0$$

since $\dfrac{a_{t_i}}{y_s^{n-i}} \in \mathcal{F}_\mu(R) \times \mathcal{F}_\phi^{-1}(R)$

Hence, $\left(\dfrac{x}{y}\right)_{t\wedge s}$ is integral over $\mathcal{F}_\mu(R) \times \mathcal{F}_\phi^{-1}(R)$.

4 Fuzzy Field of Fractions of Fuzzy Points

Let μ be a fuzzy subring of R and let ν given by

$$\nu(x) = \begin{cases} 0 & \text{if } \exists u \in R \text{ such that } x.u = 0 \\ \mu(x) & \text{elsewhere} \end{cases}$$

It's clear that $\nu \subset \mu$.

We have $0_t \notin \nu$ and $1_1 \in \nu$. Now if $xy = 0$ so $\nu(x) = \nu(y) = \nu(xy) = 0$, then $\nu(xy) \geq \nu(x) \wedge \nu(y)$, else $\nu(xy) = \mu(x, y) \geq \mu(x) \wedge \mu(y)$.

Hence, ν is a fuzzy multiplicative closed set of μ.

Definition 7 • if R is not an integral domain then $\mathcal{F}_\mu(R) \times \mathcal{F}_\nu^{-1}(R)$ is a ring.

- if R is an integral domain then $\mathcal{F}_\mu(R) \times \mathcal{F}_\nu^{-1}(R)$ is a field, named field of fractions of fuzzy points, with

$$\nu : x \rightarrow \begin{cases} 0 & \text{if } x = 0 \\ \mu(x) & \text{elsewhere} \end{cases}$$

Remark 1 • if ν is a fuzzy multiplicative closed set of μ, then $\mathcal{F}_\mu(R) \subset \mathcal{F}_\mu(R) \times \mathcal{F}_\nu^{-1}(R)$

- let R be an integral domain and μ be a fuzzy subring of R
  for all fuzzy multiplicative closed set ν of μ, we have $\nu \subset \phi$ with:

$$\phi(x) = \begin{cases} 0 & \text{if } x = 0 \\ \mu(x) & \text{elsewhere} \end{cases}$$

  Moreover $\mathcal{F}_\nu(R) \subset \mathcal{F}_\phi(R)$
  Consequently $\mathcal{F}_\mu(R) \times \mathcal{F}_\nu^{-1}(R) \subset \mathcal{F}_\mu(R) \times \mathcal{F}_\phi^{-1}(R)$.
  Then for any fuzzy multiplicative closed set of ν we have $\mathcal{F}_\mu(R) \times \mathcal{F}_\nu^{-1}(R)$ is included in the field of fractions of fuzzy points $\mathcal{F}_\mu(R)$.

Proposition 3 *Let $\mathcal{F}_\mu(R)$ be an integrally closed domain. For all S fuzzy multiplicative closed set of ν we have $\mathcal{F}_\mu(R) \times \mathcal{F}_S^{-1}(R)$ is fuzzy multiplicative closed set.*

Proof Let K be the field of fractions over $\mathcal{F}_\mu(R)$, then

$$\mathcal{F}_\mu(R) \subseteq \mathcal{F}_\mu(R) \times \mathcal{F}_S^{-1}(R) \subseteq K$$

Let $\frac{x_t}{y_s} \in K$ be a fuzzy integral over $\mathcal{F}_\mu(R) \times \mathcal{F}_S^{-1}(R)$ then $\exists P \in \mathcal{F}_\mu(R) \times \mathcal{F}_S^{-1}(R)[X]$ such that

$$P\left(\frac{x_t}{y_s}\right) = \sum_{i=0}^{i=n} \frac{a_{t_i}}{b_{s_i}}.\left(\frac{x_t}{y_s}\right)^i = 0.$$

Hence y_s^n.

$$\prod_{i=0}^{i=n} b_{\alpha_i} . P\left(\frac{x_t}{y_s}\right) = \sum_{i=0}^{i=n} C_i . x^i = 0.$$

with $C_i \in \mathcal{F}_\mu(R)$.

So x is integral over $\mathcal{F}_\mu(R)$ which is an integrally closed domain, then $x \in \mathcal{F}_\mu(R) \subseteq \mathcal{F}_\mu(R) \times \mathcal{F}_S^{-1}(R)$.

References

1. Alkhamees, Y., Mordeson, J.N.: Fuzzy localized subrings. Inf. Sci. **99**, 183–193 (1997)
2. Gathmann, A.: Einführung in die Algebra. Class Notes TU Kaiserslautern (2010). www.mathematik.uni-kl.de/~gathmann/algebra.php
3. Liu, W.: Fuzzy invariant subgroups and fuzzy ideals. Fuzzy Sets Syst. **8**, 133–139 (1982)
4. Melliani, S., Bakhadach, I., Chadli, L.S.: Fuzzy rings and fuzzy polynomial rings. In: Badawi, A. et al. (eds.) Homological and Combinatorial Methods in Algebra. Springer Proceedings in Mathematics & Statistics, vol. 228, Springer International Publishing AG, Berlin (2018). https://doi.org/10.1007/978-3-319-74195-6_8
5. Mordeson, J.N.: Fuzzy coefficient fields of fuzzy subrings. Fuzzy Sets Syst. **58**, 227–237 (1993). North-Holland
6. Mukherjee, N.P., Bhattacharya, P.: Fuzzy normal subgroups and fuzzy cosets. Inf. Sci. **34**, 225–239 (1984)
7. Pu, P.M., Liu, Y.M.: Fuzzy topology. I. Neighborhood structure of a fuzzy point and Moore-Smith convergence. J. Math. Anal. Appl. **76**, 571–599 (1980)
8. Rosenfeld, A.: Fuzzy groups. J. Math. Anal. Appl. **35**, 512–517 (1971)
9. Wang, X., Ruan, D., Kerre, E.E.: Mathematics of Fuzziness-Basic Issues. Springer, Berlin (2009)
10. Swamy, U.M., Swamy, K.L.N.: Fuzzy prime ideals of rings. J. Math. Anal. Appl. **134**, 94–103 (1988)
11. Zadeh, L.A.: Fuzzy sets. Inform. Control **8**, 338–353 (1965)

Coupled System of Nonlinear Fuzzy Volterra–Urysohn Integral Equations

Abdelati El Allaoui, Said Melliani, and Lalla Saadia Chadli

Abstract This paper investigates a coupled system of nonlinear fuzzy Volterra–Urysohn integral equations. Here, we have studied the existence of at least one continuous solution. As an application, we will consider the coupled system of nonlinear fuzzy Volterra–Hammerstein integral equations. Finally, illustrative examples are presented to validate the obtained results.

1 Introduction

The existence of solutions of the integral and the coupled systems of integral equations have been extensively investigated by a number of authors and there are many interesting results concerning this problems (see [4, 6, 7, 9, 19]), and there are several fuzzy mathematical models treated by different approaches as can be seen in [12–16].

Existence theorems for Volterra integral equations have been studied extensively in view of their applications to predator-prey models and medical diagnosis and various applied problems arising in mathematical physics, mechanics and control theory lead to multivalued analogs of the fuzzy Volterra–Urysohn integral equations, these equations have been studied in several papers and monographs (see [2, 3, 5, 8, 22]).

In this paper, we study the existence of at least one continuous solutions (u, v) of the coupled system of nonlinear fuzzy Volterra–Urysohn integral equations

A. El Allaoui · S. Melliani (✉) · L. S. Chadli
LMACS, Laboratoire de Mathématiques Appliquées & Calcul Scientifique, Sultan Moulay Slimane University, PO Box 523, 23000 Beni Mellal, Morocco
e-mail: saidmelliani@gmail.com

A. El Allaoui
e-mail: elallaoui199@gmail.com

L. S. Chadli
e-mail: sa.chadli@yahoo.fr

S. Melliani and O. Castillo (eds.), *Recent Advances in Intuitionistic Fuzzy Logic Systems and Mathematics*, Studies in Fuzziness and Soft Computing 395,
https://doi.org/10.1007/978-3-030-53929-0_3

$$u(t) = \phi(t) + \int_0^t f(t, s, v(l(s)))ds, \qquad t \in [0, 1] \tag{1}$$

$$v(t) = \psi(t) + \int_0^t g(t, s, u(k(s)))ds, \qquad t \in [0, 1] \tag{2}$$

As an application, we study the existence of at least one continuous solutions (u, v) of the coupled system of nonlinear fuzzy Volterra–Hammerstein integral equations

$$u(t) = \phi(t) + \int_0^t G(t, s) f_1(s, v(l(s)))ds, \qquad t \in [0, 1], \tag{3}$$

$$v(t) = \psi(t) + \int_0^t K(t, s) g_1(s, u(k(s)))ds, \qquad t \in [0, 1]. \tag{4}$$

2 Preliminaries

Let $\mathscr{P}_K(\mathbb{R}^n)$ denote the family of all non-empty compact convex subsets of $\mathbb{R}^n$ and define the addition and scalar multiplication in $\mathscr{P}_K(\mathbb{R}^n)$ as usual.

Let A and B be two nonempty bounded subsets of $\mathbb{R}^n$. The distance between A and B is defined by the Hausdorf metric,

$$\delta(A, B) = max \left\{ \sup_{a\in A} \inf_{b\in B} \|a - b\|, \sup_{b\in B} \inf_{a\in A} \|a - b\| \right\}$$

where $\| \|$ denotes the usual euclidean norm in $\mathbb{R}^n$. Then it is clear that $(\mathscr{P}_K(\mathbb{R}^n), \delta)$ becomes a complete and separable metric space (see [21]).

Denote

$$E^n = \left\{ u : \mathbb{R}^n \longrightarrow [0, 1] \ \mid \ \text{u satisfies (i)–(iv) below} \right\},$$

where

(i) u is normal i.e there exists an $x_0 \in \mathbb{R}^n$ such that $u(x_0) = 1$,
(ii) u is fuzzy convex,
(iii) u is upper semicontinuous,
(iv) $[u]^0 = cl\{x \in \mathbb{R}^n / u(x) > 0\}$ is compact.

For $0 < \alpha \leq 1$, denote $[u]^\alpha = \{t \in \mathbb{R}^n \ / \ u(t) \geq \alpha\}$. Then from (i)–(iv), it follows that the α-level set $[u]^\alpha \in \mathscr{P}_K(\mathbb{R}^n)$ for all $0 \leq \alpha \leq 1$.

According to Zadeh's extension principle, we have addition and scalar multiplication in fuzzy number space E^n as follows:

$$[u + v]^\alpha = [u]^\alpha + [v]^\alpha, \quad [ku]^\alpha = k[u]^\alpha$$

where $u, v \in E^n$, $k \in \mathbb{R}^n$ and $0 \leq \alpha \leq 1$.

Define $d : E^n \times E^n \to \mathbb{R}^+$ by the equation

$$d(u, v) = \sup_{0 \le \alpha \le 1} \delta\Big([u]^\alpha , [v]^\alpha\Big)$$

where δ is the Hausdorff metric for non-empty compact sets in $\mathbb{R}^n$.

Then it is easy to see that d is a metric in E^n.

Using the results in [21], we know that

(1) (E^n, d) is a complete metric space;
(2) $d(u + w, v + w) = d(u, v)$ for all $u, v, w \in E^n$;
(3) $d(k\,u, k\,v) = |k|\ d(u, v)$ for all $u, v \in E^n$ and $k \in \mathbb{R}^n$.

If we denote $\|u\| = d\left(u, \tilde{0}\right)$, $u \in E^n$, then $\|u\|$ has the properties of an usual norm on E^n (see [11, 18]),

(1) $\|u\| = 0$ iff $u = \tilde{0}$;
(2) $\|\lambda u\| = |\lambda|\|u\|$ for all $u \in E^n, \lambda \in \mathbb{R}$;
(3) $\|u + v\| \le \|u\| + \|v\|$ for all $u, v \in E^n$;
(4) $d(\alpha u, \beta u) \le |\alpha - \beta| d(u, \tilde{0})$, for all $\alpha, \beta \ge 0$ or $\alpha, \beta \le 0$, $u \in E^n$.

On E^n, we can define the substraction $-_H$, called the H-difference as follows: $u -_H v$ has sense if there exists $w \in E^n$ such that $u = v + w$.

Denote $\mathscr{C}_a = \mathscr{C}\,([0, a], E^n) = \{f : [0, a] \longrightarrow E^n;\ f \text{is continuous on}[0, a]\}$, endowed with the metric

$$D(u, v) = \sup_{t \in [0,a]} d(u(t), v(t))$$

Then $(\mathscr{C}_a, D)$ is a complete metric space.

Lets $a, b \in \mathbb{R}$, $f \in \mathscr{C}([0, a], E^n)$, if we denote $\|f\|_{\mathscr{F}} = D\left(f, \tilde{0}\right)$, then $\|f\|_{\mathscr{F}}$ has the properties of an usual norm on E^n (see [11]),

(1) $\|f\|_{\mathscr{F}} = 0$ if $f = \tilde{0}$;
(2) $\|\lambda f\|_{\mathscr{F}} = |\lambda|\|f\|_{\mathscr{F}}$ for all $f \in \mathscr{C}([0, a], E^n)$, $\lambda \in \mathbb{R}$;
(3) $\|f + g\|_{\mathscr{F}} \le \|f\|_{\mathscr{F}} + \|g\|_{\mathscr{F}}$ for all $f, g \in \mathscr{C}([0, a], E^n)$;
(4) $D(\alpha f, \beta f) \le |\alpha - \beta| D(f, \tilde{0})$, for all $\alpha, \beta \ge 0$ or $\alpha, \beta \le 0$, $f \in \mathscr{C}([0, a], E^n)$.
Let $T = [c, d] \subset \mathbb{R}$ be a compact interval.

Definition 1 A mapping $F : T \to E^n$ is Hukuhara differentiable at $t_0 \in T$ if there exists a $F'(t) \in E^n$ such that the following limits

$$\lim_{h \to 0^+} \frac{F(t + h) -_H F(t)}{h} \quad \text{and} \quad \lim_{h \to 0^+} \frac{F(t) -_H F(t - h)}{h}$$

exist and equal to $F'(t)$.

We recall some measurability, integrability properties for fuzzy set-valued mappings (see [17]).

Definition 2 A mapping $F : T \to E^n$ is strongly measurable if for all $\alpha \in [0, 1]$ the set-valued function $F_\alpha : T \to \mathscr{P}_K(\mathbb{R}^n)$ defined by $F_\alpha(t) = [F(t)]^\alpha$ is Lebesgue measurable.
A mapping $F : T \to E^n$ is called integrably bounded if there exists an integrable function k such that $||x|| \leq k(t)$ for all $x \in F_0(t)$.

Definition 3 Let $F : T \to E^n$. Then the integral of F over T denoted by $\int_T F(t)dt$ or $\int_c^d F(t)dt$, is defined by the equation

$$\left[\int_T F(t)dt\right]^\alpha = \int_T F_\alpha(t)dt = \left\{\int_T f(t)dt / f : T \to \mathbb{R}^n \text{ is a mesurable selection for } F_\alpha\right\}$$

for all $\alpha \in]0, 1]$.

Also, a strongly measurable and integrably bounded mapping $F : T \to E^n$ is said to be integrable over T if

$$\int_T F(t)dt \in E^n$$

Proposition 1 (Aumann [1]) *If $F : T \to E^n$ is strongly measurable and integrably bounded, then F is integrable.*

Proposition 2 ([17]) *Let $F, G : T \to E^n$ be integrable and $\lambda \in \mathbb{R}$. Then*

(i) $\int_T (F(t) + G(t))dt = \int_T F(t)dt + \int_T G(t)dt,$
(ii) $\int_T \lambda F(t)dt = \lambda \int_T F(t)dt,$
(iii) $d(F, G)$ *is integrable,*
(iv) $d(\int_T F(t)dt, \int_T G(t)dt) \leq \int_T d(F, G)(t)dt.$

3 Main Result

Now we present some auxiliary results that will be need in this work.

Define the space F^n by

$$F^n = E^n \times E^n.$$

Let $H : F^n \times F^n \longrightarrow [0, +\infty[$ be defined by

$$H\big((u, v), (u', v')\big) = \max\big\{d(u, u'), d(v, v')\big\}.$$

Then, $(F^n\,,\ H)$ is a complete metric space. Denote $\mathscr{C}([0,1], F^n)$ the set of all continuous maps to $[0,1]$ into F^n.

Remark 1 We can see that, if $(f, g) \in \mathscr{C}([0,1], F^n)$, then $f, g \in \mathscr{C}([0,1], E^n)$.

We define on $\mathscr{C}([0,1], F^n)$ the metric Φ given by

$$\Phi\big((f_1, g_1)\,,\ (f_2, g_2)\big) = \max\left\{D(f_1, f_2)\,,\ D(g_1, g_2)\right\}.$$

For $(u, v) \in F^n$ and $(f, g) \in \mathscr{C}([0,1], F^n)$, denote

$$|||(u, v)||| = H\big((u, v), \tilde{0}_2\big), \qquad |||(f, g)|||_{\mathscr{F}} = \Phi\big((f, g), \tilde{0}_2\big),$$

where $\tilde{0}_2 = (\tilde{0}, \tilde{0})$.

Remark 2 Note that

$$\begin{aligned} |||(u, v)||| &= H\big((u, v), \tilde{0}_2\big) \\ &= \max\{d(u, \tilde{0})\,,\ d(v, \tilde{0})\} \\ &= \max\{\|u\|\,,\ \|v\|\}. \end{aligned}$$

and

$$\begin{aligned} |||(f, g)|||_{\mathscr{F}} &= \Phi\big((f, g), \tilde{0}_2\big) \\ &= \max\{D(f, \tilde{0})\,,\ D(g, \tilde{0})\} \\ &= \max\{\|f\|_{\mathscr{F}}\,,\ \|g\|_{\mathscr{F}}\}. \end{aligned}$$

Then it is easy to prove that $|||.|||$ and $|||.|||_{\mathscr{F}}$ satisfies the same previous properties of $\|.\|$ and $\|.\|_{\mathscr{F}}$.

3.1 Coupled System of Fuzzy Volterra–Urysohn Integral Equations

Let us start by defining what we mean by a solution of the problem (1)–(2).

Definition 4 A pair (u, v) is a solution of the coupled system (1)–(2) if u and v are continuous and this ordered pair satisfies the coupled system (1)–(2).

To establish our main result concerning the existence of solutions for (1)–(2), we list the following hypotheses:

($\mathbf{H}_1$) The functions $\phi,\ \psi\ :\ [0,1] \longrightarrow E^n$ are continuous and there exist two constants $a, b > 0$ such that

$$\sup_{t\in[0,1]} \|\phi(t)\| \leq a, \qquad \sup_{t\in[0,1]} \|\psi(t)\| \leq b.$$

($\mathbf{H}_2$) The functions $l, k : [0, 1] \longrightarrow [0, 1]$ are continuous.

($\mathbf{H}_3$) The functions $f, g : [0, 1] \times [0, 1] \times E^n \longrightarrow E^n$ are continuous, and there exist two integrable functions $m, n : [0, 1] \times [0, 1] \longrightarrow \mathbb{R}$ and two positive constants $0 < \alpha, \beta < 1$ such that

$$\|f(t, s, v)\| \leq |m(t, s)| + \alpha\|v\|, \qquad \|g(t, s, u)\| \leq |n(t, s)| + \beta\|u\|,$$

and

$$\int_0^1 m(t, s)ds \leq M, \qquad \int_0^1 n(t, s)ds \leq N.$$

Define the operator Γ by

$$\Gamma(u, v) = \big(\Gamma_1 v\,,\ \Gamma_2 u\big),$$

where

$$\Gamma_1 v = \phi(t) + \int_0^t f(t, s, v(l(s)))ds, \qquad t \in [0, 1] \tag{5}$$

$$\Gamma_2 u = \psi(t) + \int_0^t g(t, s, u(k(s)))ds, \qquad t \in [0, 1] \tag{6}$$

We are now in a position to state and prove our existence result of at least one solution for the problem (1)–(2) based on Leray–Schauder degree.

Theorem 1 *Suppose that assumptions* ($\boldsymbol{H}_1$)–($\boldsymbol{H}_3$) *hold. Then the system (1)–(2) has at least one solution in* $\mathscr{C}([0, 1], F^n)$.

Proof We are going to show that there exists at least one solution to (1)–(2), by the same manner used in theorem 3.3 in [10]. Let us consider the operator $\Gamma : \mathscr{C}([0, 1], F^n) \longrightarrow \mathscr{C}([0, 1], F^n)$ defined by (5)–(6).
It is clear that the solutions of our problem are fixed points of the operator Γ. So we will look for at least $(u, v) \in \mathscr{C}([0, 1], F^n)$ such that

$$(u, v) = \Gamma(u, v). \tag{7}$$

It is easy to verify that Γ is well defined.

Consider the following family of problems associated with (7), namely

$$(u, v) = \lambda\Gamma(u, v), \qquad \lambda \in [0, 1]. \tag{8}$$

For $t \in [0, 1]$, from (8) we have

$$
\begin{aligned}
H\big((u(t), v(t)),\, \tilde{0}_2\big) &= H\big(\lambda \Gamma(u(t), v(t)),\, \tilde{0}_2\big) \\
&= H\left((\lambda \Gamma_1 v(t), \lambda \Gamma_2 u(t)),\, (\tilde{0}, \tilde{0})\right) \\
&= \max\left\{d(\lambda \Gamma_1 v(t), \tilde{0}),\, d(\lambda \Gamma_2 u(t), \tilde{0})\right\} \\
&= \max\left\{d\left(\lambda \phi(t) + \lambda \int_0^t f(t,s,v(l(s)))ds, \tilde{0}\right),\right. \\
&\qquad \left. d\left(\lambda \psi(t) + \lambda \int_0^t g(t,s,u(k(s)))ds, \tilde{0}\right)\right\},
\end{aligned}
$$

and we have

$$
\begin{aligned}
d\left(\lambda \phi(t) + \lambda \int_0^t f(t,s,v(l(s)))ds, \tilde{0}\right) &\le \lambda \|\phi(t)\| + \lambda \int_0^1 \|f(t,s,v(l(s)))\| ds \\
&\le a + \int_0^1 m(t,s)ds + \alpha \|v\|_{\mathscr{F}} \\
&\le a + M + \alpha \|v\|_{\mathscr{F}}.
\end{aligned}
$$

Also

$$
\begin{aligned}
d\left(\lambda \psi(t) + \lambda \int_0^t g(t,s,u(k(s)))ds, \tilde{0}\right) &\le \lambda \|\psi(t)\|_{\mathscr{F}} + \lambda \int_0^1 \|g(t,s,u(k(s)))\| ds \\
&\le b + \int_0^1 n(t,s)ds + \beta \|u\|_{\mathscr{F}}. \\
&\le b + N + \beta \|u\|_{\mathscr{F}},
\end{aligned}
$$

which implies that, for all $t \in [0, 1]$

$$
\begin{aligned}
H\big((u(t), v(t)),\, \tilde{0}_2\big) &\le \max\{a + M + \alpha\|v\|_{\mathscr{F}},\, b + N + \beta\|u\|_{\mathscr{F}}\} \\
&\le \max(a,b) + \max(M,N) + \max(\alpha,\beta)\max\{\|u\|_{\mathscr{F}},\, \|v\|_{\mathscr{F}}\} \\
&\le \max(a,b) + \max(M,N) + \max(\alpha,\beta)|\|(u,v)\||_{\mathscr{F}}
\end{aligned}
$$

Since

$$
H\big((u(t), v(t)),\, \tilde{0}_2\big) = \max\left\{d(u(t), \tilde{0}),\, d(v(t), \tilde{0})\right\}.
$$

then, for all $t \in [0, 1]$

$$
d(u(t), \tilde{0}) \le \max(a,b) + \max(M,N) + \max(\alpha,\beta)|\|(u,v)\||_{\mathscr{F}},
$$

and

$$
d(v(t), \tilde{0}) \le \max(a,b) + \max(M,N) + \max(\alpha,\beta)|\|(u,v)\||_{\mathscr{F}},
$$

Hence

$$
\begin{aligned}
|\|(u,v)\||_{\mathscr{F}} = \max\{\|u\|_{\mathscr{F}},\, \|v\|_{\mathscr{F}}\} &= \max\left\{\sup_{t\in[0,1]} d(u(t), \tilde{0}),\, \sup_{t\in[0,1]} d(v(t), \tilde{0})\right\} \\
&\le \frac{\max(a,b) + \max(M,N)}{1 - \max(\alpha,\beta)} =: \rho,
\end{aligned}
\tag{9}
$$

Define the open ball B_ρ with center $\tilde{0}$ and radius $\rho + 1$ by

$$B_{\rho+1} = \left\{(u, v) \in \mathscr{C}([0, 1], F^n) \; : \; |||(u, v)|||_{\mathscr{F}} \; < \rho + 1\right\}.$$

According to (9) all possible solutions to (8) verify

$$|||(u, v)|||_{\mathscr{F}} \; < \rho + 1.$$

Then, by the homotopy invariance property of the Leray–Schauder degree (see [10, 20]), we know that

$$\begin{aligned} deg\left(\mathscr{I} - \Gamma \,,\, B_{\rho+1} \,,\, \tilde{0}\right) &= deg\left(\mathscr{I} - \lambda\Gamma \,,\, B_{\rho+1} \,,\, \tilde{0}\right) \\ &= deg\left(\mathscr{I} \,,\, B_{\rho+1} \,,\, \tilde{0}\right) = 1, \end{aligned}$$

with $\mathscr{I}$ is the identity map.

Consequently, by the non-zero property of the Leray–Schauder degree (Solvability), we have

$$deg\left(\mathscr{I} - \Gamma \,,\, B_{\rho+1} \,,\, \tilde{0}\right) = 1,$$

ensures the existence of at least one solution in $B_{\rho+1}$ to the problem (7) and so to (1)–(2).

Corollary 1 *Let $u = v$, $f = g$, $\phi = \psi$ and $k = l$ in the previous Theorem. Suppose that assumptions ($\boldsymbol{H}_1$)–($\boldsymbol{H}_3$) be satisfied, then the fuzzy integral equation*

$$u(t) = \phi(t) + \int_0^t f(t, s, u(l(s)))ds, \qquad t \in [0, 1]$$

has at least one continuous solution $u \in \mathscr{C}([0, 1], E^n)$.

3.2 Coupled System of Fuzzy Volterra–Hammerstein Integral Equations

Let

$$f(t, s, v(l(s))) = G(t, s) f_1(s, v(l(s))), \qquad g(t, s, u(k(s))) = K(t, s) g_1(s, u(k(s))),$$

where $G, K \; : \; [0, 1] \times [0, 1] \longrightarrow \mathbb{R}$.

Then the system (1)–(2) becomes as the form (3)–(4).

We introduce the following assumptions:

(**H'**$_1$) The functions $f_1, g_1 \; : \; [0, 1] \times E^n \longrightarrow E^n$ are continuous, and there exist two integrable functions $m_1, n_1 \; : \; [0, 1] \longrightarrow \mathbb{R}$ and two positive constants $0 < \alpha_1, \beta_1 < 1$ such that

$$\|f_1(t, v)\| \leq |m_1(t)| + \alpha_1 \|v\|, \qquad \|g_1(t, u)\| \leq |n_1(t)| + \beta_1 \|u\|,$$

(H'$_2$) $G, K : [0,1] \times 0,1] \longrightarrow \mathbb{R}$ are continuous in $t \in [0,1]$ for every $s \in [0,1]$ and measurable in $s \in [0,1]$ for all $t \in [0,1]$ such that

$$\sup_{t\in[0,1]} \int_0^1 |G(t,s)||m_1(s)|ds \leq M_1, \qquad \sup_{t\in[0,1]} \int_0^1 |K(t,s)||n_1(s)|ds \leq N_1.$$

Corollary 2 *Suppose that assumptions* (***H***$_1$)–(***H***$_2$) *and* (***H'***$_1$)–(***H'***$_2$) *be satisfied, then the coupled system of fuzzy integral equations (3)–(4) has at least one continuous solution.*

Corollary 3 *Let* $u = v$, $G = K$, $f_1 = g_1$, $\phi = \psi$ *and* $k = l$ *in the previous Corollary. Suppose that assumptions* (***H***$_1$)–(***H***$_2$) *and* (***H'***$_1$)–(***H'***$_2$) *be satisfied, then the fuzzy integral equation*

$$u(t) = \phi(t) + \int_0^t G(t,s) f_1(s, u(l(s)))ds, \qquad t \in [0,1]$$

has at least one continuous solution $u \in \mathscr{C}([0,1], E^n)$.

3.3 Bounded Solutions

In this section, we examine some cases under which all the solutions of the coupled system of fuzzy integral equations (1)–(2) are bounded.

Define on $\mathscr{C}([0,1], F^n)$ the set B_r as follows

$$B_r = \left\{(u,v) \in \mathscr{C}([0,1], F^n) : \|u\|_{\mathscr{F}} \leq r_2,\ \|v\|_{\mathscr{F}} \leq r_1,\ r = \max(r_1, r_2)\right\},$$

where

$$r_1 = \frac{a+M}{1-\alpha}, \qquad r_2 = \frac{b+N}{1-\beta}.$$

Remark 3 Let $(u, v) \in F^n$, such that

$$\|u\|_{\mathscr{F}} \leq r_2 \text{ and } \|v\|_{\mathscr{F}} \leq r_1,$$

Then

$$\begin{aligned} |||(u,v)|||_{\mathscr{F}} &= \max\{\|u\|_{\mathscr{F}}, \|v\|_{\mathscr{F}}\} \\ &\leq \max(r_2, r_1) = r. \end{aligned}$$

Theorem 2 *Suppose that assumptions* $(\boldsymbol{H}_1)$–$(\boldsymbol{H}_3)$ *hold. Then the operator* $\Gamma : B_r \longrightarrow B_r$ *is continuous in* B_r *and the set* ΓB_r *is uniformly bounded and equicontinuous.*

Proof We distinguish in the proof several steps.

Step 1. $\Gamma : B_r \longrightarrow B_r$ and the set ΓB_r is uniformly bounded:
For $(u, v) \in B_r$, we have

$$|||\Gamma(u, v)|||_{\mathscr{F}} = |||(\Gamma_1 v, \Gamma_2 u)|||_{\mathscr{F}} = \max\{\|\Gamma_1 v\|_{\mathscr{F}}, \|\Gamma_2 u\|_{\mathscr{F}}\},$$

and we have, for all $t \in [0, 1]$

$$\begin{aligned}
\|\Gamma_1 v(t)\| &= \left\|\phi(t) + \int_0^t f(t, s, v(l(s)))ds\right\| \\
&\leq \|\phi(t)\| + \left\|\int_0^t f(t, s, v(l(s)))ds\right\| \\
&\leq \|\phi(t)\| + \int_0^t \|f(t, s, v(l(s)))\| \, ds \\
&\leq \|\phi(t)\| + \int_0^t \left(m(t, s) + \alpha\|v\|_{\mathscr{F}}\right)ds \\
&\leq \|\phi(t)\| + \int_0^t m(t, s)ds + \alpha\|v\|_{\mathscr{F}} \\
&\leq a + M + \alpha r_1 = r_1.
\end{aligned}$$

Also we have

$$\begin{aligned}
\|\Gamma_2 u(t)\| &= \left\|\psi(t) + \int_0^t g(t, s, u(k(s)))ds\right\| \\
&\leq \|\psi(t)\| + \left\|\int_0^t g(t, s, u(k(s)))ds\right\| \\
&\leq \|\psi(t)\| + \int_0^t \|g(t, s, u(k(s)))\| \, ds \\
&\leq \|\psi(t)\| + \int_0^t \left(n(t, s) + \beta\|u\|_{\mathscr{F}}\right)ds \\
&\leq \|\psi(t)\| + \int_0^t n(t, s)ds + \beta\|u\|_{\mathscr{F}} \\
&\leq b + N + \beta r_2 = r_2.
\end{aligned}$$

It follows that

$$\|\Gamma_1 v\|_{\mathscr{F}} = \sup_{t\in[0,1]} \|\Gamma_1 v(t)\| \leq r_1, \qquad \|\Gamma_2 u\|_{\mathscr{F}} = \sup_{t\in[0,1]} \|\Gamma_2 u(t)\| \leq r_2.$$

Hence

$$|||\Gamma(u, v)|||_{\mathscr{F}} \leq \max\{r_1, r_2\} = r,$$

which shows that the operator $\Gamma : B_r \longrightarrow B_r$ and the set ΓB_r is uniformly bounded.

Step 2. Γ is continuous:
Let $(u_n, v_n) \in B_r$ and $(u_0, v_0) \in B_r$ such that

$$\lim_{n\to+\infty} \Phi\big((u_n, v_n)\,,\,(u_0, v_0)\big) = 0.$$

Since

$$\Phi\big((u_n, v_n)\,,\,(u_0, v_0)\big) = \max\{D(u_n, u_0)\,,\,D(v_n, v_0)\},$$

which implies that

$$\lim_{n\to+\infty} d(u_n(t), u_0(t)) = \lim_{n\to+\infty} d(v_n(t), v_0(t)) = 0.$$

We have

$$\begin{aligned} H\big(\Gamma(u_n(t), v_n(t))\,,\,\Gamma(u_0(t), v_0(t))\big) &= H\big((\Gamma_1 v_n(t), \Gamma_2 u_n(t))\,,\,(\Gamma_1 v_0(t), \Gamma_2 u_0(t))\big) \\ &= \max\left\{d\big(\Gamma_1 v_n(t), \Gamma_1 v_0(t)\big)\,,\,d\big(\Gamma_2 u_n(t), \Gamma_2 u_0(t)\big)\right\}, \end{aligned}$$

on the one hand

$$\begin{aligned} d\big(\Gamma_1 v_n(t), \Gamma_1 v_0(t)\big) &= d\left(\phi(t) + \int_0^t f(t,s,v_n(l(s)))ds\,,\,\phi(t) + \int_0^t f(t,s,v_0(l(s)))ds\right) \\ &= d\left(\int_0^t f(t,s,v_n(l(s)))ds\,,\,\int_0^t f(t,s,v_0(l(s)))ds\right) \\ &\le \int_0^t d\left(f(t,s,v_n(l(s)))\,,\,f(t,s,v_0(l(s)))\right)ds, \end{aligned}$$

on the other hand

$$\begin{aligned} d\big(\Gamma_2 u_n(t), \Gamma_2 u_0(t)\big) &= d\left(\psi(t) + \int_0^t g(t,s,u_n(k(s)))ds\,,\,\psi(t) + \int_0^t g(t,s,u_0(k(s)))ds\right) \\ &= d\left(\int_0^t g(t,s,u_n(k(s)))ds\,,\,\int_0^t g(t,s,u_0(k(s)))ds\right) \\ &\le \int_0^t d\left(g(t,s,u_n(k(s)))\,,\,g(t,s,u_0(k(s)))\right)ds, \end{aligned}$$

which prove that Γ is continuous.

Step 3. ΓB_r is equicontinuous:
Let $t_1, t_2 \in [0, 1]$, and $|t_2 - t_1| < h$, h very small, then

$$H\big(\Gamma(u(t_2), v(t_2))\,,\,\Gamma(u(t_1), v(t_1))\big) = \max\left\{d(\Gamma_1 v(t_2), \Gamma_1 v(t_1))\,,\,d(\Gamma_2 u(t_2), \Gamma_2 u(t_1))\right\},$$

and we have

$$
\begin{aligned}
d(\Gamma_1 v(t_2), \Gamma_1 v(t_1)) &= d\left(\phi(t_2) + \int_0^{t_2} f(t_2, s, v(l(s)))ds\,,\ \phi(t_1) + \int_0^{t_1} f(t_1, s, v(l(s)))ds\right) \\
&= d\left(\phi(t_2) + \int_0^{t_1} f(t_2, s, v(l(s)))ds + \int_{t_1}^{t_2} f(t_2, s, v(l(s)))ds\,,\right. \\
&\qquad \left.\phi(t_1) + \int_0^{t_1} f(t_1, s, v(l(s)))ds\right) \\
&\leq d(\phi(t_2), \phi(t_1)) + d\left(\int_0^{t_1} f(t_2, s, v(l(s)))ds, \int_0^{t_1} f(t_1, s, v(l(s)))ds\right) \\
&\quad + d\left(\int_{t_1}^{t_2} f(t_2, s, v(l(s)))ds, \tilde{0}\right) \\
&\leq d(\phi(t_2), \phi(t_1)) + \int_0^{t_1} d\big(f(t_2, s, v(l(s))), f(t_1, s, v(l(s)))\big)ds \\
&\quad + \int_{t_1}^{t_2} d\left(f(t_2, s, v(l(s))), \tilde{0}\right) ds \\
&\leq d(\phi(t_2), \phi(t_1)) + \int_0^{t_1} d\big(f(t_2, s, v(l(s))), f(t_1, s, v(l(s)))\big)ds \\
&\quad + \int_{t_1}^{t_2} \| f(t_2, s, v(l(s)))\| ds..
\end{aligned}
$$

Also

$$
\begin{aligned}
d(\Gamma_2 u(t_2), \Gamma_2 u(t_1)) &= d\left(\psi(t_2) + \int_0^{t_2} g(t_2, s, u(k(s)))ds\,,\ \psi(t_1) + \int_0^{t_1} g(t_1, s, u(k(s)))ds\right) \\
&= d\left(\psi(t_2) + \int_0^{t_1} g(t_2, s, u(k(s)))ds + \int_{t_1}^{t_2} g(t_2, s, u(k(s)))ds\,,\right. \\
&\qquad \left.\psi(t_1) + \int_0^{t_1} g(t_1, s, u(k(s)))ds\right) \\
&\leq d(\psi(t_2), \psi(t_1)) + d\left(\int_0^{t_1} g(t_2, s, u(k(s)))ds, \int_0^{t_1} g(t_1, s, u(k(s)))ds\right) \\
&\quad + d\left(\int_{t_1}^{t_2} g(t_2, s, u(k(s)))ds, \tilde{0}\right) \\
&\leq d(\psi(t_2), \psi(t_1)) + \int_0^{t_1} d\big(g(t_2, s, u(k(s))), g(t_1, s, u(k(s)))\big)ds \\
&\quad + \int_{t_1}^{t_2} d\left(g(t_2, s, u(k(s))), \tilde{0}\right) ds \\
&\leq d(\psi(t_2), \psi(t_1)) + \int_0^{t_1} d\big(g(t_2, s, u(k(s))), g(t_1, s, u(k(s)))\big)ds \\
&\quad + \int_{t_1}^{t_2} \| g(t_2, s, u(k(s)))\| ds.
\end{aligned}
$$

Then

$$
d(\Gamma_1 v(t_2), \Gamma_1 v(t_1)) \longrightarrow 0 \text{ and } d(\Gamma_2 u(t_2), \Gamma_2 u(t_1)) \longrightarrow 0 \text{ as } t_1 \to t_2,
$$

and this means

$$
H\big(\Gamma(u(t_2), v(t_2))\,,\ \Gamma(u(t_1), v(t_1))\big) \longrightarrow 0 \text{ as } t_1 \to t_2,
$$

which shows that ΓB_r is equicontinuous.

Example 1 Let $\phi, \psi : [0, 1] \longrightarrow E^1$ defined by

$$\phi(t)(z) = \begin{cases} z(t+2) & \text{if } z \in \left[0, \frac{1}{t+2}\right], \\ 1 & \text{if } z \in \left[\frac{1}{t+2}, 1 - \frac{1}{t+2}\right], \\ (1-z)(t+2) & \text{if } z \in \left[1 - \frac{1}{t+2}, 1\right], \\ 0 & \text{otherwise.} \end{cases}$$

and

$$\psi(t)(z) = \begin{cases} \dfrac{z}{t+1} & \text{if } z \in [0, t+1], \\ 1 & \text{if } z \in [t+1, t+2], \\ \dfrac{4-z}{2-t} & \text{if } z \in [t+2, 4], \\ 0 & \text{otherwise.} \end{cases}$$

Then

$$[\phi(t)]^\alpha = \left[\frac{\alpha}{t+2}, 1 - \frac{\alpha}{t+2}\right], \quad 0 < \alpha \le 1, \qquad [\phi(t)]^0 = [0, 1],$$

and

$$[\psi(t)]^\alpha = [\alpha(t+1), 4 - \alpha(2-t))], \quad 0 < \alpha \le 1, \qquad [\psi(t)]^0 = [0, 4].$$

It is obvious that the fuzzy functions ϕ and ψ defined above are a continuous fuzzy functions.

Consider the fuzzy functions $f, g : [0, 1] \times [0, 1] \times E^1 \longrightarrow E^1$ defined by

$$f(t, s, y) = \mu \sin(t)\cos(s)y, \qquad \text{and} \qquad g(t, s, x) = \nu \frac{(t+1)(s+1)}{4} (\tilde{a} + x), \qquad 0 < \mu, \nu < 1,$$

where $\tilde{a} \in E^1$ is a fuzzy number defined by

$$\tilde{a}(z) = \begin{cases} \dfrac{1+z}{2}, & \text{if } -1 \le z \le 1 \\ \dfrac{4-z}{3}, & \text{if } 1 \le z \le 4 \\ 0 & \text{otherwise} \end{cases}$$

It is easy to check that all assumptions of the Theorem 1 are satisfied with

$$a = 1, \qquad m \equiv 0, \qquad M = 0, \qquad \alpha = \mu,$$

and

$$b = 4, \qquad n(t, s) = \frac{(t+1)(s+1)}{4}, \qquad N = 3, \qquad \beta = \nu.$$

So, by a suitable application of Theorem 1, the following coupled system of fuzzy Volterra–Urysohn integral equations

$$u(t) = \phi(t) + \int_0^t \mu \sin(t)\cos(s)v(s)ds, \qquad t \in [0, 1] \tag{10}$$

$$v(t) = \psi(t) + \int_0^t \nu \frac{(t+1)(s+1)}{4} (\tilde{a} + u(s))\, ds, \qquad t \in [0, 1] \tag{11}$$

has at least one solution in $\mathscr{C}([0, 1], F^1)$.

Example 2 We can see also the coupled system (10)–(11) as a coupled system of fuzzy Volterra–Hammerstein integral equations with
$G, K : [0, 1] \times [0, 1] \longrightarrow \mathbb{R}$ and $f_1, g_1 : [0, 1] \times E^1 \longrightarrow E^1$ are given by

$$G(t, s) = \sin(t)\cos(s), \qquad K(t, s) = \frac{(t+1)(s+1)}{4},$$

and

$$f_1(t, y) = \mu y, \qquad g_1(t, x) = \nu\,(\tilde{a} + u(s))\,.$$

It is clear that all assumptions of Corollary 2 are satisfied, and we have the existence of at least one solution for the problem (10)–(11).

4 Conclusion

In this paper, we proved the existence of at least one solution of coupled system of fuzzy Volterra–Urysohn integral equations. The coupled system of fuzzy Volterra–Hammerstein integral equations as a particular case. In addition, future work includes expanding the idea signalized in this work and introducing some examples of application. This is a fertile field with vast research projects, which can lead to numerous theories and applications. We plan to devote significant attention to this direction. And we intend to investigate the applications which are based on experimental data (real world problems)of the proposed theory.

Acknowledgements The authors express their sincere thanks to the anonymous referees for numerous helpful and constructive suggestions which have improved the manuscript.

References

1. Aumann, R.J.: Integrals of set-valued functions. J. Math. Anal. Appl. **12**, 1–12 (1965)
2. Balachandran, K., Kanagarajan, K.: Existence of solutions of general nonlinear fuzzy Volterra-Fredholm integral equations. Int. J. Stoch. Anal. **3**, 333–343 (2005)

3. Balachandran, K., Prakash, P.: Existence of solutions of nonlinear fuzzy Volterra-Fredholm integral equations. Indian J. Pure Appl. Math. **33**, 329–343 (2002)
4. Banas, J.: Inegrable solutions of Hammerstein and Urysohn integral equations. J. Aust. Math. Soc. **4**(6), 61–68 (1989)
5. Bica, A.M.: Error estimation in the approximation of the solution of nonlinear fuzzy Fredholm integral equations. Inf. Sci. **178**, 1279–1292 (2008)
6. Bugajewski, D.: On the existence of weak solutions of integral equations in Banach spaces. Comment. Math. Univ. Carol. **35**, 35–41 (1994)
7. Cichon, M., Kubiaczyk, I.: Existence theorem for the Hammerstien integral equation. Discuss. Math. Differ. Incl. Control Optim. **16**, 171–177 (1996)
8. Darwish, M.A., Kashkari, B.S.: On a fuzzy Urysohn integral equation. Int. Math. Forum **8**, 1955–1962 (2013)
9. El-Sayed, A.M.A., Hashem, H.H.G.: Coupled systems of Hammerstien and Urysohn integral equations in reflexive Banach spaces. Differ. Equ. Control Process. **1**, 85–96 (2012)
10. Esfahani, A., Fard, O.S., Bidgoli, T.A.: On the existence and uniqueness of solutions to fuzzy boundary value problems. Ann. Fuzzy Math. Inform. **7**, 15–29 (2014)
11. Gal, C.G., Gal, S.G.: Semigroups of operators on spaces of fuzzy-number-valued functions with applications to fuzzy differential equations. arXiv:1306.3928v1 (2013). Accessed 17 Jun 2013
12. Garai, T., Chakraborty, D., Roy, T.K.: A multi-item periodic review probabilistic fuzzy inventory model with possibility and necessity constraints. Int. J. Bus. Forecast. Mark. Intell. **2**, 175–189 (2016)
13. Garai, T., Chakraborty, D., Roy, T.K.: A multi-item inventory model with fuzzy rough coefficients via fuzzy rough expectation. In: Conference Proceeding Mathematics & Statistics of "Frotiers in Optimization: Theory and Application", vol. 225, pp. 377–394 (2016)
14. Garai, T., Chakraborty, D., Roy, T.K.: Possibility-necessity-credibility measures on generalized intuitionistic fuzzy number and their applications to multi-product manufacturing system. Granul. Comput. **2**, 1–15 (2017)
15. Garai, T., Chakraborty, D., Roy, T.K.: Expected value of exponential fuzzy number and its application to multi-item deterministic inventory model for deteriorating items. J. Uncertain. Anal. Appl. **5**, 1–8 (2017)
16. Garai, T., Chakraborty, D., Roy, T.K.: A fuzzy rough multi-objective multi-item inventory model with both stock-dependent demand and holding cost rate. Granul. Comput. **3**, 1–18 (2018)
17. Kaleva, O.: Fuzzy differential equations. Fuzzy Sets Syst. **24**, 301–317 (1987)
18. Melliani, S., El Allaoui, A., Chadli, L.S.: Relation between fuzzy semigroups and fuzzy dynamical systems. Nonlinear Dyn. Syst. Theory **17**(1), 60–69 (2017)
19. O'Regan, D.: Volterra and Urysohn integral equations in Banach spaces. Int. J. Stoch. Anal. **11**, 449–464 (1998)
20. O'Regan, D., Cho, Y.J., Chen, Y.Q.: Topological Degree Theory and Applications. Series in Mathematical Analysis and Applications, vol. 10. Chapman & Hall/CRC, Boca Raton (2006)
21. Puri, M.L., Ralescu, D.A.: Fuzzy random variables. J. Math. Anal. Appl. **114**, 409–422 (1986)
22. Subrahmanyam, P.V., Sudarsanam, S.K.: A note on fuzzy Volterra integral equations. Fuzzy Sets Syst. **81**, 237–240 (1996)

Optimal Bounded Control on the Velocity Term of a Bilinear Plate Equation

Abderrahman Ait Aadi and El Hassan Zerrik

Abstract This paper investigates a regional optimal control problem of a plate equation described by a bilinear system evolving in a spacial domain Ω. The control is distributed, bounded and acts on the velocity term. Then, we minimize a functional cost constituted of the deviation between a desired state and the reached one only on a subregion ω of Ω and the energy term. The purpose of this study is to prove that a control solution of such problem exists, and characterized as a solution to an optimality system. Numerical approach is given and successfully illustrated by simulations.

1 Introduction

Bilinear systems constitute an important subclass of nonlinear systems. The nonlinearity in mathematical models of bilinear systems appears in the multiplication of state and control in the dynamical process. The motivation for studying the bilinear systems is that numerous real-world problems have bilinear structures (see for instance [9–11]). In addition, bilinear systems can well approximate a lot of nonlinear systems. Furthermore, it is in general more accurate to use a bilinear model to represent the dynamics of a nonlinear system than to use a linear model, and there are several characteristics that render bilinear systems appealing also from the theoretical point of view.

The controllability for distributed bilinear systems has been investigated by numerous authors using different control techniques. In [3], authors studied approximate controllability of bilinear beam and rod equations in the mono-dimensional case. In [12], author proved the multiplicative controllability of various parabolic

A. A. Aadi · E. H. Zerrik (✉)
MACS Team, Department of Mathematics, Moulay Ismail University,
Meknes, Morocco
e-mail: zerrik3@yahho.fr

A. A. Aadi
e-mail: abderrahman.aitaadi@gmail.com

S. Melliani and O. Castillo (eds.), *Recent Advances in Intuitionistic Fuzzy Logic Systems and Mathematics*, Studies in Fuzziness and Soft Computing 395,
https://doi.org/10.1007/978-3-030-53929-0_4

and hyperbolic equations of semilinear type using asymptotic qualitative methods. He investigated the controllability of bilinear parabolic equations with the reaction-diffusion term satisfying Newton's law [13]. For linear plate model, in [14], author studied exact controllability of vibrating plate with boundary control using HUM method, and in [20], author considered exact controllability of vibrating plate for arbitrarily small time.

Optimal control for a class of distributed bilinear systems have been developed in many works: in [6], author proved the existence and gave characterization of an optimal control of a convective-diffusive fluid problem. In [1], authors studied unbounded optimal control for a bilinear system governed by a fourth-order parabolic operator. In [15], author developed optimal control of the convective velocity coefficient of a bilinear heat equation. In [7], authors considered optimal control of heat transfer problem using boundary control. In [8], authors studied optimal control of a wave equation with boundary control. In [4], authors considered the optimal control of Kirchhoff plate equation by controls acting on the state position.

The concept of regional controllability for a distributed linear system evolving on a spacial domain Ω concerns the study of the classical notion of controllability only on a subregion ω of Ω [5]. The main reasons for considering this notion is that it is close to real applications. For example, in the problem of a tunnel furnace when one has to maintain a prescribed temperature only in a subregion of the furnace. Also, it becomes possible to control a system on a subregion of its evolution domain acting out of the subregion. Besides there exist systems which are not controllable on the whole domain Ω but controllable on some subregion. Moreover, controlling a system on a subregion is cheaper than controlling it globally [5]. Regional optimal control of parabolic distributed bilinear systems with unbounded and bounded controls involving the minimization of the final state error and the energy, was considered by [18, 19], they established the existence and gave characterization of an optimal control. In [2], authors studied regional optimal control of a bilinear plate equation by controls without constraints and acting on the state position. In the present work, we examine the regional optimal control of a bilinear plate equation with bounded controls and acting on the velocity term. Then, we prove the existence and we give characterization of an optimal control. Moreover, we develop a numerical approach that leads to an algorithm that we illustrate by simulations. More precisely, let Ω be an open bounded of $\mathbb{R}^2$, with a smooth boundary $\partial\Omega$, for $T > 0$, we denote by $Q = \Omega \times [0, T]$ and $\Sigma = \partial\Omega \times [0, T]$, we consider the following bilinear plate equation

$$\begin{cases} y_{tt}(x,t) + \Delta^2 y(x,t) = u(x,t) y_t(x,t) & \text{on } Q \\ y(x,0) = y_0(x), \;\; y_t(x,0) = y_1(x) & \text{in } \Omega \\ y(x,t) = \dfrac{\partial y}{\partial \nu}(x,t) = 0 & \text{in } \Sigma, \end{cases} \tag{1}$$

where Δ^2 is the bilaplacian operator, ν is the unit outer normal to $\partial\Omega$ and $u \in \mathscr{U}_\rho := \{u \in L^\infty(Q) \mid -\rho \leq u(x,t) \leq \rho\}$ is a scalar control function.

Let us define the following product Hilbert spaces

$$H_0^2(\Omega) = \{y \in H^2(\Omega) \; : \; y = \frac{\partial y}{\partial \nu} = 0 \text{ on } \partial\Omega\}$$

and

$$\mathscr{X} = H_0^2(\Omega) \times L^2(\Omega).$$

For $u \in \mathscr{U}_\rho$, $\tilde{y} = \tilde{y}(u) = (y, y_t)$ is a weak solution of system (1) if $\tilde{y} \in \mathscr{C}([0, T], \mathscr{X})$, $\tilde{y}(0) = (y_0, y_1)$, and $\tilde{y}$ satisfies

$$\langle y_{tt}, \varphi \rangle + B(y, \varphi) = \int_\Omega u y_t \varphi dx, \quad \text{for all } \varphi \in H_0^2(\Omega), \text{ and a.e } 0 \le t \le T, \tag{2}$$

where $B(y, \varphi) = \int_\Omega \Delta y \Delta \varphi dx$ and $\langle ., . \rangle$ denotes the duality pairing of $H^{-2}(\Omega)$ and $H_0^2(\Omega)$.

The system (1) can be written as

$$\begin{cases} \dfrac{d}{dt}\tilde{y}(t) = \mathbb{A}\tilde{y}(t) + \mathbb{B}\tilde{y}(t) \\ \tilde{y}(0) = (y_0, y_1), \end{cases} \tag{3}$$

where $\mathbb{A} : H^4(\Omega) \times H_0^2(\Omega) \longrightarrow \mathscr{X}$

$$\mathbb{A}\tilde{y}(t) = \begin{pmatrix} 0 & I \\ -\Delta^2 & 0 \end{pmatrix} \tilde{y}(t)$$

with domain $\mathscr{D}(\mathbb{A}) = (H^4(\Omega) \cap H_0^2(\Omega)) \times H_0^2(\Omega)$ and the operator $\mathbb{B}$ is given by

$$\mathbb{B}\tilde{y}(t) = \begin{pmatrix} 0 \\ u(t)y_t(t) \end{pmatrix}.$$

Since the operator $\mathbb{A}$ generates a strongly continuous unitary group on $\mathscr{X}$ and $\mathbb{B}$ is bounded on $\mathscr{X}$. Then system (3) has a unique weak solution $\tilde{y}(t) \in \mathscr{C}([0, T], \mathscr{X})$ (see [16]).

Let us consider a non-empty subset $\omega \subset \Omega$, with a positive Lebesgue measure, we define $\chi_\omega : L^2(\Omega) \longrightarrow L^2(\omega)$ the restriction operator to ω, and χ_ω^* is the adjoint operator of χ_ω given by

$$(\chi_\omega^* y)(x) = \begin{cases} y(x) & \text{if } x \in \omega \\ 0 & \text{else } x \in \Omega \backslash \omega. \end{cases}$$

We consider the optimal control problem

$$\min_{u \in \mathscr{U}_\rho} J(u), \tag{4}$$

where

$$J(u) = \frac{1}{2}\int_{\omega}\int_{0}^{T}\Big(\chi_{\omega}y(x,t) - y_d(x)\Big)dtdx + \frac{\beta}{2}\int_{Q}u^2(x,t)dxdt, \tag{5}$$

with $y_d \in L^2(\omega)$ is a desired state and β is a positive constant.

The paper is organized as follows: in Sect. 2, we prove the existence of an optimal control solution of problem (4). In Sect. 3, we give characterization of control solution of problem (4). In Sect. 4, we develop a numerical approach that leads to an algorithm we illustrate by numerical simulations.

2 Existence of an Optimal Control

This section is devoted to the existence of an optimal control solution of problem (4).

First, we give an apriori estimate needed for the existence of an optimal control.

Lemma 1 *For $\tilde{y}_0 = (y_0, y_1) \in \mathscr{X}$ and $u \in \mathscr{U}_\rho$, then the weak solution $\tilde{y}$ of system (1) satisfies the estimate*

$$\|\tilde{y}\|_{\mathscr{C}([0,T],\, \mathscr{X})} \leq M\big(1+\rho T\big)^{\frac{1}{2}}e^{\rho CT}, \tag{6}$$

where $M = \|\tilde{y}_0\|_{\mathscr{X}}$ and C is a positive constant.

Proof Let $(y_0, y_1) \in \mathscr{X}$, since $\mathscr{D}(\mathbb{A})$ is dense in $\mathscr{X}$, there exist sequences (y_0^n, y_1^n) in $\mathscr{D}(\mathbb{A})$ and $u^n \in \mathscr{U}_\rho \cap \mathscr{C}^2(Q)$ such that

$$\begin{aligned} (y_0^n, y_1^n) &\longrightarrow (y_0, y_1) \;\text{ strongly in }\; \mathscr{X}, \\ u^n &\longrightarrow u \;\text{ strongly in }\; L^2(Q). \end{aligned}$$

Denote by $\tilde{y}^n$ the weak solution of system (1) corresponding to the initial data (y_0^n, y_1^n) with control u^n.

Multiplying the Eq. (1) by y_t and integrating over $\Omega \times (0, \tau)$, we obtient

$$\begin{aligned} 0 &= \int_0^\tau\int_\Omega\Big(y_{tt}^n\, y_t^n + \Delta^2 y^n\, y_t^n - u^n(y_t^n)^2\Big)dxdt \\ &= \int_0^\tau\int_\Omega\Big(\frac{1}{2}\frac{d}{dt}\Big[(y_t^n)^2 + (\Delta y^n)^2\Big] - u^n(y_t^n)^2\Big)dxdt \\ &= \int_0^\tau\int_\Omega\frac{1}{2}\frac{d}{dt}(y_t^n)^2dxdt + \frac{1}{2}\int_0^\tau\frac{d}{dt}B(y^n, y^n)dt - \int_0^\tau\int_\Omega u^n(y_t^n)^2dxdt. \end{aligned}$$

Thus, we have

$$\frac{1}{2}\int_{\Omega}(y_t^n)^2(x,\tau)dx + B(y^n,y^n)(\tau) = \frac{1}{2}\|y_1^n\|^2_{L^2(\Omega)} + \frac{1}{2}B(y_0^n,y_0^n) + \int_0^{\tau}\int_{\Omega}u^n(y_t^n)^2dxdt$$
$$\leq \frac{1}{2}\|\tilde{y}(0)^n\|^2_{\mathscr{X}} + \rho\int_0^{\tau}\|\tilde{y}(t)^n\|^2_{\mathscr{X}}\,dt.$$

Gronwall's Inequality gives

$$\sup_{0\leq t\leq T}\left(\int_{\Omega}\left(y_t^n\right)^2(x,\tau)dx + B(y^n,y^n)(\tau)\right) \leq \|\tilde{y}(0)^n\|^2_{\mathscr{X}}\left(1+2\rho T\right)e^{\rho CT}. \quad (7)$$

We pass to the limit and obtain (6) for $\tilde{y}(t)$.

Now, we obtain the existence of an optimal control.

Theorem 1 *There exists an optimal control $u^* \in \mathscr{U}_\rho$, solution of problem (4).*

Proof Let u^n be a minimizing sequence in $\mathscr{U}_\rho$, such that

$$\lim_{n\to+\infty} J(u^n) = \inf_{u\in\mathscr{U}_\rho} J(u). \quad (8)$$

By Lemma 1, we have the estimate

$$\|y^n\|^2_{H_0^2(\Omega)} + \|y_t^n\|^2_{L^2(\Omega)} \leq Me^{\rho CT}. \quad (9)$$

Using the weak solution form (2) and (9), we conclude that

$$\|y_{tt}^n\|^2_{H^{-2}(\Omega)} \leq Me^{\rho CT}. \quad (10)$$

From (9) and (10), we deduce the following convergence properties

$$y^n \rightharpoonup y^* \quad \text{weakly}^* \text{ in } L^\infty([0,T],H_0^2(\Omega)) \quad (11)$$
$$y_t^n \to y_t^* \quad \text{weakly}^* \text{ in } L^\infty([0,T],L^2(\Omega)) \quad (12)$$
$$y_{tt}^n \rightharpoonup y_{tt}^* \quad \text{weakly}^* \text{ in } L^\infty([0,T],H^{-2}(\Omega))$$
$$u^n \rightharpoonup u^* \quad \text{weakly in } L^2(Q). \quad (13)$$

Since $\mathscr{U}_\rho$ is a closed and convex subset of $L^\infty(Q) \subset L^2(Q)$, $\mathscr{U}_\rho$ is weakly closed in $L^2(Q)$.

Then $u^* \in \mathscr{U}_\rho \subset L^2(Q)$. On the other hand, since $-\rho \leq u^n(x,t) \leq \rho$ for all n, $u^n \rightharpoonup u^{**}$ weakly* in $L^\infty(Q)$, and hence $u^n \rightharpoonup u^{**}$ weakly in $L^2(Q)$.

By the uniqueness of the weak limit, we obtain $u^* = u^{**}$ and $u^* \in \mathscr{U}_\rho \subset L^\infty(Q)$.

In other hand, we have y^n satisfies the weak form

$$\int_0^T\langle y_{tt}^n,\phi\rangle dt + \int_0^T B(y^n,\phi)dt = \int_0^T\langle u^n y_t^n,\phi\rangle dt, \quad \text{for all } \phi\in H_0^2(\Omega). \quad (14)$$

Using a compactness result from [17], we have

$$y^n \to y^* \text{ strongly in } \mathscr{C}([0,T], L^2(\Omega)). \tag{15}$$

For $\phi \in H_0^2(\Omega) \subset \mathscr{C}(\Omega)$, by (12), (13) and (15), we obtain

$$\lim_{n\to+\infty} \int_0^T \langle u^n y_t^n, \phi\rangle dt = \int_0^T \langle u^* y_t^*, \phi\rangle dt \tag{16}$$

Taking the limit as $n \to +\infty$ in (14) and using (11), (12), (13) and (16), we conclude the formula

$$\int_0^T \langle y_{tt}^*, \phi\rangle dt + \int_0^T B(y^*, \phi) dt = \int_0^T \langle u^* y_t^*, \phi\rangle dt, \quad \text{for all } \phi \in H_0^2(\Omega).$$

Thus $y^* = y(u^*)$ is the solution of state equation (1) with control u^*.
Since

$$J(u^*) = \frac{1}{2}\int_\omega \int_0^T \Big(\chi_\omega y^*(x,t) - y_d(x)\Big) dt dx + \frac{\beta}{2}\int_Q (u^*)^2(x,t) dx dt,$$

using lower-semicontinuity of L^2 norm with respect to weak convergence, we have

$$\begin{aligned} J(u^*) &\le \frac{1}{2}\lim_{n\to+\infty}\int_\omega \int_0^T \Big(\chi_\omega y^n(x,t) - y_d(x)\Big) dt dx + \frac{\beta}{2}\liminf_{n\to+\infty}\int_Q (u^n)^2(x,t) dx dt \\ &\le \liminf_{n\to+\infty} J(u^n) \\ &= \inf_{u\in\mathscr{U}_\rho} J(u). \end{aligned}$$

Finally, we conclude that u^* is an optimal control.

3 Optimal Control Formulas

In this section, we give characterization of an optimal control solution of problem (4).

Let now examine the differentiability of the mapping $u \to \tilde{y}(u)$.

Lemma 2 *The mapping $u \in \mathscr{U}_\rho \to \tilde{y}(u) \in \mathscr{C}([0,T], \mathscr{X})$ is differentiable in the following sense*

$$\frac{\tilde{y}(u+\varepsilon h) - \tilde{y}(u)}{\varepsilon} \rightharpoonup \tilde{\lambda} \text{ weakly in } L^\infty([0,T], \mathscr{X}) \text{ as } \varepsilon \to 0, \text{ for any } u,\ u+\varepsilon h \in \mathscr{U}_\rho.$$

Moreover, $\tilde{\lambda} = (\lambda, \lambda_t)$ is a weak solution of the following system

$$\begin{cases} \lambda_{tt}(x,t) + \Delta^2\lambda(x,t) = u(x,t)\lambda_t(x,t) + h(x,t)y_t(x,t) & on \;\; Q \\ \lambda(x,0) = \lambda_t(x,0) = 0 & in \;\; \Omega \\ \lambda(x,t) = \dfrac{\partial \lambda}{\partial \nu}(x,t) = 0 & in \;\; \Sigma \end{cases} \tag{17}$$

Proof Denote $\tilde{y}^\varepsilon = \tilde{y}(u+\varepsilon h) = (y^\varepsilon, y_t^\varepsilon)$ and $\tilde{y} = \tilde{y}(u)$. Then $\left(\frac{\tilde{y}^\varepsilon - \tilde{y}}{\varepsilon}\right)$ is a weak solution of

$$\begin{cases} \left(\dfrac{y^\varepsilon - y}{\varepsilon}\right)_{tt} + \Delta^2\left(\dfrac{y^\varepsilon - y}{\varepsilon}\right) = u\left(\dfrac{y^\varepsilon - y}{\varepsilon}\right)_t + hy_t^\varepsilon & \text{on } Q \\ \left(\dfrac{y^\varepsilon - y}{\varepsilon}\right)(x,0) = \left(\dfrac{y^\varepsilon - y}{\varepsilon}\right)_t (x,0) = 0 & \text{in } \Omega \\ \left(\dfrac{y^\varepsilon - y}{\varepsilon}\right) = \dfrac{\partial}{\partial \nu}\left(\dfrac{y^\varepsilon - y}{\varepsilon}\right) = 0 & \text{in } \Sigma \end{cases}$$

Using Lemma 1 with source term hy_t^ε, we obtain

$$\| \frac{\tilde{y}^\varepsilon - \tilde{y}}{\varepsilon} \|_{\mathscr{C}([0,T],\mathscr{X})} \leq \|hy_t^\varepsilon\|_{L^2(Q)} e^{\rho CT}.$$

y_t^ε satisfies the estimate

$$\|hy_t^\varepsilon\|_{L^2(Q)} \leq T\|h\|_\infty \|\tilde{y}^\varepsilon\|_{\mathscr{C}([0,T],\mathscr{X})} \leq (1+\rho T)^{1/2} e^{\rho CT} \|\tilde{y}(0)\|_{\mathscr{X}}.$$

Hence, we have

$$\frac{\tilde{y}^\varepsilon - \tilde{y}}{\varepsilon} \rightharpoonup \tilde{\lambda} \text{ weakly in } L^\infty([0,T],\mathscr{X}) \;\; as \;\; \varepsilon \to 0.$$

We conclude that λ is a weak solution of system (17).

3.1 *Time and Space Control Dependent*

Here, we give characterization of an optimal control that depend on time and space.

Theorem 2 *An optimal control solution of problem (4) is given by the formula*

$$u^*(x,t) = \max\left(-\rho, \min\left(-\frac{1}{\beta}\chi_\omega^* \chi_\omega y_t^*(x,t) p(x,t), \rho\right)\right), \tag{18}$$

where $(p, p_t) \in \mathscr{C}([0,T],\mathscr{X})$ *is the weak solution of the adjoint system*

$$\begin{cases} p_{tt}(x,t) + \Delta^2 p(x,t) = u^*(x,t)p_t(x,t) + y^*(x,t) - \chi_\omega^* y_d(x) & on \ Q \\ p(x,T) = p_t(x,T) = 0 & in \ \Omega \\ p(x,t) = \dfrac{\partial p}{\partial \nu}(x,t) = 0 & in \ \Sigma. \end{cases} \tag{19}$$

Proof The proof of existence of the solution to the adjoint system is similar to the proof of existence of solution of the state equation (1) since the source term $(y^* - \chi_\omega^* y_d) \in \mathscr{C}([0,T], L^2(\Omega))$.

We now proceed to characterize the optimal control in terms of the state (y, y_t) of system (1) and the one (p, p_t) of the adjoint system (19). Let $u^* \in \mathscr{U}_\rho$ be an optimal control and $\tilde{y} = \tilde{y}(u^*)$ be the corresponding optimal solution. Let $u^* + \varepsilon h \in \mathscr{U}_\rho$ for $\varepsilon > 0$ and $\tilde{y}^\varepsilon = \tilde{y}(u^* + \varepsilon h)$ be the corresponding weak solution of system (1). We compute the directional derivative of the cost functional J with respect to u^* in the direction of h.

Since J reaches its minimum at u^*, we have

$$\begin{aligned} 0 \le & \lim_{\varepsilon \to 0^+} \frac{J(u^* + \varepsilon h) - J(u^*)}{\varepsilon} \\ = & \lim_{\varepsilon \to 0^+} \frac{1}{2} \int_\omega \int_0^T \frac{(\chi_\omega y^\varepsilon - y_d)^2 - (\chi_\omega y^* - y_d)^2}{\varepsilon} dt dx \\ + & \lim_{\varepsilon \to 0^+} \frac{\beta}{2} \int_Q \frac{(u^* + \varepsilon h)^2 - u^{*2}}{\varepsilon} dx dt \\ = & \lim_{\varepsilon \to 0^+} \frac{1}{2} \int_\omega \int_0^T \chi_\omega \frac{(y^\varepsilon - y^*)}{\varepsilon} (\chi_\omega y^\varepsilon + \chi_\omega y^* - 2 y_d) dt dx \\ + & \lim_{\varepsilon \to 0^+} \frac{\beta}{2} \int_Q (2hu^* + \varepsilon h^2) dx dt. \end{aligned}$$

Then,

$$\lim_{\varepsilon \to 0^+} \frac{J(u^* + \varepsilon h) - J(u^*)}{\varepsilon} = \int_Q \chi_\omega^* \chi_\omega \lambda \, (y^* - \chi_\omega^* y_d) dx dt + \beta \int_Q h u^* dx dt,$$

where λ is solution of system (17).

Using the adjoint system (19), and the solution of system (17), we obtain

$$\begin{aligned} 0 \le & \int_\Omega \chi_\omega^* \chi_\omega \int_0^T \lambda(x,t)(p_{tt}(x,t) + \Delta^2 p(x,t) - u^*(x,t) p_t(x,t)) dt dx \\ + & \beta \int_Q h(x,t) u^*(x,t) dx dt \\ = & \int_\Omega \chi_\omega^* \chi_\omega \int_0^T (\lambda_{tt}(x,t) + \Delta^2 \lambda(x,t) - u^*(x,t) \lambda_t(x,t)) p(x,t) \, dt dx \\ + & \beta \int_Q h(x,t) u^*(x,t) \, dx dt \end{aligned}$$

$$= \int_\Omega \chi_\omega^* \chi_\omega \int_0^T h(x,t) y_t^*(x,t)\, p(x,t) dt dx + \beta \int_Q h(x,t) u^*(x,t) dt dx$$
$$= \int_Q h(x,t) \left(\beta u^*(x,t) + \chi_\omega^* \chi_\omega y_t^*(x,t)\, p(x,t) \right) dx dt.$$

Using a standard control argument based on the choices for the variation $h(x,t)$, an optimal control is given by

$$u^*(x,t) = \max\left(-\rho, \min\left(-\frac{1}{\beta} \chi_\omega^* \chi_\omega y_t^*(x,t) p(x,t), \rho \right) \right).$$

3.2 *Time or Space Control Dependent*

In this part, we will treat two cases that the controls are only functions of time or space.

First, we deal with the time dependent case.

• **Case 1:** $u = u(t)$

We take the admissible controls set

$$\mathcal{U}_\rho = \{ u \in L^\infty(0,T) \mid -\rho \le u \le \rho \} \tag{20}$$

and the functional cost

$$J(u) = \frac{1}{2} \int_\omega \int_0^T \left(\chi_\omega y(x,t) - y_d(x) \right)^2 dt dx + \frac{\beta}{2} \int_0^T u^2(t) dt. \tag{21}$$

Corollary 1 *Under conditions (20) and (21), an optimal control is given by*

$$u(t) = \max\left(-\rho, \min\left(-\frac{1}{\beta} \int_\Omega \chi_\omega^* \chi_\omega y_t(x,t) p(x,t) dx, \rho \right) \right), \tag{22}$$

where (y, y_t) *is the solution of*

$$\begin{cases} y_{tt}(x,t) + \Delta^2 y(x,t) = u(t) y_t(x,t) & on\ Q \\ y(x,0) = y_0(x), \;\; y_t(x,0) = y_1(x) & in\ \Omega \\ y(x,t) = \dfrac{\partial y}{\partial \nu}(x,t) = 0 & in\ \Sigma, \end{cases}$$

and (p, p_t) *is the solution of*

$$\begin{cases} p_{tt}(x,t) + \Delta^2 p(x,t) = u(t)p_t(x,t) + y(x,t) - \chi_\omega^* y_d(x) & on\ Q \\ p(x,T) = p_t(x,T) = 0 & in\ \Omega \\ p(x,t) = \dfrac{\partial p}{\partial \nu}(x,t) = 0 & in\ \Sigma. \end{cases}$$

Proof Using the same notations as in the proof of Theorem 2. Let $h = h(t)$ be an arbitrary function with $u + \varepsilon h \in \mathcal{U}_\rho$ for small ε.

We have

$$\int_0^T h(t)\Big(\int_\Omega \chi_\omega^* \chi_\omega y_t(x,t)p(x,t)dx + \beta u(t)\Big)dt \geq 0.$$

By using a standard control argument concerning the sign of the variation h, we obtain that

$$u(t) = \max\Big(-\rho, \min\Big(-\frac{1}{\beta}\int_\Omega \chi_\omega^*\chi_\omega y_t(x,t)p(x,t)dx, \rho\Big)\Big).$$

• **Case 2:** $u = u(x)$

We take the admissible controls set

$$\mathcal{U}_\rho = \{u \in L^\infty(\Omega) \mid -\rho \leq u \leq \rho\} \tag{23}$$

and the functional cost

$$J(u) = \frac{1}{2}\int_\omega \int_0^T \Big(\chi_\omega y(x,t) - y_d(x)\Big)^2 dxdt + \frac{\beta}{2}\int_\Omega u^2(x)dx. \tag{24}$$

Corollary 2 *Under conditions (23) and (24), an optimal control satisfies*

$$u(x) = \max\Big(-\rho, \min\Big(-\frac{1}{\beta}\int_0^T \chi_\omega^*\chi_\omega y_t(x,t)p(x,t)dt, \rho\Big)\Big), \tag{25}$$

where (y, y_t) *is the solution of*

$$\begin{cases} y_{tt}(x,t) + \Delta^2 y(x,t) = u(x)y_t(x,t) & on\ Q \\ y(x,0) = y_0(x),\ y_t(x,0) = y_1(x) & in\ \Omega \\ y(x,t) = \dfrac{\partial y}{\partial \nu}(x,t) = 0 & in\ \Sigma, \end{cases}$$

and (p, p_t) *is the solution of*

$$\begin{cases} p_{tt}(x,t) + \Delta^2 p(x,t) = u(x)p_t(x,t) + y(x,t) - \chi_\omega^* y_d(x) & on \;\; Q \\ p(x,T) = p_t(x,T) = 0 & in \;\; \Omega \\ p(x,t) = \dfrac{\partial p}{\partial \nu}(x,t) = 0 & in \;\; \Sigma \end{cases}$$

Proof Using the same notations as in the proof of Theorem 2. Let $h = h(x)$ be an arbitrary function with $u + \varepsilon h \in \mathcal{U}_\rho$ for small ε.

We have

$$\int_\Omega h(x)\Big(\int_0^T \chi_\omega^* \chi_\omega y_t(x,t)p(x,t)dt + \beta u(x)\Big)dx \geq 0.$$

A standard control argument gives

$$u(x) = \max\Big(- \rho, \min\Big(- \frac{1}{\beta}\int_0^T \chi_\omega^* \chi_\omega y_t(x,t)p(x,t)dt, \rho\Big)\Big).$$

4 Numerical Approach and Simulations

We have seen that the solution of problem (4) is given by

$$u^*(x,t) = \max\Big(-\rho, \min\Big(-\frac{1}{\beta}\chi_\omega^* \chi_\omega y_t^*(x,t)p(x,t), \rho\Big)\Big),$$

where y^* is the solution of system (1) associated with the control u^* and p is the solution of the adjoint system (19).

The computation of such control can be realised by the following formula

$$\begin{cases} u_{n+1}^*(x,t) = \max\Big(-\rho, \min\Big(-\frac{1}{\beta}\chi_\omega^* \chi_\omega (y_t^*)_n(x,t)p_n(x,t), \rho\Big)\Big) \\ u_0^* = 0, \end{cases} \quad (26)$$

where y_n^* is the solution of system (1) associated to the control u_n^* and p_n is the solution of the adjoint system (19). This allows to consider the following algorithm:

Step 1: Initials system data.

- $\odot$ Initial state y_0, y_1 and u_0^*.
- $\odot$ Desired state y_d.
- $\odot$ Threshold accuracy ϵ, subregion ω and time T.

Step 2 :

- $\odot$ Solving the equation (1) gives y_n^*.
- $\odot$ Solving the equation (19) gives p_n.
- $\odot$ Calculate u_{n+1}^* by the formula (26).

Until $\|u_{n+1}^* - u_n^*\|_{L^\infty(Q)} \leq \epsilon$ stop, else $n = n + 1$ go to step 2.
Step 3 : The control u_n^* is optimal.

Simulations
On $\Omega =]0, 1[\times]0, 1[$, consider a bilinear plate equation

$$\begin{cases} y_{tt}(x_1, x_2, t) + \Delta^2 y(x_1, x_2, t) = u(t)y_t(x_1, x_2, t) & \text{on } Q \\ y(x_1, x_2, 0) = y_0(x_1, x_2), \quad y_t(x_1, x_2, 0) = y_1(x_1, x_2) & \text{in } \Omega \\ y(x_1, x_2, t) = \dfrac{\partial y}{\partial \nu}(x_1, x_2, t) = 0 & \text{in } \Sigma, \end{cases} \tag{27}$$

and consider problem (4) with the control set $\mathscr{U}_\rho = \{u \in L^\infty(0, T) \ : \ -\rho \leq u(t) \leq \rho\}$.

An optimal control solution of problem (4) is given by the following formula

$$u^*(t) = \max\left(-\rho, \min\left(-\frac{1}{\beta}\int_0^1\int_0^1 \chi_\omega^* \chi_\omega y_t^*(x_1, x_2, t) p_n(x_1, x_2, t) dx_1 dx_2, \rho\right)\right),$$

where y^* is solution of system (27) associated to the control u^* and p is the solution of the following adjoint system

$$\begin{cases} p_{tt}(x_1, x_2, t) + \Delta^2 p(x_1, x_2, t) = u^*(t)p_t(x_1, x_2, t) + y^*(x_1, x_2, t) - \chi_\omega^* y_d(x_1, x_2) & \text{on } Q \\ p(x_1, x_2, T) = p_t(x_1, x_2, T) = 0 & \text{in } \Omega \\ p(x_1, x_2, t) = \dfrac{\partial p}{\partial \nu}(x_1, x_2, t) = 0 & \text{in } \Sigma \end{cases}$$

We take $T = 1, \rho = 1, \beta = 0.1, y_0(x_1, x_2) = x_1x_2(1 - x_1)(1 - x_2), y_1(x_1, x_2) = 0$, and the desired state $y_d(x_1, x_2) = 0$ on $\omega \subset \Omega$.

Applying the previous algorithm, with $\varepsilon = 10^{-4}$ we obtain.

- For $\omega =]0.7, 1[\times]0, 1[$

Figure 1 shows that the reached state is very close to the desired state on ω and the evolution of control function is given by the Fig. 2. The desired state is obtained with error $\|\chi_\omega y^*(., T)\|^2_{L^2(\omega)} = 4.2 \times 10^{-4}$, and cost $J(u^*) = 3.8 \times 10^{-3}$.

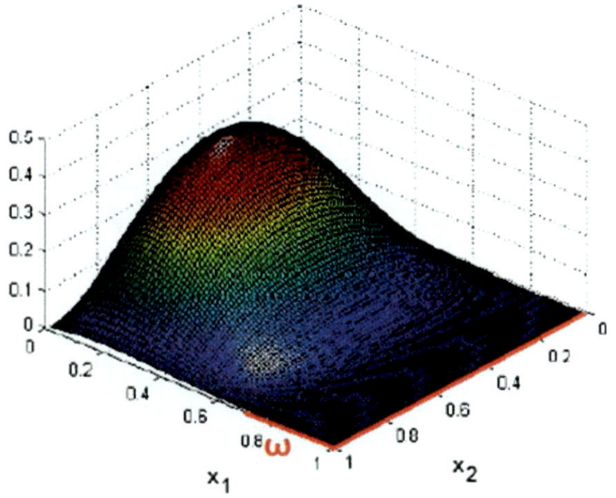

Fig. 1 Desired and final state on Ω

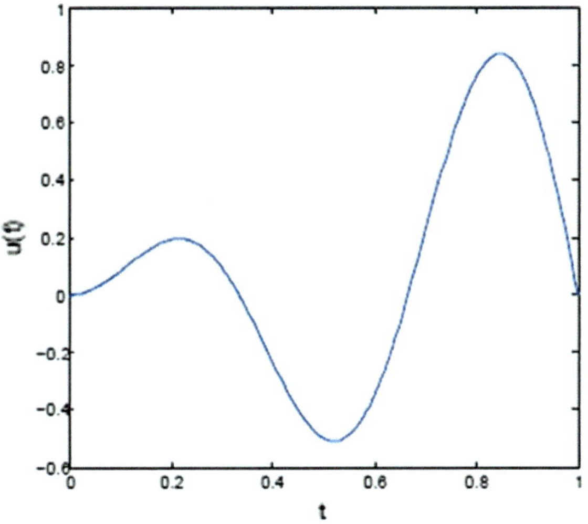

Fig. 2 Evolution of control function

- For $\omega =]0.5, 1[\times]0, 1[$

Figure 3 shows that the reached state is very close to the desired state on ω and the evolution of control function is given by the Fig. 4. The desired state is obtained with error $\|\chi_\omega y^*(., T)\|^2_{L^2(\omega)} = 7.5 \times 10^{-4}$, and cost $J(u^*) = 6.4 \times 10^{-3}$.

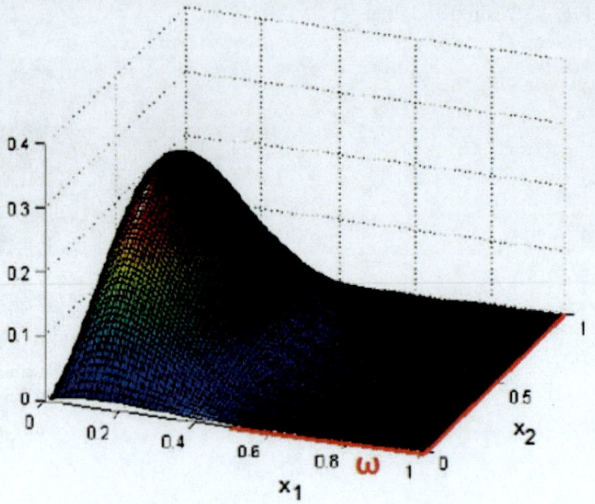

Fig. 3 Desired and final state on Ω

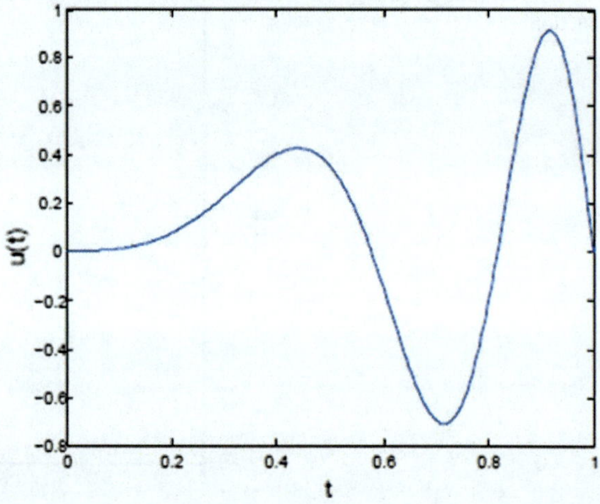

Fig. 4 Evolution of control function

5 Conclusion

We consider a regional optimal control problem of a bilinear plate equation with bounded controls and acting on the velocity term. The existence of an optimal control is proved and characterised as a solution of an optimality system. The obtained results are successfully tested through numerical examples. Questions are still open, as is the case of boundary optimal control of plate equation.

Acknowledgements This work was carried out with the help of the Academy Hassan II of Sciences and Technology.

References

1. Addou, A., Benbrik, A.: Existence and uniqueness of optimal control for a distributed parameter bilinear systems. J. Dyn. Control Syst. **8**, 141–152 (2002)
2. Ait Aadi, A., Zerrik, E.: Regional optimal control for a bilinear plate equation. In: Proceedings IEEE Xplore. https://doi.org/10.1109/CoDIT.2019.8820379
3. Ball, J.M., Marsden, J.E., Slemrod, M.: Controllability for distributed bilinear systems. SIAM J. Control Opt. **20**, 575–597 (1982)
4. Bradley, M.E., Lenhart, S.: Bilinear optimal control of a Kirchhoff plate. J. Syst. Control Lett. **22**, 27–38 (1994)
5. El Jai, A., Simon, M.C., Zerrik, E., Pritchard, A.J.: Regional controllability of distributed parameter systems. Int. J. Control **62**, 1351–1365 (1995)
6. Lenhart, S.: Optimal control of a convective-diffuusive fluid problem. J. Math. Models Methods Appl. Sci. **5**, 225–237 (1995)
7. Lenhart, S., Wilson, D.G.: Optimal control of a heat transfer problem with convective boundary condition. J. Opt. Theory Appl. **79**, 581–597 (1993)
8. Lenhart, S., Protopopescu, V., Yong, J.: Optimal control of a reflection boundary coefficient in an acoustic wave equation. J. Appl. Anal. **68**, 179–194 (1998)
9. Mohler, R.R.: Natural bilinear control processes. IEEE Trans. Syst. Sci. Cybern. **6**, 192–197 (1970)
10. Mohler, R.R.: Bilinear Control Processes: With Applications to Engineering, Ecology, and Medicine. Academic, New York (1973)
11. Mohler, R.R., Ruberti, A.: Theory and Applications of Variable Structure Systems. Academic, New York (1972)
12. Khapalov, A.Y.: Controllability of Partial Differential Equations Governed by Multiplicative Controls. Springer, Berlin (2010)
13. Khapalov, A.Y.: On bilinear controllability of the parabolic equation with the reaction-diffusion term satisfying Newton's law. J. Comp. Appl. Math. **21**, 1–23 (2002)
14. Lions, J.L.: Contrôlabilité exacte, perturbations et stabilisation des systèmes distribués. Masson, Paris (1988)
15. Joshi, H.R.: Optimal control of the convective velocity coefficient in a parabolic problem. J. Nonlinear Anal. **63**, 1383–1390 (2005)
16. Pazy, A.: Semi-groups of Linear Operators and Applications to Partial Differential Equations. Springer, New York (1983)
17. Simon, J.: Compact sets in the space in $L^p(0, T; B)$. Ann. Mat. Pura Appl. **146**, 65–96 (1987)
18. Ztot, K., Zerrik, E., Borray, H.: Regional control problem for distributed bilinear systems. Int. J. Appl. Math. Comput. Sci. **21**, 499–508 (2011)
19. Zerrik, E., Ould Sidi, M.: Regional controllability for infinite dimensional distributed bilinear systems. Int. J. Control **84**, 2108–2116 (2011)
20. Zuazua, E.: Contrôlabilité exacte de quelques modèles de plaques en un temps arbitrairement petit. C. R. A. S, Parie. serie I. Math (1988)

Numerical Solution of Intuitionistic Fuzzy Differential Equations by Runge–Kutta Verner Method

Bouchra Ben Amma, Said Melliani, and Lalla Saadia Chadli

Abstract In this paper we present a numerical algorithm for solving intuitionistic fuzzy differential equations. We discuss in detail a numerical method based on a Runge–Kutta Verner method. Finally, numerical example is presented to illustrate this method and compared with the numerical methods in [9, 12]. It is observed that the Runge–Kutta Verner method yield more accurate results than the existing methods.

Keywords Numerical solution · Intuitionistic fuzzy differential equations · Runge–kutta verner method

1 Introduction

Generalizations of fuzzy sets theory [23] is considered to be one of intuitionistic fuzzy set (IFS). Later on Atanassov generalized the concept of fuzzy set and introduced the idea of intuitionistic fuzzy set [1, 3]. Atanassov [2] explored the concept of fuzzy set theory by intuitionistic fuzzy set (IFS) theory. Now-a-days, IFSs are being studied extensively and being used in different disciplines of Science and Technology. Amongst the all research works mainly on IFS we can include in [13, 16]. They are very necessary and powerful tool in modeling imprecision, valuable applications of IFSs have been flourished in many different fields [14, 19–22].

As far as we know, the progress of the research on intuitionistic fuzzy differential equations are very slow, since the first publication on intuitionistic fuzzy differen-

B. Ben Amma · S. Melliani (✉) · L. S. Chadli
Laboratory of Applied Mathematics and Scientific Computing, Faculty of Sciences and Technologies, Sultan Moulay Slimane University, BP 523, 23000 Beni Mellal, Morocco
e-mail: saidmelliani@gmail.com

B. Ben Amma
e-mail: bouchrabenamma@gmail.com

L. S. Chadli
e-mail: sa.chadli@yahoo.fr

S. Melliani and O. Castillo (eds.), *Recent Advances in Intuitionistic Fuzzy Logic Systems and Mathematics*, Studies in Fuzziness and Soft Computing 395,
https://doi.org/10.1007/978-3-030-53929-0_5

tial equations was introduced by Melliani and Chadli in [15], recently the authors established, the Cauchy problem for intuitionistic fuzzy differential equations [7], intuitionistic fuzzy functional differential equations [8], the existence and uniqueness of intuitionistic fuzzy solutions for intuitionistic fuzzy partial functional differential equations [10], integral boundary value problem for intuitionistic fuzzy partial hyperbolic differential equations [11]. They proved the existence and uniqueness of the intuitionistic fuzzy solution for these intuitionistic fuzzy differential equations using different concepts.

The still-standing problem in the theory of intuitionistic fuzzy differential equations is to find implementable numerical methods. Much more effort has been made in this direction as well. There are some applications of numerical methods such as the intuitionistic fuzzy Euler and Taylor methods [4], Runge–Kutta of Order Four [6], Runge–Kutta Gill [9], Runge–Kutta Nyström [12], Variational iteration method [18], Adams-Bashforth, Adams-Moulton and Predictor-Corrector methods [5]. In this paper, intuitionistic fuzzy Cauchy problem is solved numerically by Runge–Kutta Verner method and the numerical results are compared with the existing methods in [9, 12].
The paper is organized as follows: In Sect. 2, some basic definitions and results are brought. Section 3 contains intuitionistic fuzzy differential equation whose numerical solution is the main interest of this paper. Solving numerically the intuitionistic fuzzy differential equation by Runge–Kutta Verner method in Sect. 4. An example is presented in Sect. 5, and finally conclusion is drawn.

2 Preliminaries

2.1 Notations and Definitions

Consider the initial value problem:

$$\begin{cases} x'(t) = f(t, x(t)), & t \in [t_0, T] \\ x(t_0) = x_0 \end{cases} \tag{1}$$

The basis of all Runge–Kutta method is to express the difference between the value of y at t_{n+1} and t_n as

$$x_{n+1} - x_n = \sum_{i=1}^{i=m} w_i k_i, \tag{2}$$

where for $i = 1, 2, 3, \dots, m$ the w_i's are constants and

$$k_i = hf\big(t_n + c_i h, x_n + \sum_{j=1}^{i-1} \beta_{ij} k_j\big) \tag{3}$$

Equation (2) is to be exact for powers of h through h^m, because it is to be coincident with Taylor series of order m. Therefore, the truncation error T_m can be written as

$$T_m = \gamma_m h^{m+1} + O(h^{m+2})$$

The true value of γ_m will generally be much less than the bound of Theorem 1. Thus, if the $O(h^{m+2})$ term is small compared to $\gamma_m h^{m+1}$ for small h, then the bound on $\gamma_m h^{m+1}$ will usually be a bound on the error as a whole. The famous nonzero constants c_i, β_{ij} in the Runge Kutta Verner Method are:

$$\begin{aligned}
c_1 = 0, c_2 = \frac{1}{6}, c_3 = \frac{4}{15}, c_4 = \frac{2}{3}, c_5 = \frac{5}{6}, c_6 = 1, c_7 = \frac{1}{15}, c_8 = 1,\\
\beta_{21} = \frac{1}{6}, \beta_{31} = \frac{4}{15}, \beta_{32} = \frac{16}{75}, \beta_{41} = \frac{5}{6}, \beta_{42} = -\frac{8}{3}, \beta_{43} = \frac{5}{2},\\
\beta_{51} = -\frac{165}{64}, \beta_{52} = \frac{55}{6}, \beta_{53} = -\frac{425}{64}, \beta_{54} = \frac{85}{96},\\
\beta_{61} = \frac{12}{5}, \beta_{62} = -8, \beta_{63} = \frac{4015}{612}, \beta_{64} = -\frac{11}{36}, \beta_{65} = \frac{86}{225},\\
\beta_{71} = \frac{-8263}{15000}, \beta_{72} = \frac{124}{75}, \beta_{73} = \frac{643}{680}, \beta_{74} = \frac{84}{250}, \beta_{75} = \frac{2484}{10625},\\
\beta_{81} = \frac{3501}{1720}, \beta_{82} = \frac{300}{43}, \beta_{83} = \frac{297278}{52632}, \beta_{84} = \frac{319}{2322}, \beta_{85} = \frac{24068}{84065}, \beta_{87} = \frac{3850}{26703},
\end{aligned}$$

where $m = 8$. Hence we have

$$\begin{aligned}
x(t_0) &= x_0\\
k_1 &= hf(t_i, x_i)\\
k_2 &= hf(t_i + \frac{1}{6}h, x_i + \frac{1}{6}k_1)\\
k_3 &= hf(t_i + \frac{4}{15}h, x_i + \frac{4}{75}k_1 + \frac{16}{75}k_2)\\
k_4 &= hf(t_i + \frac{2}{3}h, x_i + \frac{5}{6}k_1 - \frac{8}{3}k_2 + \frac{5}{2}k_3)\\
k_5 &= hf(t_i + \frac{5}{6}h, x_i - \frac{165}{64}k_1 + \frac{55}{6}k_2 - \frac{425}{64}k_3 + \frac{85}{96}k_4)\\
k_6 &= hf(t_i + h, x_i + \frac{12}{5}k_1 - 8k_2 + \frac{4015}{612}k_3 - \frac{11}{36}k_4 + \frac{88}{255}k_5)\\
k_7 &= hf(t_i + \frac{1}{15}h, x_i - \frac{8263}{15000}k_1 + \frac{124}{75}k_2 - \frac{643}{680}k_3 - \frac{84}{250}k_4 + \frac{2484}{10625}k_5)\\
k_8 &= hf\Big(t_i + h, x_i + \frac{3501}{1720}k_1 - \frac{300}{43}k_2 + \frac{297278}{52632}k_3 - \frac{319}{2322}k_4 + \frac{24068}{84065}k_5 + \frac{3850}{26703}k_7\Big)
\end{aligned}$$

$$x_{i+1} = x_i + \frac{3}{40}k_1 + \frac{875}{2244}k_3 + \frac{23}{72}k_4 + \frac{264}{1955}k_5 + \frac{125}{11592}k_7 + \frac{43}{616}k_8$$

where

$$t_0 < t_1 < t_2 < \cdots < t_N = T, \quad h = \frac{T - t_0}{N}, \quad t_i = t_0 + ih, \quad i = 0, 1, \ldots N \qquad (4)$$

Theorem 1 *Let $f(t,x)$ belong to $C^8[a,b]$ and let it's partial derivatives are bounded and assume there exists, L,M positive numbers, such that*

$$|f(t,x)| < M, |\frac{\partial^{i+j} f}{\partial t^i \partial x^j}| < \frac{L^{i+j}}{M^{j-i}} \tag{5}$$

then in the Runge–Kutta Verner method, $x(t_{i+1}) - x_{i+1} \approx \frac{4}{9} h^8 M L^7 + O(h^9), i + j \leq m$.

Throughout this paper, $(\mathbb{R}, \mathbb{B}(\mathbb{R}), \mu)$ denotes a complete finite measure space.
Let us $P_k(\mathbb{R})$ the set of all nonempty compact convex subsets of $\mathbb{R}$.
we denote by

$$IF_1 = \mathrm{IF}(\mathbb{R}) = \{\langle u, v\rangle : \mathbb{R} \to [0,1]^2, |\forall\, x \in \mathbb{R}\ 0 \leq u(x) + v(x) \leq 1\}$$

An element $\langle u, v\rangle$ of IF_1 is said an intuitionistic fuzzy number if it satisfies the following conditions

1. $\langle u, v\rangle$ is normal i.e there exists $x_0, x_1 \in \mathbb{R}$ such that $u(x_0) = 1$ and $v(x_1) = 1$.
2. u is fuzzy convex and v is fuzzy concave.
3. u is upper semi-continuous and v is lower semi-continuous
4. $supp\langle u, v\rangle = \mathrm{cl}\{x \in \mathbb{R} : | v(x) < 1\}$ is bounded.

So we denote the collection of all intuitionistic fuzzy number by IF_1.
For $\alpha \in [0,1]$ and $\langle u, v\rangle \in \mathrm{IF}_1$, the upper and lower α-cuts of $\langle u, v\rangle$ are defined by

$$[\langle u, v\rangle]^\alpha = \{x \in \mathbb{R} : v(x) \leq 1 - \alpha\}$$

and

$$[\langle u, v\rangle]_\alpha = \{x \in \mathbb{R} : u(x) \geq \alpha\}$$

Remark 1 If $\langle u, v\rangle \in IF_1$, so we can see $[\langle u, v\rangle]_\alpha$ as $[u]^\alpha$ and $[\langle u, v\rangle]^\alpha$ as $[1-v]^\alpha$ in the fuzzy case.

We define $0_{(1,0)} \in IF_1$ as

$$0_{(1,0)}(t) = \begin{cases} (1,0) & t = 0 \\ (0,1) & t \neq 0 \end{cases}$$

Let $\langle u, v\rangle$,$\langle u', v'\rangle \in \mathrm{IF}_1$ and $\lambda \in \mathbb{R}$, we define the following operations by:

$$\Big(\langle u, v\rangle + \langle u', v'\rangle\Big)(z) = \Big(\sup_{z=x+y} \min\big(u(x), u'(y)\big),\ \inf_{z=x+y} \max\big(v(x), v'(y)\big)\Big)$$

$$\lambda\,\langle u, v\rangle = \begin{cases} \langle \lambda u, \lambda v\rangle & \text{if } \lambda \neq 0 \\ 0_{(1,0)} & \text{if } \lambda = 0 \end{cases}$$

For $\langle u, v\rangle, \langle z, w\rangle \in \mathrm{IF}_1$ and $\lambda \in \mathbb{R}$, the addition and scaler-multiplication are defined as follows

$$\Big[\langle u, v\rangle + \langle z, w\rangle\Big]^{\alpha} = \Big[\langle u, v\rangle\Big]^{\alpha} + \Big[\langle z, w\rangle\Big]^{\alpha}, \quad \Big[\lambda\langle z, w\rangle\Big]^{\alpha} = \lambda\Big[\langle z, w\rangle\Big]^{\alpha}$$
$$\Big[\langle u, v\rangle + \langle z, w\rangle\Big]_{\alpha} = \Big[\langle u, v\rangle\Big]_{\alpha} + \Big[\langle z, w\rangle\Big]_{\alpha}, \quad \Big[\lambda\langle z, w\rangle\Big]_{\alpha} = \lambda\Big[\langle z, w\rangle\Big]_{\alpha}$$

Definition 1 Let $\langle u, v\rangle$ an element of IF_1 and $\alpha \in [0, 1]$, we define the following sets:

$$\Big[\langle u, v\rangle\Big]_l^+(\alpha) = \inf\{x \in \mathbb{R} \mid u(x) \geq \alpha\}, \quad \Big[\langle u, v\rangle\Big]_r^+(\alpha) = \sup\{x \in \mathbb{R} \mid u(x) \geq \alpha\}$$
$$\Big[\langle u, v\rangle\Big]_l^-(\alpha) = \inf\{x \in \mathbb{R} \mid v(x) \leq 1-\alpha\}, \quad \Big[\langle u, v\rangle\Big]_r^-(\alpha) = \sup\{x \in \mathbb{R} \mid v(x) \leq 1-\alpha\}$$

Remark 2

$$\Big[\langle u, v\rangle\Big]_{\alpha} = \Big[\Big[\langle u, v\rangle\Big]_l^+(\alpha), \Big[\langle u, v\rangle\Big]_r^+(\alpha)\Big]$$
$$\Big[\langle u, v\rangle\Big]^{\alpha} = \Big[\Big[\langle u, v\rangle\Big]_l^-(\alpha), \Big[\langle u, v\rangle\Big]_r^-(\alpha)\Big]$$

A Triangular Intuitionistic Fuzzy Number (TIFN) $\langle u, v\rangle$ is an intuitionistic fuzzy set in $\mathbb{R}$ with the following membership function u and non-membership function v:

$$u(x) = \begin{cases} \frac{x-a_1}{a_2-a_1} & \text{if } a_1 \leq x \leq a_2 \\ \frac{a_3-x}{a_3-a_2} & \text{if } a_2 \leq x \leq a_3 \\ 0 & \text{otherwise} \end{cases}$$

$$v(x) = \begin{cases} \frac{a_2-x}{a_2-a_1'} & \text{if } a_1' \leq x \leq a_2 \\ \frac{x-a_2}{a_3'-a_2} & \text{if } a_2 \leq x \leq a_3' \\ 1 & \text{otherwise} \end{cases}$$

where $a_1' \leq a_1 \leq a_2 \leq a_3 \leq a_3'$ and $u(x), v(x) \leq 0.5$ for $u(x) = v(x), \quad \forall x \in \mathbb{R}$.

This TIFN is denoted by $\langle u, v\rangle = \langle a_1, a_2, a_3; a_1', a_2, a_3'\rangle$ where,

$$[\langle u, v\rangle]_{\alpha} = [a_1 + \alpha(a_2 - a_1), a_3 - \alpha(a_3 - a_2)] \tag{6}$$

$$[\langle u, v\rangle]^{\alpha} = [a_1' + \alpha(a_2 - a_1'), a_3' - \alpha(a_3' - a_2)] \tag{7}$$

Theorem 2 ([17]) *d_∞ define a metric on IF_1.*

Theorem 3 ([17]) *The metric space (IF_1, d_∞) is complete.*

Remark 3 If $F : [a, b] \to \mathrm{IF}_1$ is Hukuhara differentiable and its Hukuhara derivative F' is integrable over $[0, 1]$ then

$$F(t) = F(t_0) + \int_{t_0}^{t} F'(s)ds$$

Definition 2 Let $\langle u, v\rangle$ and $\langle u', v'\rangle \in IF_1$, the H-difference is the IFN $\langle z, w\rangle \in IF_1$, if it exists, such that

$$\langle u, v\rangle - \langle u', v'\rangle = \langle z, w\rangle \Longleftrightarrow \langle u, v\rangle = \langle u', v'\rangle + \langle z, w\rangle$$

Definition 3 A mapping $F : [a, b] \to IF_1$ is said to be Hukuhara derivable at t_0 if there exist $F'(t_0) \in IF_1$ such that both limits:

$$\lim_{\Delta t \to 0^+} \frac{F(t_0 + \Delta t) - F(t_0)}{\Delta t}$$

and

$$\lim_{\Delta t \to 0^+} \frac{F(t_0) - F(t_0 - \Delta t)}{\Delta t}$$

exist and they are equal to $F'(t_0) = \langle u'(t_0), v'(t_0)\rangle$, which is called the Hukuhara derivative of F at t_0.

3 Intuitionistic Fuzzy Cauchy Problem

In this section we consider the initial value problem for the intuitionistic fuzzy differential equation

$$\begin{cases} x'(t) = f(t, x(t)), \ t \in I \\ x(t_0) = x_0 \in IF_1 \end{cases} \tag{8}$$

where $x \in IF_1$ is unknown $I = [t_0, T]$ and $f : I \times IF_1 \to IF_1$ and $x(t_0)$ is intuitionistic fuzzy number.

Denote the $\alpha-$ level set

$$[x(t)]_\alpha = \Big[[x(t)]_l^+(\alpha), [x(t)]_r^+(\alpha)\Big]$$

$$[x(t)]^\alpha = \Big[x(t)]_l^-(\alpha), [x(t)]_r^-(\alpha)\Big]$$

and

$$[x(t_0)]_\alpha = \left[[x(t_0)]_l^+(\alpha), [x(t_0)]_r^+(\alpha)\right]$$

$$[x(t_0)]^\alpha = \left[x(t_0)]_l^-(\alpha), [x(t_0)]_r^-(\alpha)\right]$$

$$[f(t, x(t))]_\alpha = \left[f_l^+(t, x(t); \alpha), [f_r^+(t, x(t); \alpha)\right]$$

$$[f(t, x(t))]^\alpha = \left[f_l^-(t, x(t); \alpha), [f_r^-(t, x(t); \alpha)\right]$$

where

$$\begin{aligned}
f_1^+(t, x(t); \alpha) &= \min\left\{f(t, u) | u \in \left[[x(t)]_l^+(\alpha), [x(t)]_r^+(\alpha)\right]\right\} \\
f_2^+(t, x(t); \alpha) &= \max\left\{f(t, u) | u \in \left[[x(t)]_l^+(\alpha), [x(t)]_r^+(\alpha)\right]\right\} \\
f_3^-(t, x(t); \alpha) &= \min\left\{f(t, u) | u \in \left[x(t)]_l^-(\alpha), [x(t)]_r^-(\alpha)\right]\right\} \\
f_4^-(t, x(t); \alpha) &= \max\left\{f(t, u) | u \in \left[x(t)]_l^-(\alpha), [x(t)]_r^-(\alpha)\right]\right\}
\end{aligned} \tag{9}$$

Denote

$$\begin{aligned}
f_1^+(t, x(t); \alpha) &= G\big(t, [x(t)]_l^+(\alpha), [x(t)]_r^+(\alpha)\big) \\
f_2^+(t, x(t); \alpha) &= H\big(t, [x(t)]_l^+(\alpha), [x(t)]_r^+(\alpha)\big) \\
f_3^-(t, x(t); \alpha) &= L\big(t, [x(t)]_l^-(\alpha), [x(t)]_r^-(\alpha)\big) \\
f_4^-(t, x(t); \alpha) &= K\big(t, [x(t)]_l^-(\alpha), [x(t)]_r^-(\alpha)\big)
\end{aligned} \tag{10}$$

Sufficient conditions for the existence of an unique solution to Eq. (8) are:

1. Continuity of f
2. Lipschitz condition:
   For any pair $\big(t, x\big), \big(t, y\big) \in I \times \mathrm{IF}_1$, we have

$$d_\infty(f(t, x), f(t, y)) \leq K d_\infty(x, y) \tag{11}$$

   where $K > 0$ is a given constant.

Theorem 4 ([7]) *Let us suppose that the following conditions hold:*

1. Let $R_0 = [t_0, t_0 + p] \times \overline{B}(x_0, q)$, $p, q \geq 0$, $x_0 \in IF_1$ where $\overline{B}(x_0, q) = \{x \in IF_1 : d_\infty(x, x_0) \leq q\}$ denote a closed ball in IF_1 and let $f : R_0 \longrightarrow IF_1$ be a continuous function such that $d_\infty\big(f(t, x), 0_{(1,0)}\big) \leq M$ for all $(t, x) \in R_0$.

2. *Let* $g : [t_0, t_0 + p] \times [0, q] \longrightarrow \mathbb{R}$ *such that* $g(t, 0) \equiv 0$ *and* $0 \leq g(t, z) \leq M_1$, $\forall t \in [t_0, t_0 + p], 0 \leq z \leq q$ *such that* $g(t, z)$ *is non-decreasing in* z *and* g *is such that the initial value problem*

$$z'(t) = g(t, z(t)), z(t_0) = x_0. \tag{12}$$

has only the solution $z(t) \equiv 0$ *on* $[t_0, t_0 + p]$
3. *We have* $d_\infty(f(t, x), f(t, y)) \leq g(t, d_\infty(x, y)), \forall (t, x), (t, y) \in R_0$ *and* $d_\infty(x, y) \leq q$. *Then the intuitionistic fuzzy initial value problem*

$$\begin{cases} x(t)' = f(t, x(t)), \\ x(t_0) = x_0 \end{cases} \tag{13}$$

has an unique solution $x \in C^1[[t_0, t_0 + r], B(x_0, q)]$ *on* $[t_0, t_0 + r]$ *where* $r = \min\{p, \frac{q}{M}, \frac{q}{M_1}, d\}$ *and the successive iterations*

$$x(t_0) = x_0, \quad x_{n+1}(t) = x_0 + \int_{t_0}^{t} f(s, x_n(s))ds \tag{14}$$

converge to $x(t)$ *on* $[t_0, t_0 + r]$.

4 The Runge–Kutta Verner Method

Let the exact solutions

$$[X(t_n)]_\alpha = \Big[[X(t_n)]_l^+(\alpha), [X(t_n)]_r^+(\alpha)\Big], \quad [X(t_n)]^\alpha = \Big[[X(t_n)]_l^-(\alpha), [X(t_n)]_r^-(\alpha)\Big]$$

be approximated by

$$[x(t_n)]_\alpha = \Big[[x(t_n)]_l^+(\alpha), [x(t_n)]_r^+(\alpha)\Big], \quad [x(t_n)]^\alpha = \Big[[x(t_n)]_l^-(\alpha), [x(t_n)]_r^-(\alpha)\Big]$$

at t_n, $0 \leq n \leq N$.

The solutions are calculated by grid points at

$$t_0 < t_1 < t_2 < \cdots < t_N = T, \quad h = \frac{T - t_0}{N}, \quad t_n = t_0 + nh, \quad n = 0, 1, \ldots N \quad (15)$$

From (2) and (3) we define

$$[x(t_{n+1})]_l^+(\alpha) - [x(t_n)]_l^+(\alpha) = \sum_{i=1}^{i=8} w_i [k_i]_l^+(\alpha) \quad (16)$$

$$[x(t_{n+1})]_r^+(\alpha) - [x(t_n)]_r^+(\alpha) = \sum_{i=1}^{i=8} w_i [k_i]_r^+(\alpha) \quad (17)$$

$$[x(t_{n+1})]_l^-(\alpha) - [x(t_n)]_l^-(\alpha) = \sum_{i=1}^{i=8} w_i [k_i]_l^-(\alpha) \quad (18)$$

$$[x(t_{n+1})]_r^-(\alpha) - [x(t_n)]_r^-(\alpha) = \sum_{i=1}^{i=8} w_i [k_i]_r^-(\alpha) \quad (19)$$

where the $w_i's$ are constants and

$$\begin{cases} [k_i]_\alpha &= \left[[k_i]_l^+(\alpha), [k_i]_r^+(\alpha)\right], \quad i = 1, 2, 3, 4, 5, 6, 7, 8 \\ [k_i]^\alpha &= \left[[k_i]_l^-(\alpha), [k_i]_r^-(\alpha)\right] \\ [k_i]_l^+(\alpha) &= hG\left(t_n + c_i h, [x(t_n)]_l^+ + \sum_{j=1}^{i-1} \beta_{ij}[k_j]_l^+(\alpha), [x(t_n)]_r^+ + \sum_{j=1}^{i-1} \beta_{ij}[k_j]_r^+(\alpha)\right) \\ [k_i]_r^+(\alpha) &= hH\left(t_n + c_i h, [x(t_n)]_l^+ + \sum_{j=1}^{i-1} \beta_{ij}[k_j]_l^+(\alpha), [x(t_n)]_r^+ + \sum_{j=1}^{i-1} \beta_{ij}[k_j]_r^+(\alpha)\right) \\ [k_i]_l^-(\alpha) &= hL\left(t_n + c_i h, [x(t_n)]_l^- + \sum_{j=1}^{i-1} \beta_{ij}[k_j]_l^-(\alpha), [x(t_n)]_r^- + \sum_{j=1}^{i-1} \beta_{ij}[k_j]_r^-(\alpha)\right) \\ [k_i]_r^-(\alpha) &= hK\left(t_n + c_i h, [x(t_n)]_l^- + \sum_{j=1}^{i-1} \beta_{ij}[k_j]_l^-(\alpha), [x(t_n)]_r^- + \sum_{j=1}^{i-1} \beta_{ij}[k_j]_r^-(\alpha)\right) \end{cases}$$

The Runge–Kutta Verner method define as follows:

$$\begin{aligned} [x(t_{n+1})]_l^+(\alpha) =& [x(t_n)]_l^+(\alpha) + \frac{3}{40}[k_1]_l^+(\alpha) + \frac{875}{2244}[k_3]_l^+(\alpha) + \frac{23}{72}[k_4]_l^+(\alpha) + \frac{264}{1955}[k_5]_l^+(\alpha) \\ &+ \frac{125}{11592}[k_7]_l^+(\alpha) + \frac{43}{616}[k_8]_l^+(\alpha) \end{aligned} \quad (20)$$

where

$$\begin{cases} [k_1]_l^+(\alpha) &= hG\big(t_n, [x(t_n)]_l^+(\alpha), [x(t_n)]_r^+(\alpha)\big) \\ [k_2]_l^+(\alpha) &= hG\big(t_n + \frac{1}{6}h, [z_1]_l^+(\alpha), [z_1]_r^+(\alpha)\big) \\ [k_3]_l^+(\alpha) &= hG\big(t_n + \frac{4}{15}h, [z_2]_l^+(\alpha), [z_2]_r^+(\alpha)\big) \\ [k_4]_l^+(\alpha) &= hG\big(t_n + \frac{2}{3}h, [z_3]_l^+(\alpha), [z_3]_r^+(\alpha)\big) \\ [k_5]_l^+(\alpha) &= hG\big(t_n + \frac{5}{6}h, [z_4]_l^+(\alpha), [z_4]_r^+(\alpha)\big) \\ [k_6]_l^+(\alpha) &= hG\big(t_n + h, [z_5]_l^+(\alpha), [z_5]_r^+(\alpha)\big) \\ [k_7]_l^+(\alpha) &= hG\big(t_n + \frac{1}{15}h, [z_6]_l^+(\alpha), [z_6]_r^+(\alpha)\big) \\ [k_8]_l^+(\alpha) &= hG\big(t_n + h, [z_7]_l^+(\alpha), [z_7]_r^+(\alpha)\big) \\ [z_1]_i^+(\alpha) &= [x(t_n)]_i^+(\alpha) + \frac{1}{6}[k_1]_i^+(\alpha) \\ [z_2]_i^+(\alpha) &= [x(t_n)]_i^+(\alpha) + \frac{4}{75}[k_1]_i^+(\alpha) + \frac{16}{75}[k_2]_i^+(\alpha) \\ [z_3]_i^+(\alpha) &= [x(t_n)]_i^+(\alpha) + \frac{5}{6}[k_1]_i^+(\alpha) - \frac{8}{3}[k_2]_i^+(\alpha) + \frac{5}{2}[k_3]_i^+(\alpha) \\ [z_4]_i^+(\alpha) &= [x(t_n)]_i^+(\alpha) - \frac{165}{64}[k_1]_i^+(\alpha) + \frac{55}{6}[k_2]_i^+(\alpha) - \frac{425}{64}[k_3]_i^+(\alpha) + \frac{85}{96}[k_4]_i^+(\alpha) \\ [z_5]_i^+(\alpha) &= [x(t_n)]_i^+(\alpha) + \frac{12}{5}[k_1]_i^+(\alpha) - 8[k_2]_i^+(\alpha) + \frac{4015}{612}[k_3]_i^+(\alpha) - \frac{11}{36}[k_4]_i^+(\alpha) + \frac{88}{255}[k_5]_i^+(\alpha) \\ [z_6]_i^+(\alpha) &= [x(t_n)]_i^+(\alpha) - \frac{8263}{15000}[k_1]_i^+(\alpha) + \frac{124}{75}[k_2]_i^+(\alpha) - \frac{643}{680}[k_3]_i^+(\alpha) - \frac{81}{250}[k_4]_i^+(\alpha) + \frac{2484}{10625}[k_5]_i^+(\alpha) \\ [z_7]_i^+(\alpha) &= [x(t_n)]_i^+(\alpha) + \frac{3501}{1720}[k_1]_i^+(\alpha) - \frac{300}{43}[k_2]_i^+(\alpha) + \frac{297275}{52632}[k_3]_i^+(\alpha) - \frac{319}{2322}[k_4]_i^+(\alpha) \\ &\quad + \frac{24068}{84065}[k_5]_i^+(\alpha) + \frac{3850}{26703}[k_7]_i^+(\alpha) \end{cases}$$

and

$$\begin{aligned} [x(t_{n+1})]_r^+(\alpha) =& [x(t_n)]_r^+(\alpha) + \frac{3}{40}[k_1]_r^+(\alpha) + \frac{875}{2244}[k_3]_r^+(\alpha) + \frac{23}{72}[k_4]_r^+(\alpha) + \frac{264}{1955}[k_5]_r^+(\alpha) \\ &+ \frac{125}{11592}[k_7]_r^+(\alpha) + \frac{43}{616}[k_8]_r^+(\alpha) \end{aligned} \tag{21}$$

where

$$
\begin{cases}
[k_1]_r^+(\alpha) & = hH\big(t_n, [x(t_n)]_l^+(\alpha), [x(t_n)]_r^+(\alpha)\big) \\
[k_2]_r^+(\alpha) & = hH\big(t_n + \frac{1}{6}h, [z_1]_l^+(\alpha), [z_1]_r^+(\alpha)\big) \\
[k_3]_r^+(\alpha) & = hH\big(t_n + \frac{4}{15}h, [z_2]_l^+(\alpha), [z_2]_r^+(\alpha)\big) \\
[k_4]_r^+(\alpha) & = hH\big(t_n + \frac{2}{3}h, [z_3]_l^+(\alpha), [z_3]_r^+(\alpha)\big) \\
[k_5]_r^+(\alpha) & = hH\big(t_n + \frac{5}{6}h, [z_4]_l^+(\alpha), [z_4]_r^+(\alpha)\big) \\
[k_6]_r^+(\alpha) & = hH\big(t_n + h, [z_5]_l^+(\alpha), [z_5]_r^+(\alpha)\big) \\
[k_7]_r^+(\alpha) & = hH\big(t_n + \frac{1}{15}h, [z_6]_l^+(\alpha), [z_6]_r^+(\alpha)\big) \\
[k_8]_r^+(\alpha) & = hH\big(t_n + h, [z_7]_l^+(\alpha), [z_7]_r^+(\alpha)\big) \\
[z_1]_i^+(\alpha) & = [x(t_n)]_i^+(\alpha) + \frac{1}{6}[k_1]_i^+(\alpha) \\
[z_2]_i^+(\alpha) & = [x(t_n)]_i^+(\alpha) + \frac{4}{75}[k_1]_i^+(\alpha) + \frac{16}{75}[k_2]_i^+(\alpha) \\
[z_3]_i^+(\alpha) & = [x(t_n)]_i^+(\alpha) + \frac{5}{6}[k_1]_i^+(\alpha) - \frac{8}{3}[k_2]_i^+(\alpha) + \frac{5}{2}[k_3]_i^+(\alpha) \\
[z_4]_i^+(\alpha) & = [x(t_n)]_i^+(\alpha) - \frac{165}{64}[k_1]_i^+(\alpha) + \frac{55}{6}[k_2]_i^+(\alpha) - \frac{425}{64}[k_3]_i^+(\alpha) + \frac{85}{96}[k_4]_i^+(\alpha) \\
[z_5]_i^+(\alpha) & = [x(t_n)]_i^+(\alpha) + \frac{12}{5}[k_1]_i^+(\alpha) - 8[k_2]_i^+(\alpha) + \frac{4015}{612}[k_3]_i^+(\alpha) - \frac{11}{36}[k_4]_i^+(\alpha) + \frac{88}{255}[k_5]_i^+(\alpha) \\
[z_6]_i^+(\alpha) & = [x(t_n)]_i^+(\alpha) - \frac{8263}{15000}[k_1]_i^+(\alpha) + \frac{124}{75}[k_2]_i^+(\alpha) - \frac{643}{680}[k_3]_i^+(\alpha) - \frac{81}{250}[k_4]_i^+(\alpha) + \frac{2484}{10625}[k_5]_i^+(\alpha) \\
[z_7]_i^+(\alpha) & = [x(t_n)]_i^+(\alpha) + \frac{3501}{1720}[k_1]_i^+(\alpha) - \frac{300}{43}[k_2]_i^+(\alpha) + \frac{297275}{52632}[k_3]_i^+(\alpha) - \frac{319}{2322}[k_4]_i^+(\alpha) \\
& \quad + \frac{24068}{84065}[k_5]_i^+(\alpha) + \frac{3850}{26703}[k_7]_i^+(\alpha)
\end{cases}
$$

and

$$
\begin{aligned}
[x(t_{n+1})]_l^-(\alpha) = & [x(t_n)]_l^-(\alpha) + \frac{3}{40}[k_1]_l^-(\alpha) + \frac{875}{2244}[k_3]_l^-(\alpha) + \frac{23}{72}[k_4]_l^-(\alpha) + \frac{264}{1955}[k_5]_l^-(\alpha) \\
& + \frac{125}{11592}[k_7]_l^-(\alpha) + \frac{43}{616}[k_8]_l^-(\alpha)
\end{aligned} \tag{22}
$$

where

$$\begin{cases} [k_1]_l^-(\alpha) &= hL\big(t_n, [x(t_n)]_l^-(\alpha), [x(t_n)]_r^-(\alpha)\big) \\ [k_2]_l^-(\alpha) &= hL\big(t_n + \frac{1}{6}h, [z_1]_l^-(\alpha), [z_1]_r^-(\alpha)\big) \\ [k_3]_l^-(\alpha) &= hL\big(t_n + \frac{4}{15}h, [z_2]_l^-(\alpha), [z_2]_r^-(\alpha)\big) \\ [k_4]_l^-(\alpha) &= hL\big(t_n + \frac{2}{3}h, [z_3]_l^-(\alpha), [z_3]_r^-(\alpha)\big) \\ [k_5]_l^-(\alpha) &= hL\big(t_n + \frac{5}{6}h, [z_4]_l^-(\alpha), [z_4]_r^-(\alpha)\big) \\ [k_6]_l^-(\alpha) &= hL\big(t_n + h, [z_5]_l^-(\alpha), [z_5]_r^-(\alpha)\big) \\ [k_7]_l^-(\alpha) &= hL\big(t_n + \frac{1}{15}h, [z_6]_l^-(\alpha), [z_6]_r^-(\alpha)\big) \\ [k_8]_l^-(\alpha) &= hL\big(t_n + h, [z_7]_l^-(\alpha), [z_7]_r^-(\alpha)\big) \\ [z_1]_i^-(\alpha) &= [x(t_n)]_i^-(\alpha) + \frac{1}{6}[k_1]_i^-(\alpha) \\ [z_2]_i^-(\alpha) &= [x(t_n)]_i^-(\alpha) + \frac{4}{75}[k_1]_i^-(\alpha) + \frac{16}{75}[k_2]_i^-(\alpha) \\ [z_3]_i^-(\alpha) &= [x(t_n)]_i^-(\alpha) + \frac{5}{6}[k_1]_i^-(\alpha) - \frac{8}{3}[k_2]_i^-(\alpha) + \frac{5}{2}[k_3]_i^-(\alpha) \\ [z_4]_i^-(\alpha) &= [x(t_n)]_i^-(\alpha) - \frac{165}{64}[k_1]_i^-(\alpha) + \frac{55}{6}[k_2]_i^-(\alpha) - \frac{425}{64}[k_3]_i^-(\alpha) + \frac{85}{96}[k_4]_i^-(\alpha) \\ [z_5]_i^-(\alpha) &= [x(t_n)]_i^-(\alpha) + \frac{12}{5}[k_1]_i^-(\alpha) - 8[k_2]_i^-(\alpha) + \frac{4015}{612}[k_3]_i^-(\alpha) - \frac{11}{36}[k_4]_i^-(\alpha) + \frac{88}{255}[k_5]_i^-(\alpha) \\ [z_6]_i^-(\alpha) &= [x(t_n)]_i^-(\alpha) - \frac{8263}{15000}[k_1]_i^-(\alpha) + \frac{124}{75}[k_2]_i^-(\alpha) - \frac{643}{680}[k_3]_i^-(\alpha) - \frac{81}{250}[k_4]_i^-(\alpha) + \frac{2484}{10625}[k_5]_i^-(\alpha) \\ [z_7]_i^-(\alpha) &= [x(t_n)]_i^-(\alpha) + \frac{3501}{1720}[k_1]_i^-(\alpha) - \frac{300}{43}[k_2]_i^-(\alpha) + \frac{297275}{52632}[k_3]_i^-(\alpha) - \frac{319}{2322}[k_4]_i^-(\alpha) \\ &\quad + \frac{24068}{84065}[k_5]_i^-(\alpha) + \frac{3850}{26703}[k_7]_i^-(\alpha) \end{cases}$$

and

$$[x(t_{n+1})]_r^-(\alpha) = [x(t_n)]_r^-(\alpha) + \frac{3}{40}[k_1]_r^-(\alpha) + \frac{875}{2244}[k_3]_r^-(\alpha) + \frac{23}{72}[k_4]_r^-(\alpha) + \frac{264}{1955}[k_5]_r^-(\alpha) + \frac{125}{11592}[k_7]_r^-(\alpha) + \frac{43}{616}[k_8]_r^-(\alpha) \tag{23}$$

where

$$\begin{cases}
[k_1]_r^-(\alpha) &= hK\big(t_n, [x(t_n)]_l^-(\alpha), [x(t_n)]_r^-(\alpha)\big)\\
[k_2]_r^-(\alpha) &= hK\big(t_n + \frac{1}{6}h, [z_1]_l^-(\alpha), [z_1]_r^-(\alpha)\big)\\
[k_3]_r^-(\alpha) &= hK\big(t_n + \frac{4}{15}h, [z_2]_l^-(\alpha), [z_2]_r^-(\alpha)\big)\\
[k_4]_r^-(\alpha) &= hK\big(t_n + \frac{2}{3}h, [z_3]_l^-(\alpha), [z_3]_r^-(\alpha)\big)\\
[k_5]_r^-(\alpha) &= hK\big(t_n + \frac{5}{6}h, [z_4]_l^-(\alpha), [z_4]_r^-(\alpha)\big)\\
[k_6]_r^-(\alpha) &= hK\big(t_n + h, [z_5]_l^-(\alpha), [z_5]_r^-(\alpha)\big)\\
[k_7]_r^-(\alpha) &= hK\big(t_n + \frac{1}{15}h, [z_6]_l^-(\alpha), [z_6]_r^-(\alpha)\big)\\
[k_8]_r^-(\alpha) &= hK\big(t_n + h, [z_7]_l^-(\alpha), [z_7]_r^-(\alpha)\big)\\
[z_1]_i^-(\alpha) &= [x(t_n)]_i^-(\alpha) + \frac{1}{6}[k_1]_i^-(\alpha)\\
[z_2]_i^-(\alpha) &= [x(t_n)]_i^-(\alpha) + \frac{4}{75}[k_1]_i^-(\alpha) + \frac{16}{75}[k_2]_i^-(\alpha)\\
[z_3]_i^-(\alpha) &= [x(t_n)]_i^-(\alpha) + \frac{5}{6}[k_1]_i^-(\alpha) - \frac{8}{3}[k_2]_i^-(\alpha) + \frac{5}{2}[k_3]_i^-(\alpha)\\
[z_4]_i^-(\alpha) &= [x(t_n)]_i^-(\alpha) - \frac{165}{64}[k_1]_i^-(\alpha) + \frac{55}{6}[k_2]_i^-(\alpha) - \frac{425}{64}[k_3]_i^-(\alpha) + \frac{85}{96}[k_4]_i^-(\alpha)\\
[z_5]_i^-(\alpha) &= [x(t_n)]_i^-(\alpha) + \frac{12}{5}[k_1]_i^-(\alpha) - 8[k_2]_i^-(\alpha) + \frac{4015}{612}[k_3]_i^-(\alpha) - \frac{11}{36}[k_4]_i^-(\alpha) + \frac{88}{255}[k_5]_i^-(\alpha)\\
[z_6]_i^-(\alpha) &= [x(t_n)]_i^-(\alpha) - \frac{8263}{15000}[k_1]_i^-(\alpha) + \frac{124}{75}[k_2]_i^-(\alpha) - \frac{643}{680}[k_3]_i^-(\alpha) - \frac{81}{250}[k_4]_i^-(\alpha) + \frac{2484}{10625}[k_5]_i^-(\alpha)\\
[z_7]_i^-(\alpha) &= [x(t_n)]_i^-(\alpha) + \frac{3501}{1720}[k_1]_i^-(\alpha) - \frac{300}{43}[k_2]_i^-(\alpha) + \frac{297275}{52632}[k_3]_i^-(\alpha) - \frac{319}{2322}[k_4]_i^-(\alpha)\\
&\quad + \frac{24068}{84065}[k_5]_i^-(\alpha) + \frac{3850}{26703}[k_7]_i^-(\alpha)
\end{cases}$$

Let $G(t, u^+, v^+)$, $H(t, u^+, v)^+$, $L(t, u^-, v)^-$ and $K(t, u^-, v^-)$ be the functions of (10), where u^+, v^+, u^- and v^- are the constants and $u^+ \leq v^+$ and $u^- \leq v^-$.
The domain of G and H is

$$M_1 = \{(t, u^+, v^+)\backslash\ t_0 \leq t \leq T,\ \ \infty < u^+ \leq v^+,\ \ -\infty < v^+ < +\infty\}$$

and the domain of L and K is

$$M_2 = \{(t, u^-, v^-)\backslash\ t_0 \leq t \leq T,\ \ \infty < u^- \leq v^-,\ \ -\infty < v^- < +\infty\}$$

where $M_1 \subseteq M_2$

Theorem 5 *Let $G(t, u^+, v^+)$, $H(t, u^+, v^+)$ belong to $C^8(M_1)$ and $L(t, u^-, v^-)$, $K(t, u^-, v^-)$ belong to $C^8(M_2)$ and the partial derivatives of G, H and L, K be bounded over M_1 and M_2 respectively. Then, for arbitrarily fixed $0 \leq \alpha \leq 1$, the* *numerical solutions of (20), (21), (22) and (23) converge to the exact solutions* $[X(t)]_l^+(\alpha)$, $[X(t)]_r^+(\alpha)$, $[X(t)]_l^-(\alpha)$ *and* $[X(t)]_r^-(\alpha)$ *uniformly in* t.

Proof see [4].

5 Example

Example 5.1 Consider the intuitionistic fuzzy initial value problem

$$\begin{cases} x'(t) = x(t) \text{ for all } t \in [0, T] \\ x_0 \quad = \Big((\alpha - 1, 1 - \alpha), (-2\alpha, 2\alpha)\Big) \end{cases} \tag{24}$$

Then, we have the following parametrized differential system:

$$\begin{cases} [x(t)]_l^+(\alpha) = (\alpha - 1)\exp(t) \\ [x(t)]_r^+(\alpha) = (1 - \alpha)\exp(t) \\ [x(t)]_l^-(\alpha) = -2\alpha\exp(t) \\ [x(t)]_r^-(\alpha) = 2\alpha\exp(t) \end{cases}$$

Therefore the exact solutions are given by

$$[X(t)]_\alpha = \Big[(\alpha - 1)\exp(t), (1 - \alpha)\exp(t)\Big]$$

$$[X(t)]^\alpha = \Big[-2\alpha\exp(t), 2\alpha\exp(t)\Big]$$

which at $t = 1$ are

$$[X(1)]_\alpha = \Big[(\alpha - 1)\exp(1), (1 - \alpha)\exp(1)\Big]$$

$$[X(1)]^\alpha = \Big[(-2\alpha\exp(1), 2\alpha\exp(1)\Big]$$

Comparison of results of the Runge–Kutta Verner method, Runge–Kutta Nyström and Runge–Kutta Gill method in [9, 12] for $h = 0.2$ and $t = 1$:

	Exact		RK-Verner	
α	$([X]_l^+, [X]_r^+)$	$([X]_l^-, [X]_r^-)$	$([x]_l^+, [x]_r^+)$	$([x]_l^-, [x]_r^-)$
0	(-2.718281828,2.718281828)	(0,0)	(-2.718281825,2.718281825)	(0,0)
0.2	(-2.174625462,2.174625462)	(-1.087312731,1.087312731)	(-2.174625460,2.174625460)	(-1.087312730,1.087312730)
0.4	(-1.630969097,1.630969097)	(-2.174625462,2.174625462)	(-1.630969095,1.630969095)	(-2.174625460,2.174625460)
0.6	(-1.087312731,1.087312731)	(-3.261938194,3.261938194)	(-1.087312730,1.087312730)	(-3.261938191,3.261938191)
0.8	(-0.543656365,0.543656365)	(-4.349250925,4.349250925)	(0.543656365,0.543656365)	(-4.349250921,4.349250921)
1	(0,0)	(-5.436563656,5.436563656)	(0,0)	(5.436563651,5.436563651)

	RK-Gill		RK-Nyström	
α	$([x]_l^+, [x]_r^+)$	$([x]_l^-, [x]_r^-)$	$([x]_l^+, [x]_r^+)$	$([x]_l^-, [x]_r^-)$
0	(-2.718251136,2.718251136)	(0,0)	(-2.717509377,2.717509377)	(0,0)
0.2	(-2.174600909,2.174600909)	(-1.087300454,1.087300454)	(-2.174007501,2.174007501)	(-1.087003750,1.087003750)
0.4	(-1.630950681,1.630950681)	(-2.174600909,2.174600909)	(-1.630505626,1.630505626)	(-2.174007501,2.174007501)
0.6	(-1.087300454,1.087300454)	(-3.261901363,3.261901363)	(-1.087003750,1.087003750)	(-3.261011252,3.261011252)
0.8	(-0.543650227,0.543650227)	(-4.349201818,4.349201818)	(-0.543501875,0.543501875)	(-4.348015003,4.348015003)
1	(0,0)	(-5.436502273,5.436502273)	(0,0)	(-5.435018754,5.435018754)

α	Error in RK-Nyström	Error in RK-Gill	Error in RK-Verner
0	3.8622×10^{-4}	1.5345×10^{-5}	1.3032×10^{-9}
0.2	4.6347×10^{-4}	1.8415×10^{-5}	1.5639×10^{-9}
0.4	5.4071×10^{-4}	2.1484×10^{-5}	1.8246×10^{-9}
0.6	6.1796×10^{-4}	2.4553×10^{-5}	2.0852×10^{-9}
0.8	6.9520×10^{-4}	2.7622×10^{-5}	2.3459×10^{-9}
1	7.7245×10^{-4}	3.0691×10^{-5}	2.6065×10^{-9}

The exact and approximate solutions obtained by the Runge–Kutta Nyström method, Runge–Kutta Gill method and the Runge–Kutta Verner method are compared and plotted at $t = 1$ and $h = 0.2$ in Fig. 1.

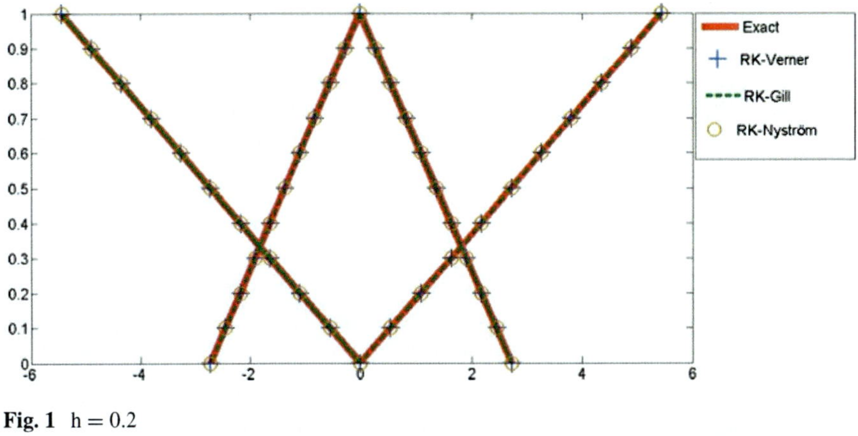

Fig. 1 h = 0.2

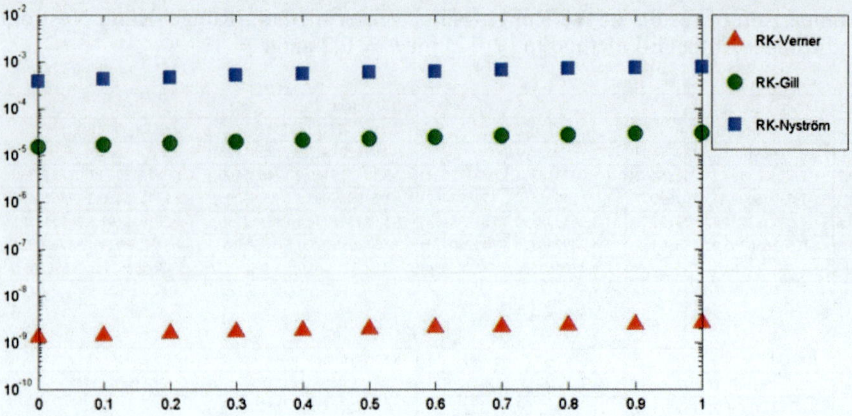

Fig. 2 h = 0.2

The error between the Runge–Kutta Nyström method, Runge–Kutta Gill method and the Runge–Kutta Verner method is plotted in Fig. 2.

6 Conclusion

In this work we have applied iterative solution of Runge-Kutta Verner method for numerical solution of intuitionistic fuzzy differential equations. The numerical results indicate that the Runge–Kutta Verner method is much more efficient than the other well-known Runge Kutta methods in [9, 12].

References

1. Atanassov, K.T.: Intuitionistic fuzzy sets. VII ITKR's session, Sofia (deposited in Central Science and Technical Library of the Bulgarian Academy of Sciences 1697/84)(1983)
2. Atanassov, K.T.: Intuitionistic fuzzy sets. Fuzzy Sets Syst **20**, 87–96 (1986)
3. Atanassov, K.T.: Operators over interval valued intuitionistic fuzzy sets. Fuzzy Sets Syst. **64**(2), 159–174 (1994)
4. Ben Amma, B., Melliani, S., Chadli, L.S.: Numerical solution of intuitionistic fuzzy differential equations by Euler and Taylor methods. Notes Intuitionistic Fuzzy Sets **22**(2), 71–86 (2016)
5. Ben Amma, B., Melliani, S., Chadli, L.S.: Numerical solution of intuitionistic fuzzy differential equations by Adams three order predictor-corrector method. Notes Intuitionistic Fuzzy Sets **22**(3), 47–69 (2016)
6. Ben Amma, B., Melliani, S., Chadli, L.S.: Numerical solution of intuitionistic fuzzy differential equations by Runge-Kutta Method of order four. Notes Intuitionistic Fuzzy Sets **22**(4), 42–52 (2016)
7. Ben Amma, B., Melliani, S., Chadli, L.S.: The cauchy problem of intuitionistic fuzzy differential equations. Notes Intuitionistic Fuzzy Sets **24**(1), 37–47 (2018)

8. Ben Amma, B., Melliani, S., Chadli, L.S.: Intuitionistic fuzzy functional differential equations. In: Fuzzy Logic in Intelligent System Design: Theory and Applications, pp. 335–357. Springer International Publishing, Cham (2018)
9. Ben Amma, B., Melliani, S., Chadli, L.S.: A fourth order Runge-Kutta gill method for the numerical solution of intuitionistic Fuzzy differential equations. Recent Advances in Intuitionistic Fuzzy Logic Systems Studies in Fuzziness and Soft Computing, vol. 372. Springer, Berlin (2019)
10. Ben, Amma B., Melliani, S., Chadli, L.S.: The existence and uniqueness of intuitionistic fuzzy solutions for intuitionistic fuzzy partial functional differential equations. Int. J. Diff. Equ. **2019**(2019), 1–13 (2019)
11. Ben Amma, B., Melliani , S., Chadli, L.S.: Integral boundary value problem for intuitionistic fuzzy partial hyperbolic differential equations. Nonlinear Analysis and Boundary Value Problems. Springer Proceedings in Mathematics & Statistics, vol. 292 (2019)
12. Ben Amma, B., Melliani, S., Saadia Chadli, L.: The numerical solution of intuitionistic fuzzy differential equations by the third order Runge-Kutta Nyström method, Intuitionistic and Type-2 Fuzzy Logic Enhancements in Neural and Optimization Algorithms: Theory and Applications, vol. 862. Springer International Publishing (2020)
13. Castillo, O., Melin, P.: Short remark on fuzzy sets, interval type-2 fuzzy sets, general type-2 fuzzy sets and intuitionistic fuzzy sets. IEEE Int. Conf. Intell. Syst. **2014**(1), 183–190 (2014)
14. De, S.K., Biswas, R. Roy, A.R.: An application of intuitionistic fuzzy sets in medical diagnosis. Fuzzy Sets Syst. **117**, 209–213 (2001)
15. Melliani, S., Chadli, L.S.: Intuitionistic fuzzy differential equation. Notes Intuitionistic Fuzzy Sets **6**, 37–41 (2000)
16. Mahapatra, G.S., Roy, T.K.: Reliability evaluation using triangular intuitionistic fuzzy numbers arithmetic operations. In: Proceedings of World Academy of Science Engineering and Technology, vol. 38, pp. 587–595 (2009)
17. Melliani, S., Elomari, M., Chadli, L.S., Ettoussi, R.: Intuitionistic Fuzzy metric space. Notes Intuitionistic Fuzzy sets **21**(1), 43–53 (2015)
18. Melliani, S., Atti, H., Ben Amma, B. Chadli, L.S.: Solution of n-th order intuitionistic fuzzy differential equation by variational iteration method. Notes Intuitionistic Fuzzy sets **24**(3), 92–105 (2018)
19. Shu, M.H., Cheng, C.H., Chang, J.R.: Using intuitionistic fuzzy sets for fault-tree analysis on printed circuit board assembly. Microelectron. Reliab. **46**(12), 2139–2148 (2006)
20. Sotirov, S., Sotirova, E., Melin, P., Castillo, O., Atanassov: Modular Neural Network Preprocessing Procedure with Intuitionistic Fuzzy InterCriteria Analysis Method. Flexible Query Answering Systems, pp. 175–186 (2015)
21. Sotirov, S., Sotirova, E., Atanassova, V., Atanassov, K., Castillo, O., Melin, P., Petkov, T., Surchev, S.: A hybrid approach for modular neural network design using intercriteria analysis and intuitionistic fuzzy logic. Complexity **2018** (2018)
22. Ye, J.: Multicriteria fuzzy decision-making method based on a novel accuracy function under interval valued intuitionistic fuzzy environment. Expert Syst. Appl. **36**, 6899–6902 (2009)
23. Zadeh, L.A.: Fuzzy sets. Inf. Control **8**(3), 338–353 (1965)

Fuzzy Hidden Markov Model in Images Segmentation

Meryem Ameur, Cherki Daoui, and Najlae Idrissi

Abstract In this work, we segment some fuzzy color and grey level images using the fuzzy model of Markov: Fuzzy Hidden Markov Chain, we use three algorithms EM, SEM and ICE for estimating the parameters to this model. We compare these estimators under some criteria of evaluation such as: segmentation quality, running time and convergence and complexity. The results of segmentation show that Fuzzy Hidden Markov Chain provides the same encouragement results under EM, SEM and ICE estimators. Moreover, EM algorithm has the less running time comparing to SEM and ICE, but SEM and ICE converge quickly than EM. They keep the same level of complexity, it is linear $O(N)$.

Keywords Image segmentation · HMC · FHMC · EM · SEM · ICE · MPM

1 Introduction

Hidden Markov Model(HMM) [1] is an statistical unsupervised method of image segmentation. It has a lot of representations in this latter such as: classical [2], pairwise [3, 4] and triplet [5–7], each representation represents a generalization to other, pairwise is the generalization of classical [8], and triplet is most general than these latter. Using a Markov model depends on the posed problem, classical categories of this model are used to treat the filtered stationaries data [9]. The pairwise categories are destinated to treat the problem of noised stationaries data [10], and the triplet categories are proposed to treat the problem of non-stationarity of data [11].

M. Ameur (✉) · C. Daoui · N. Idrissi
Faculty of Sciences and techniques, BP 523, Beni Mellal, Morocco
e-mail: m.ameur@usms.ma

C. Daoui
e-mail: c.daoui@usms.ma

N. Idrissi
e-mail: n.idrissi@usms.ma

S. Melliani and O. Castillo (eds.), *Recent Advances in Intuitionistic Fuzzy Logic Systems and Mathematics*, Studies in Fuzziness and Soft Computing 395,
https://doi.org/10.1007/978-3-030-53929-0_6

There exist another category of Markov models, that can treat the thematic and fuzzy data in the same time. This category is called Fuzzy Hidden Markov Model(FHMM) [12]. FHMM uses a mixture between probabilistic [13] and fuzzy set theories [14] to model the problem of fuzzy data [15].

HMM used in image segmentation considers the input image as an observed process Y, and the image result of segmentation as a hidden process X, and it estimates the process X from the process Y by calculating the a posteriori probability of X knowing Y basing on Bayes theory [13]. Following an estimation procedure. Starting by initialization, iterative estimation until convergence, finishing by obtained the final result. Theoretically, FHMM follows the same procedure but, it adds the fuzzy set theory [14] to estimate the fuzzy data.

In this work, we are interested to fuzzy model of Markov to segmenting some fuzzy color and grey level images. We consider a problem of image segmentation that contains two thematic classes between them there exist several fuzzy levels. Basing on discretization the continuous data.

In the phase of parameters estimation, we propose a comparative study between three estimators such as: EM(Expectation-Maximization) [16], SEM (Stochastic Expectation-Maximization) [17] and ICE(iterative Conditional Estimation) [18]. We are looking for conclude the estimator most effective, that can segment images fast with high quality taking account into fuzzy information. The final process X has been estimated by MPM(Marginal Posterior Mode) algorithm [19].

The rest of this paper is organized as follows: Sect. 2 presents the Fuzzy Hidden Markov Chain model and the different algorithms used in this computation. Section 3 shows the realized experiments and the obtained results. Last section gives a conclusion.

2 Fuzzy Hidden Markov Chain Model

This section begins by a brief presentation of classical model of Markov Hidden Markov Chain(HMC), also, it shows its fuzzy version, EM, SEM and ICE, algorithms suitable to Fuzzy Hidden Markov Chain(FHMC), finally, it closes by MPM algorithm.

2.1 Hidden Markov Chain with Independent Noise

Let consider $Z = (X, Y)$ where $X = (X_n)_{n=1}^N \in \Omega = \{\omega_1, \dots, \omega_K\}, \forall n \in N, \forall k \in K$, where K is number of membership classes and $Y = (Y_n)_{n=1}^N \in \mathbf{R}$ where N is the totale size of image.

The process $Z = (X, Y)$ is a Hidden Markov Chain(HMC) model, if it satisfied these three conditions:

1. The hidden process X is a Markov chain, its law is:

$$P(x) = P(x_1) \prod_{n=1}^{N-1} P(x_{n+1}|x_n) \tag{1}$$

2. $Y = (Y_n)_{n=1}^{N}$ are conditionally independent to $X = (X_n)_{n=1}^{N}$.
3. Each y_n depends only on its hidden state x_n.

$$P(Y_n = y_n|X) = P(Y_n = y_n|X_n = x_n) \tag{2}$$

This model is defined by two probability laws: the a priori law of process X and the gaussian law of observations Y. Each law has its parameters. The process X has two parameters θ_x: The initial law of X $PI(i)$ and the matrix of transition $A(i, j)$. The observations law admits two parameters θ_y: The mean m and the variance σ^2.

Parameters estimation requierts three phases of estimation: The first phase is to initialize parameters $\theta^0 = (\theta_x^0, \theta_y^0)$, the second phase is to estimate these parameters iteratively for a number of iterations Q using iterative algorithms like EM, SEM......The final phase is to decide the final configuration of process X.

To estimate parameters, it should use the Baum Welch algorithm [20]. Baum Welch is a recursive algorithm, it proceeds as:

1. Calculating the Forward and the Backward probabilities α β.
2. Calculating the Marginala posteriori probabilities ξ, and the Joint a posteriori probabilities γ basing on Forward Backward probabilities.

- Forward and Backward probabilities $\alpha = p(y_1, \ldots, y_n, x_n)$ and $\beta = p(y_{n+1}, \ldots, y_N|x_n)$ are calculated by:

  1. Initialization: for $n = 1$

$$\alpha_1(x_1) = p(x_1).p(y_1|x_1) \tag{3}$$

  for $n = N$

$$\beta_N(x_N) = 1 \tag{4}$$

  2. Induction: for $n > 1$

$$\alpha_n(x_n) = \sum_{x_{n-1}} \alpha_{n-1}(x_{n-1}) p(x_n|x_{n-1}).p(y_n|x_n) \tag{5}$$

for $n < N$

$$\beta_n(x_n) = \sum_{x_{n+1}} \beta_{n+1}(x_{n+1}) p(x_{n+1}|x_n).p(y_{n+1}|x_{n+1}) \tag{6}$$

- The joint a posteriori probabilities $\gamma_n(x_n, x_{n+1}) = p(x_n, x_{n+1}|y)$ are calculated by:

$$\gamma_n(x_n, x_{n+1}) = \alpha_n(x_n).p(x_{n+1}|x_n).p(y_{n+1}|x_{n+1}).\beta_{n+1}(x_{n+1}) \tag{7}$$

- The Marginal a posteriori probabilities $\xi_n(x_n) = p(x_n|y)$ are calculated using: $\forall n \in N$

$$\xi_n(x_n) = \sum_{x_{n+1}} \gamma_n(x_n, x_{n+1}) \tag{8}$$

or using:

$$\xi_n(x_n) = \alpha_n(x_n).\beta_n(x_n) \tag{9}$$

2.2 *Fuzzy Hidden Markov Chain with Independent Noise*

This model is introduced in [21] it is capable to represent the uncertain and the imprecision of observations in the same time using a mixture between Bayes and fuzzy set theories. It uses the probabilistic approach to represent the uncertain and the fuzzy approach to modelize the imprecision.

FHMC model keeps the same proprieties as classical model HMC.

Let $Z = (X, Y)$ Where $Y = (Y_n)_{n=1}^N \in \mathbf{R}$ and $X = (X_n)_{n=1}^N \in [0, 1]$ Z is a FHMC, if and else if it satisfied the same proprieties of HMC cited in the Sect. 2.1.

Contrary to HMC. To define the membership classes of pixels, FHMC assumes that:

- if $x_n = 0$, x_n belongs to thematic class 0.
- if $x_n = 1$, x_n belongs to thematic class 1.
- if $x_n = \zeta_f$, x_n belongs to fuzzy class of membership degree equal ζ_f.

To simplify:

- if $x_n \in \{0, 1\}$, x_n is a pixel of the thematic classes 0 or 1.
- if $x_n \in]0, 1[$, x_n is a pixel of the fuzzy classes.

The membership class of x_n is included on the interval $\Omega = [0, 1]$.

The law of X can be defined by the density measure ν where ν is a mixture of discrete and continuous components. This measure represents the discrete components by Dirac measure δ include on 0, 1 such as:

- δ_0 to represent the thematic class 0.
- δ_1 to represent the thematic class 1.

The fuzzy components are defined by Lesbegue measure μ, this measure defined under the open interval]0, 1[. Finally, the measure density ν is:

$$\nu = \delta_0 + \delta_1 + \mu \tag{10}$$

Also, we can use this:

$$p(0) + p(1) + \int_0^1 p(\zeta)\,\mathrm{d}\zeta = 1 \tag{11}$$

where:

- $p(0)$ and $p(1)$ represent the probability of thematic classes 0 and 1.
- The integral $\int_0^1 p(\zeta)\,\mathrm{d}\zeta$ represents the density of fuzzy components.

Numerically, FHMC discretizes the continuous part of fuzzy data approximately dividing the interval]0, 1[on a number of subintervals F such as: $[a_0 = 0, a_1 = \frac{1}{F}], [a_1 = \frac{1}{F}, a_2 = \frac{2}{F}], \ldots, [a_{F-1} = \frac{F-1}{F}, a_F = 1]$. The degrees of membership of each fuzzy level ζ_f is obtained by calculating the median value of each subinterval [22]. F is the number of fuzzy classes.

FHMC has the same laws and parameters as classical model HMC. The initial law PI, the transition matrix A, the mean m_0 and the variance σ_0^2 of thematic class 0, the mean m_1 and the variance σ_1^2 of thematic class 1 and the mean m_f and the variance σ_f of each classes $f \in]0, 1[$.

The mean m_f and the variance σ_f of fuzzy classes are calculated from the mean and the variance of the thematic classes.

$$m_f = \zeta_f.m_0 + \zeta_f.m_1 \forall \in]0, 1[. \tag{12}$$

$$\sigma_f^2 = (\zeta_f)^2.\sigma_0^2 + (1 - \zeta_f)^2.\sigma_1^2 \forall \in]0, 1[. \tag{13}$$

2.3 Parameters Estimation

Iterative Algorithms are used to estimate the parameters θ of HMC iteratively until ensuring the convergence. Each algorithm has its procedure to calculate parameters. Parameters estimation proceeds in two steps, the first step is for calculating the probabilities using Baum Welch algorithm [20], the second phase is to estimate the parameters of HMC.

Iterative algorithms execute these steps for a number of iterations until achieve convergence. The first step is the same for all estimators but, the second step differs from an estimator to another, each algorithm uses its owner estimation strategies.

In the first phase the algorithm calculates some probabilities for each pixel component of the input image thematic or fuzzy using Baum Welch algorithm cited in Sect. 2.1 these probabilities will be used to calculate the parameters of FHMC for the second step of estimation.

2.3.1 EM Algorithm

EM algorithm [23] estimates the parameters by maximizing the likelihood $P(x, y|\theta)$. it uses the deterministic strategy to calculate the parameters $\theta = (\theta_x, \theta_y)$ of FHMC. The convergence of this algorithm depends on parameters initialization. The following algorithm explains EM procedure:

Algorithm 1 EM Algorithm

Require: $q = 0$
Require: $\theta^0 = (PI^0, A^0, m_i^0, (\sigma_i^0)^2, g_i^0)$
Ensure: $\theta^Q = (PI^Q, A^Q, m_i^Q, (\sigma_i^Q)^2)$
for Each iteration $q \in Q$: **do**
Calculating the probabilities α^q,β^q,γ^q and ξ^q using Baum Welch algorithm.
Estimating the parameters θ_x^{q+1} of hidden process X:

$$PI^{q+1}(i) = \xi_1^q(i) \forall i \in [0, 1] \tag{14}$$

$$A^{q+1}(i, j) = \frac{\sum_{n=1}^{N} \gamma_n^q(i, j)}{\sum_{n=1}^{N} \xi_n^q(i)} \forall i, j \in [0, 1] \tag{15}$$

Estimating the parameters θ_y^{q+1} of thematic observations Y:

$$m_i^{q+1} = \frac{\sum_{n=1}^{N} y_n . \xi_n^q(i)}{\sum_{n=1}^{N} \xi_n^q(i)} \forall i \in \{0, 1\} \tag{16}$$

$$(\sigma_i^{q+1})^2 = \frac{\sum_{n=1}^{N} (y_n - m_i^q)^2 . \xi_n^q(i)}{\sum_{n=1}^{N} \xi_n^q(i)} \forall i \in \{0, 1\} \tag{17}$$

Estimating the parameters θ_y^{q+1} of fuzzy observations Y using the Eqs. 12 and 13 $\forall i \in]0, 1[$
Calculating the gaussian density g
$q = q + 1$
end for

2.3.2 SEM Algorithm

SEM algorithm is a stochastic version of EM [24], it replaces the likelihood $P(x, y|\theta)$ by the empire average $\rho = \frac{1}{\mathcal{T}}.\sum_{\tau=1}^{\mathcal{T}}.P(x, y|\theta)$, it uses the stochastic strategy to calculate the parameters $\theta = (\theta_x, \theta_y)$ of FHMC. The following algorithm explains SEM procedure:

Algorithm 2 SEM Algorithm

Require: $q = 0$
Require: $\theta^0 = (PI^0, A^0, m_i^0, (\sigma_i^0)^2, g_i^0)$
Ensure: $\theta^Q = (PI^Q, A^Q, m_i^Q, (\sigma_i^Q)^2)$
1: **for** Each iteration $q \in Q$: **do**
2: Calculating the probabilities α^q,β^q,γ^q and ξ^q using Baum Welch algorithm.
3: Simulating the process X^q for one random simulation
4: Estimating the parameters θ_x^{q+1} of hidden process $X \in [0, 1]$:

$$PI^{q+1}(i) = \frac{1}{N-1}\sum_{n=1}^{N} 1[x_n = i] \forall i \in [0, 1] \tag{18}$$

$$A^{q+1}(i, j) = \frac{\sum_{n=2}^{N} 1[x_n = j, x_{n-1} = i]}{N-1} \forall i, j \in [0, 1] \tag{19}$$

5: Estimating the parameters θ_y^{q+1} of thematic observations Y:

$$m_i^{q+1} = \frac{\sum_{n=1}^{N} y_n 1[x_n = i]}{\sum_{n=1}^{N} 1[x_n = i]} \forall i \in \{0, 1\} \tag{20}$$

$$(\sigma_i^{q+1})^2 = \frac{\sum_{n=1}^{N} (y_n - \mu_i^q)^2 1[x_n = i]}{\sum_{n=1}^{N} 1[x_n = i]} \forall i \in \{0, 1\} \tag{21}$$

6: Estimating the parameters θ_y^{q+1} of fuzzy observations Y using the Eqs. 12 and 13 $\forall i \in]0, 1[$
7: Calculating the gaussian density g
8: $q = q + 1$
9: **end for**

2.3.3 ICE Algorithm

ICE algorithm has the same principle as SEM, but it uses the deterministic and stochastic strategies at the same time to calculate parameters of FHMC, it estimates the parameters of X by deterministic strategy and those of observations Y using stochastic strategy. The following algorithm explains ICE procedure:

Algorithm 3 ICE Algorithm

Require: $q = 0$
Require: $\theta^0 = (PI^0, A^0, m_i^0, (\sigma_i^0)^2, g_i^0)$
Ensure: $\theta^Q = (PI^Q, A^Q, m_i^Q, (\sigma_i^Q)^2)$
1: **for** Each iteration $q \in Q$: **do**
2: Calculating the probabilities α^q,β^q,γ^q and ξ^q using Baum Welch algorithm.
3: Simulating the process X^q for one random simulation
4: Estimating the parameters θ_x^{q+1} of hidden process X:

$$PI^{q+1}(i) = \xi_1^q(i) \forall i \in [0, 1] \tag{22}$$

$$A^{q+1}(i, j) = \frac{\sum_{n=1}^{N} \gamma_n^q(i, j)}{\sum_{n=1}^{N} \xi_n^q(i)} \forall i, j \in [0, 1] \tag{23}$$

5: Estimating the parameters θ_y^{q+1} of thematic observations Y:

$$m_i^{q+1} = \frac{\sum_{n=1}^{N} y_n 1[x_n = i]}{\sum_{n=1}^{N} 1[x_n = i]} \forall i \in \{0, 1\} \tag{24}$$

$$(\sigma_i^{q+1})^2 = \frac{\sum_{n=1}^{N} (y_n - \mu_i^q)^2 1[x_n = i]}{\sum_{n=1}^{N} 1[x_n = i]} \forall i \in \{0, 1\} \tag{25}$$

6: Estimating the parameters θ_y^{q+1} of fuzzy observations Y using the Eqs. 12 and 13 $\forall i \in]0, 1[$
7: Calculating the gaussian density g
8: $q = q + 1$
9: **end for**

2.3.4 MPM Algorithm

MPM algorithm is used to estimating the configuration of final process X^Q. MPM estimator estimates for each pixel y_n its suitable membership thematic or fuzzy class by maximizing the a posteriori probability $P(X_n = x_n|Y_n = y_n)$. It applies the following formula:

$$\bar{x}_n = argmax_{i \in [0,1]}(\alpha_i(n).\beta_i(n)) = argmax_{i \in [0,1]}((\xi_i(n))). \tag{26}$$

3 Experimental Results

This section presents the different realized experiments and obtained results, we have carried out four experiments, in the first and the second one we have segmented color images, in the third and the fourth experiment we have grey level images. We consider the case of the number of thematic classes $K = 2$. We have varied the number of fuzzy classes F beginning by $F = 1$ until $F = 8$. We have defined each experiment by its visual results and the values of PSNR index, SSIM index, error rate, running time and convergence. The parameters of FHMC have been initialized by: The initial law of X PI^0 and the matrix of transition A^0 are initialized using these algorithms:

Algorithm 4 Initialization PI^0

Require: $K + F$: number of membership class
Ensure: Initial law PI^0
for $(i = 1; i = K + F; i++)$ **do**
  $PI^0(i) = 1./K + F$
end for

Algorithm 5 Initialization A^0

Require: $K + F$: number of membership class
Ensure: Transition matrix A^0
for $(i = 1; i = K + F; i++)$ **do**
  for $(j = 1, j = K + F; j++)$ **do**
    if $(i == j)$ **then**
      $A^0(i, j) = 1/2$
    else
      $A^0(i, j) = 1/2((K + F) - 1)$
    end if
  end for
end for

The mean m^0 and the variance $(\sigma_i^0)^2$ of thematic classes are calculated from the initial process X^0, this latter has been obtained by K-means [25]. The fuzzy initial mean m_f^0 and variance $(\sigma_f^0)^2$ are calculated from the mean and the variance of thematic classes using the Eqs. 12 and 13.

3.1 *Visual Results*

We have summarized the visual results in the following figures (Figs. 1, 2, 3 and 4):

(a) Input image Y

(b) Initial Configuration X^0

(c) Final Process X EM F=1

(d) Final Process X SEM F=1

(e) Final Process X ICE F=1

(f) Final Process X EM F=2

(g) Final Process X SEM F=2

(h) Final Process XICE F=2

(i) Final Process X EM F=3

Fig. 1 Results of fuzzy grey level image segmentation

(j) Final Process X SEM F=3

(k) Final Process X ICE F=3

(l) Final Process XEM F=4

(m) Final Process XSEM F=4

(n) Final Process X ICE F=4

(o) Final Process X EM F=5

(p) Final Process X SEM F=5

(q) Final Process X ICE F=5

(r) Final Process X EM F=6

Fig. 1 (continued)

(s) Final Process X SEM F=6

(t) Final Process X ICE F=6

(u) Final Process X EM F=7

(v) Final Process X SEM F=7

(w) Final Process X ICE F=7

(x) Final Process X EM F=8

(y) Final Process X SEM F=8

(z) Final Process X ICE F=8

Fig. 1 (continued)

(a) Input image Y

(b) Initial Configuration X^0

(c) Final Process X EM F=1

(d) Final Process X SEM F=1

(e) Final Process X ICE F=1

(f) Final Process X EM F=2

(g) Final Process X SEM F=2

(h) Final Process X ICE F=2

(i) Final Process X EM F=3

(j) Final Process X SEM F=3

(k) Final Process X ICE F=3

(l) Final Process X EM F=4

Fig. 2 Results of fuzzy grey level image segmentation

(m) Final Process X SEM F=4

(n) Final Process X ICE F=4

(o) Final Process X EM F=5

(p) Final Process X SEM F=5

(q) Final Process X ICE F=5

(r) Final Process X EM F=6

(s) Final Process X SEM F=6

(t) Final Process X ICE F=6

(u) Final Process X EM F=7

(v) Final Process X SEM F=7

(w) Final Process X ICE F=7

(x) Final Process X EM F=8

(y) Final Process X SEM F=8

(z) Final Process X ICE F=8

Fig. 2 (continued)

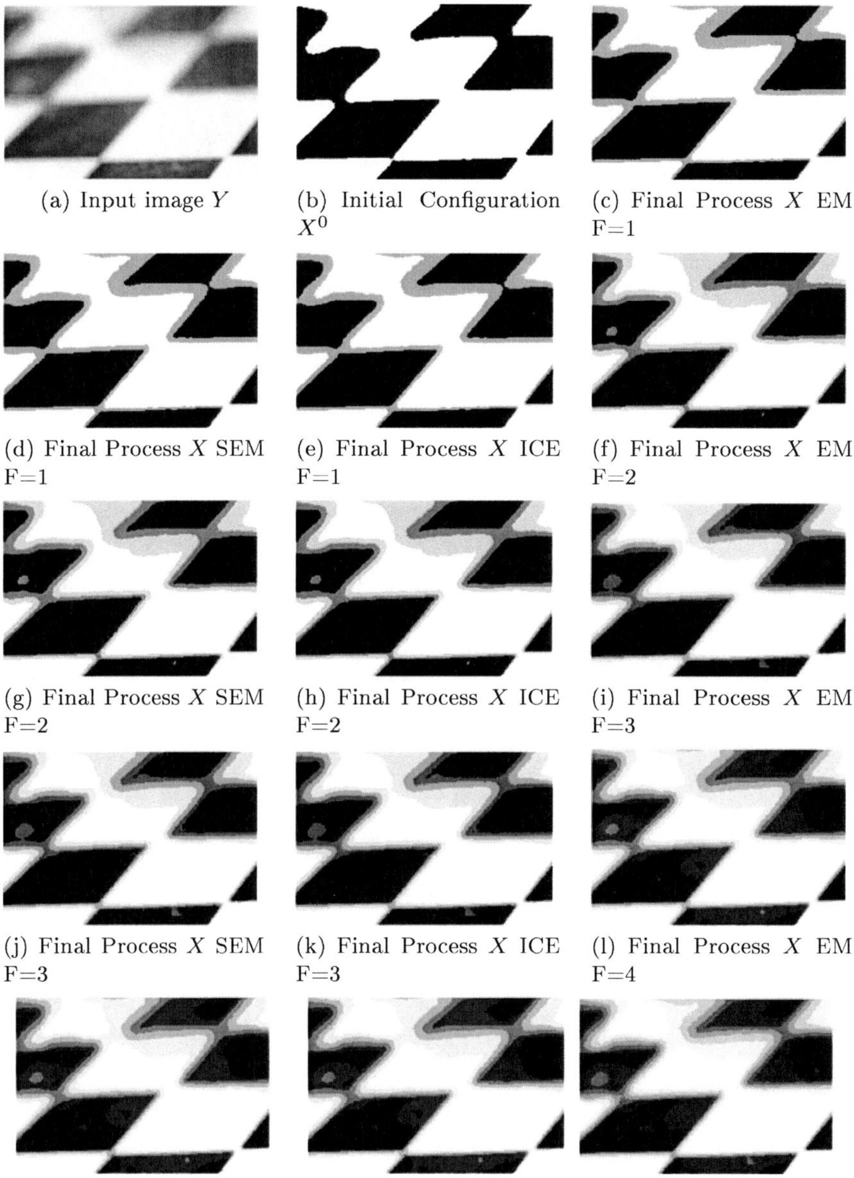

(a) Input image Y

(b) Initial Configuration X^0

(c) Final Process X EM F=1

(d) Final Process X SEM F=1

(e) Final Process X ICE F=1

(f) Final Process X EM F=2

(g) Final Process X SEM F=2

(h) Final Process X ICE F=2

(i) Final Process X EM F=3

(j) Final Process X SEM F=3

(k) Final Process X ICE F=3

(l) Final Process X EM F=4

(m) Final Process X SEM F=4

(n) Final Process X ICE F=4

(o) Final Process X EM F=5

Fig. 3 Results of fuzzy grey level image segmentation

(p) Final Process X SEM F=5

(q) Final Process X ICE F=5

(r) Final Process X EM F=6

(s) Final Process X SEM F=6

(t) Final Process X ICE F=6

(u) Final Process X EM F=7

(v) Final Process X SEM F=7

(w) Final Process X ICE F=7

(x) Final Process X EM F=8

(y) Final Process X SEM F=8

(z) Final Process X ICE F=8

Fig. 3 (continued)

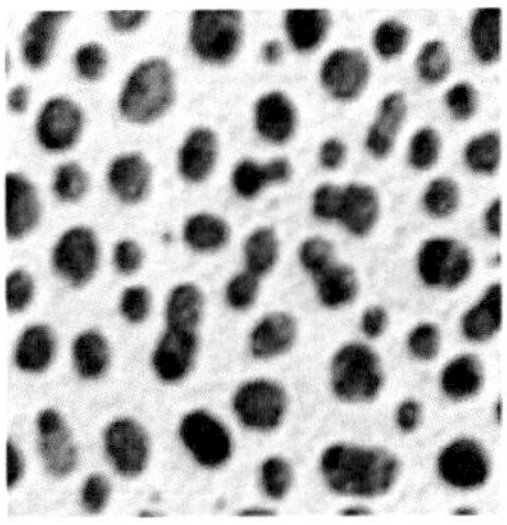
(a) Input image Y

(b) Initial Configuration X^0

(c) Final Process X EM F=1

(d) Final Process X SEM F=1

(e) Final Process X ICE F=1

(f) Final Process X EM F=2

(g) Final Process X SEM F=2

(h) Final Process X ICE F=2

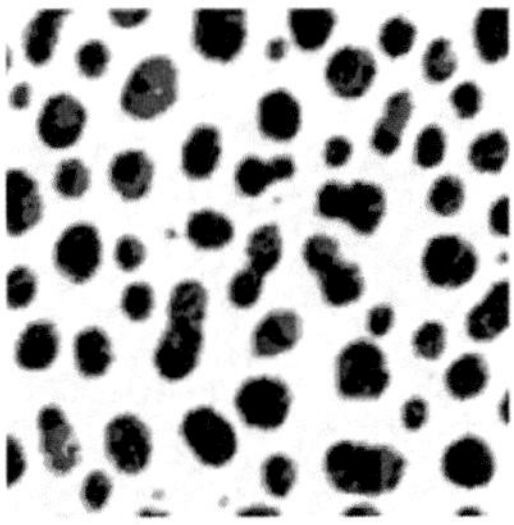
(i) Final Process X EM F=3

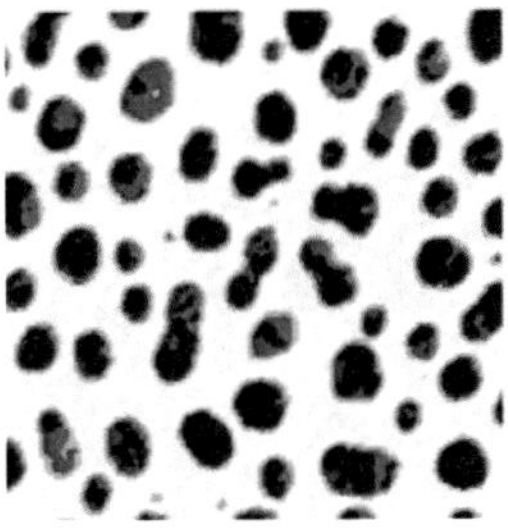
(j) Final Process X SEM F=3

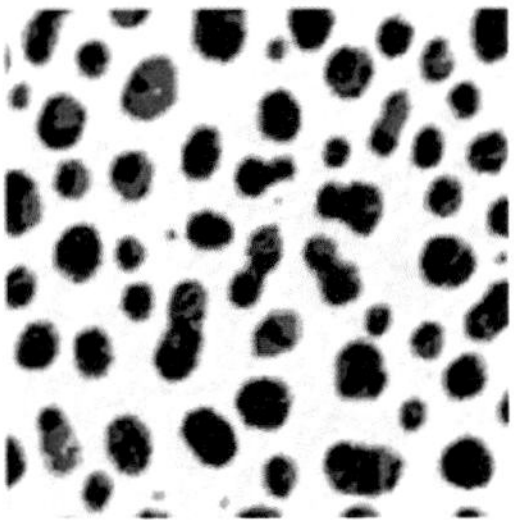
(k) Final Process X ICE F=3

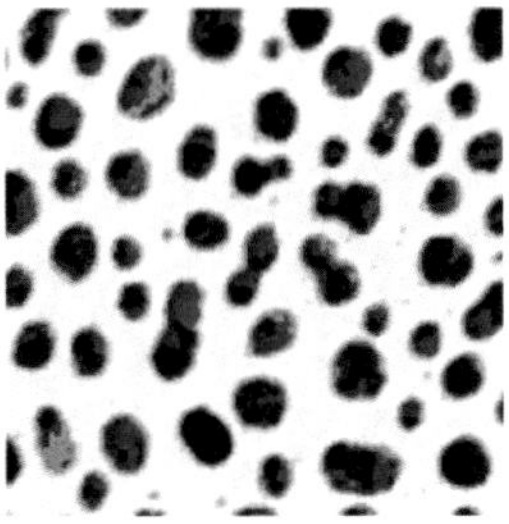
(l) Final Process X EM F=4

Fig. 4 Results of fuzzy grey level image segmentation

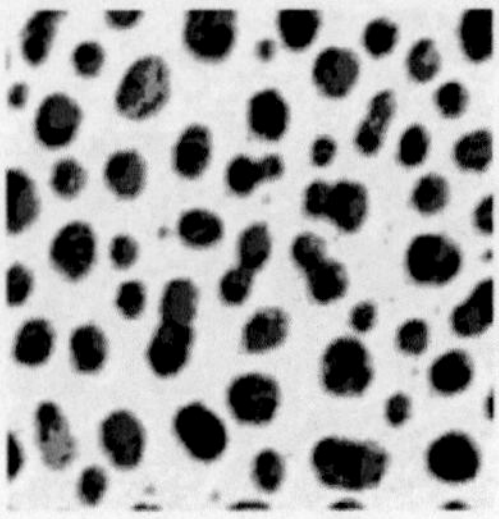

(m) Final Process X SEM F=4

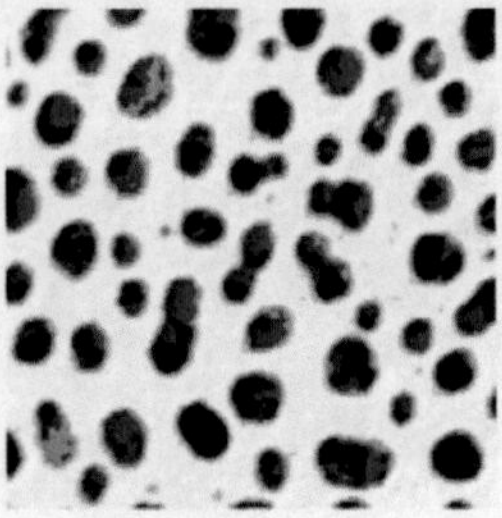

(n) Final Process X ICE F=4

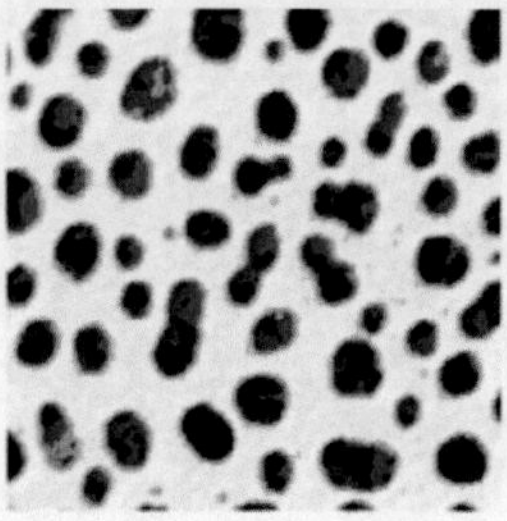

(o) Final Process X EM F=5

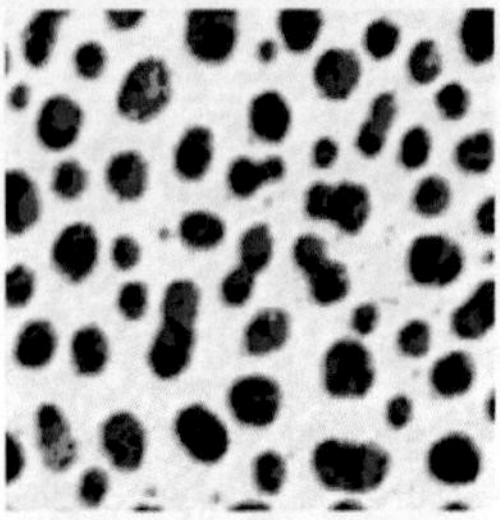

(p) Final Process X SEM F=5

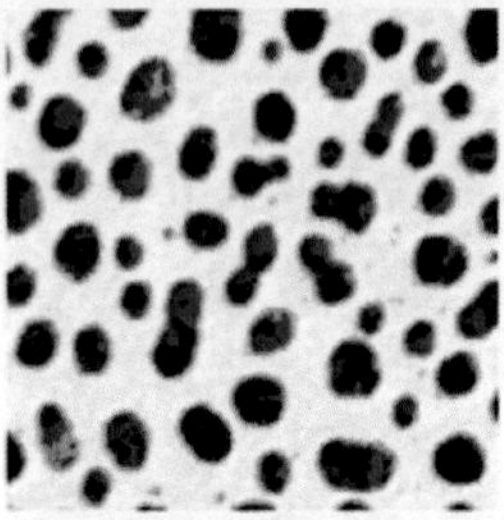

(q) Final Process X ICE F=5

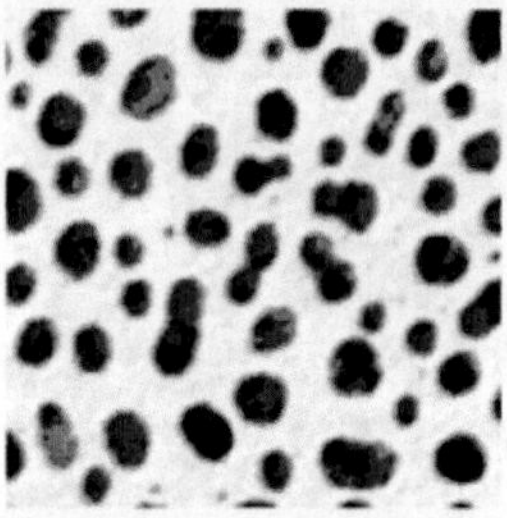

(r) Final Process X EM F=6

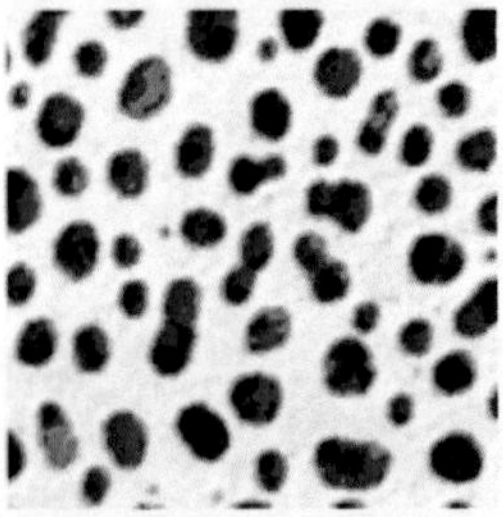

(s) Final Process X SEM F=6

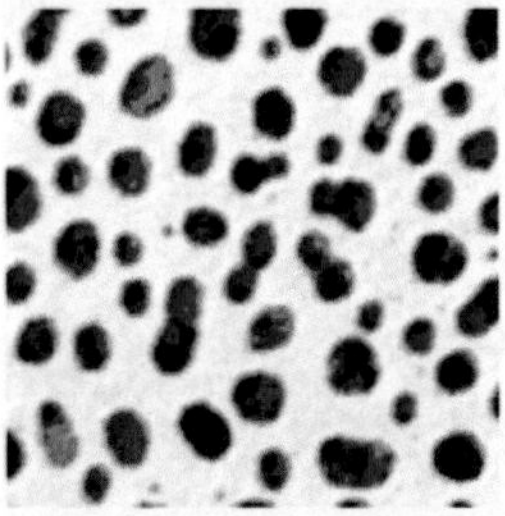

(t) Final Process X ICE F=6

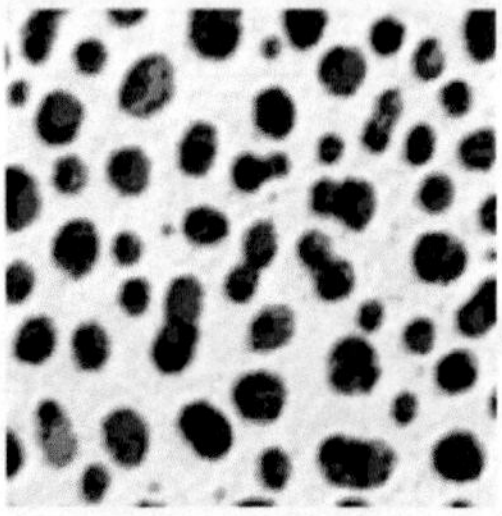

(u) Final Process X EM F=7

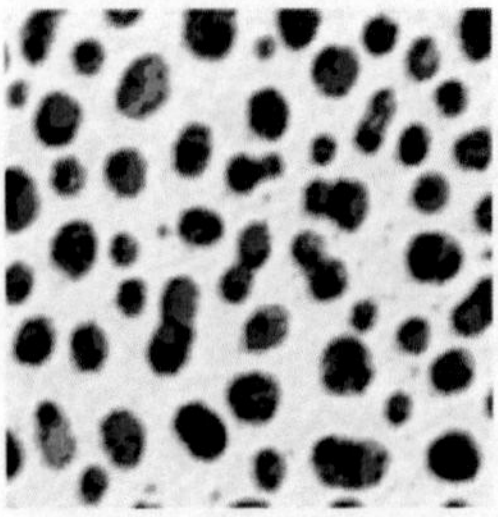

(v) Final Process X SEM F=7

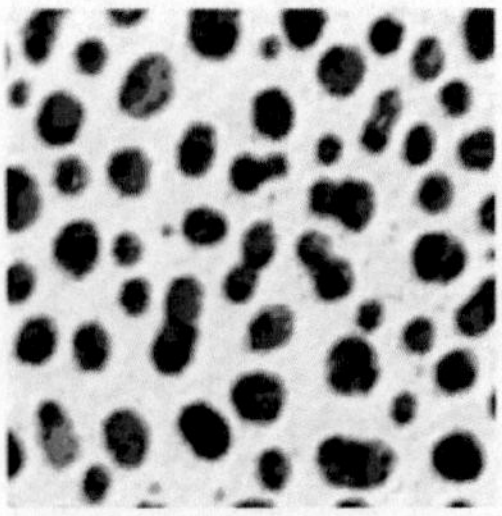

(w) Final Process X ICE F=7

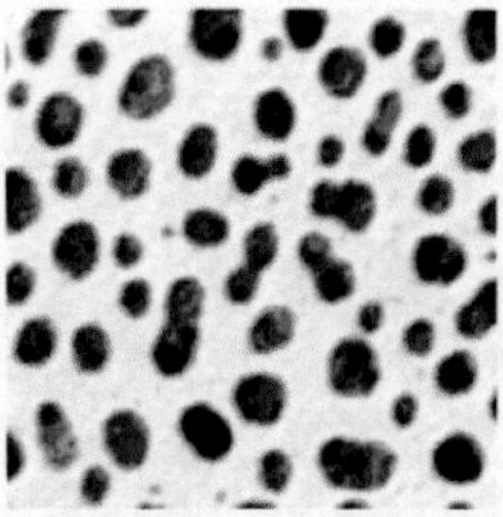

(x) Final Process X EM F=8

Fig. 4 (continued)

Fig. 4 (continued)

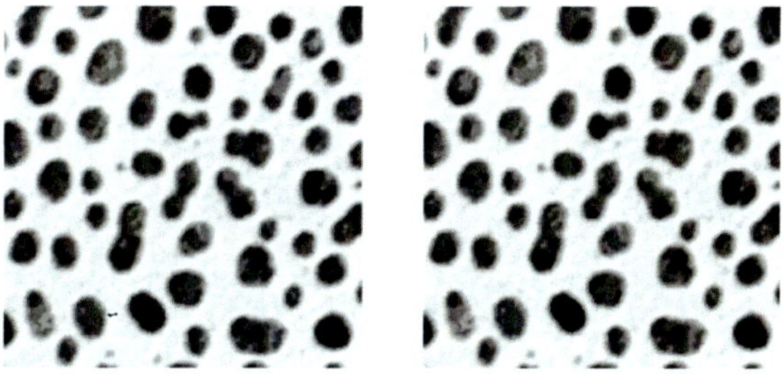

(y) Final Process X SEM F=8

(z) Final Process X ICE F=8

From the obtained resulted images we can remark that, when we increase the number of fuzzy levels, the fuzzy information appear well. There do not exist a difference between estimators. To verify that we have calculated some evaluated measures, the next Sect. 3.2 shows the obtained quantitative results of each experiment by estimator.

3.2 *Quantitative Results*

To validate the visual results. We have calculated for each estimator: the index of PSNR(Peak Signal-to-Noise Ratio), SSIM(Structural Similarity Index Measure) [26] and error rate. The Tables 1, 2, 3, 4, 5, 6, 7 and 8 represent respectively the obtained values by experiment 1, 2, 3 and 4.

Table 1 PSNR index and SSIM index values of experiment 1

Fuzzy level	EM	SEM	ICE	EM	SEM	ICE
F = 1	32.7822	32.7822	32.7822	0.8875	0.8875	0.8875
F = 2	35.211	35.211	35.211	0.8949	0.8949	0.8949
F = 3	36.9479	36.9479	36.9479	0.9129	0.9129	0.9129
F = 4	38.6609	38.6609	38.6609	0.9208	0.9208	0.9208
F = 5	39.7903	39.7903	39.7903	0.9312	0.9312	0.9312
F = 6	41.0536	41.0536	41.0536	0.9442	0.9442	0.9442
F = 7	42.0451	42.0451	42.0451	0.9534	0.9534	0.9534
F = 8	42.8346	42.8346	42.8346	0.9589	0.9589	0.9589

Table 2 Error rate values of experiment 1

Fuzzy level	EM	SEM	ICE
F = 1	8.3851	8.3851	8.3851
F = 2	7.6520	7.6520	7.6520
F = 3	5.129	5.129	5.129
F = 4	3.5735	3.5735	3.5735
F = 5	2.6564	2.6564	2.6564
F = 6	1.9287	1.9287	1.9287
F = 7	1.8609	1.8609	1.8609
F = 8	1.2245	1.2245	1.2245

Table 3 PSNR index and SSIM index values of experiment 2

Fuzzy level	EM	SEM	ICE	EM	SEM	ICE
F = 1	28.9551	28.9551	28.9551	0.8697	0.8697	0.8697
F = 2	29.6579	29.6579	29.6579	0.8715	0.8715	0.8715
F = 3	30.4530	30.4530	30.4530	0.8767	0.8767	0.8767
F = 4	31.0568	31.0568	31.0568	0.8855	0.8855	0.8855
F = 5	32.4166	32.4166	32.4166	0.8933	0.8933	0.8933
F = 6	33.2867	33.2867	33.2867	0.8999	0.8999	0.8999
F = 7	34.3880	34.3880	34.3880	0.9094	0.9094	0.9094
F = 8	35.5017	35.5017	35.5017	0.9227	0.9227	0.9227

Table 4 Error rate values of experiment 2

Fuzzy level	EM	SEM	ICE
F = 1	8.9102	8.9102	8.9102
F = 2	5.1631	5.1631	5.1631
F = 3	3.6205	3.6205	3.6205
F = 4	2.6007	2.6007	2.6007
F = 5	1.8835	1.8835	1.8835
F = 6	1.4897	1.4897	1.4897
F = 7	1.1607	1.1607	1.1607
F = 8	0.0882	0.0882	0.0882

Table 5 PSNR index and SSIM index values of experiment 3

Fuzzy level	EM	SEM	ICE	EM	SEM	ICE
F = 1	28.4167	28.4167	28.4167	0.7425	0.7425	0.7425
F = 2	29.6297	29.6297	29.6297	0.7950	0.7950	0.7950
F = 3	30.0566	30.0566	30.0566	0.8330	0.8330	0.8330
F = 4	31.0789	31.0789	31.0789	0.8488	0.8488	0.8488
F = 5	31.6134	31.6134	31.6134	0.8686	0.8686	0.8686
F = 6	32.2726	32.2726	32.2726	0.8908	0.8908	0.8908
F = 7	32.9150	32.9150	32.9150	0.8972	0.8972	0.8972
F = 8	33.2435	33.2435	33.2435	0.9115	0.9115	0.9115

Table 6 Error rate values of experiment 3

Fuzzy level	EM	SEM	ICE
F = 1	5.9939	5.9939	5.9939
F = 2	3.2610	3.2610	3.2610
F = 3	2.1363	2.1363	2.1363
F = 4	1.4876	1.4876	1.4876
F = 5	1.1025	1.1025	1.1025
F = 6	0.8304	0.8304	0.8304
F = 7	0.7191	0.7191	0.7191
F = 8	0.6122	0.6122	0.6122

Table 7 PSNR index and SSIM index values of experiment 4

Fuzzy level	EM	SEM	ICE	EM	SEM	ICE
F = 1	29.0691	29.0691	29.0691	0.6935	0.6935	0.6935
F = 2	29.3664	29.3664	29.3664	0.7562	0.7562	0.7562
F = 3	29.9264	29.9264	29.9264	0.8356	0.8356	0.8356
F = 4	30.3177	30.3177	30.3177	0.8642	0.8642	0.8642
F = 5	31.0392	31.0392	31.0392	0.8901	0.8901	0.8901
F = 6	31.4969	31.4969	31.4969	0.9115	0.9115	0.9115
F = 7	31.8957	31.8957	31.8957	0.9223	0.9223	0.9223
F = 8	32.1925	32.1925	32.1925	0.9381	0.9381	0.9381

Table 8 Error rate values of experiment 4

Fuzzy level	EM	SEM	ICE
F = 1	4.7335	4.7335	4.7335
F = 2	3.0220	3.0220	3.0220
F = 3	2.1496	2.1496	2.1496
F = 4	1.5313	1.5313	1.5313
F = 5	1.3740	1.3740	1.3740
F = 6	0.9490	0.9490	0.9490
F = 7	0.9139	0.9139	0.9139
F = 8	0.6614	0.6614	0.6614

The obtained values of PSNR, SSIM and error rate confirm the visual results, EM, SEM and ICE provide the same results of segmentation. Despite of the strategies of estimation used are differents. When, the number of fuzzy levels increases the values of PSNR and SSIM increase, and the error rate values decrease. The quality of fuzzy segmentation increases when the fuzzy levels increase.

Additionally to that, we have compared these estimators in level of running time and convergence. We have calculated the average running time by taking the sum of execution time of all fuzzy levels dividing by the total number of fuzzy levels $F = 8$ for EM, SEM and ICE in each experiment. By the same manner, we have calculated the averge of iteration number ensuring convergence for each estimator. The Figs. 5 and 6 represent respectively the running time and convergence.

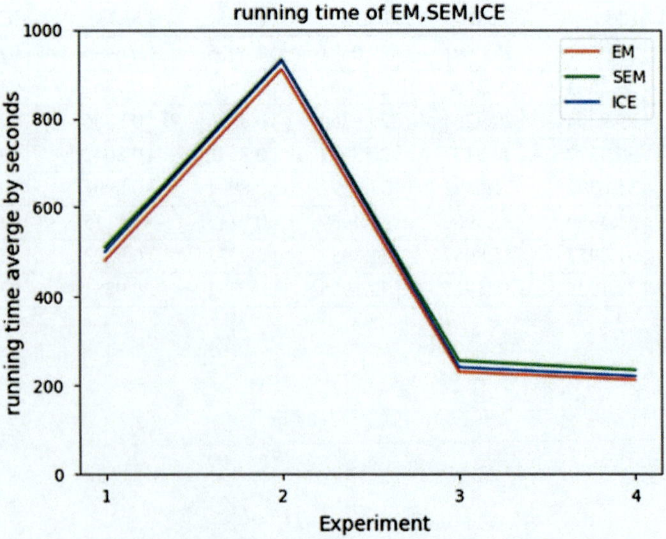

Fig. 5 Running time by algorithm per seconds

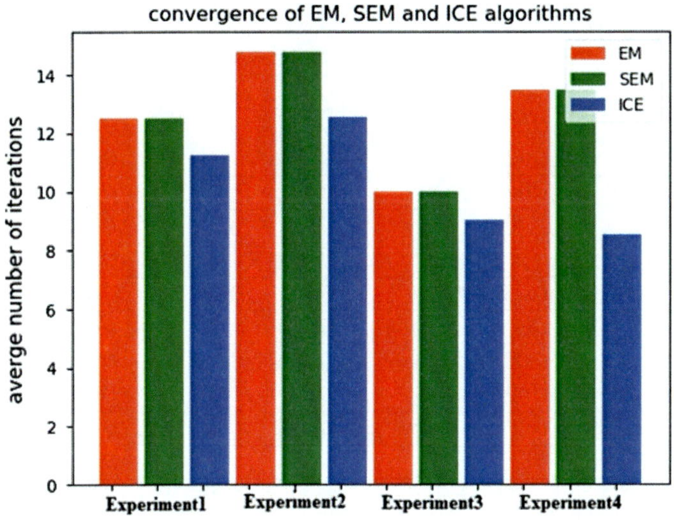

Fig. 6 Convergence by algorithm

From the graphic 1, EM is faster than SEM and ICE, but the difference between them is not important. EM is faster approximately by 15% comparing to SEM and by 10% comparing to ICE. The running time of ICE is close to SEM running time. ICE is faster than SEM by around 5%. We explain that by the estimation strategies used. SEM and ICE use the stochastic strategy to estimate parameters, this strategy can be slow than deterministic strategy that EM uses, because, stochastic strategy needs a phase of simulation process for several times. Remarkably, When the size and the type of treated data increase, the running time increases.

The graphic 2 shows that ICE and SEM converge faster than EM.

ICE and SEM have a speed of convergence faster than EM. But, EM has an execution speed rapide than ICE and SEM. It is difficult to conclude the most effective algorithm estimator, in level of quality, running time and convergence.

Additionally to that, we have compared these estimators in level of complexity, for that, we have calculated the complexity of each task executed by each algorithm, the following table illustrates the complexity of EM, SEM and ICE by task executed.

Let:

- N: is the total size of input image.
- K: is the number of thematic classes.
- F: is the number of fuzzy classes.

Table 9 Comparison of the complexity

Task	EM	SEM	ICE
Forward α	$O((K+F)^2N)$	$O((K+F)^2N)$	$O((K+F)^2N)$
Backward β	$O((K+F)^2N)$	$O((K+F)^2N)$	$O((K+F)^2N)$
Joint a posteriori γ	$O((K+F)^2N)$	$O((K+F)^2N)$	$O((K+F)^2N)$
Marginal a posteriori ξ	$O((K+F)N)$	$O((K+F)^2N)$	$O((K+F)^2N)$
Simulating X	–	$O((K+F)N)$	$O((K+F)N)$
Initial law PI	$O(K+F)$	$O(K+F)$	$O(K+F)$
Matrix of transition A	$O((K+F)^2N)$	$O((K+F)^2N)$	$O((K+F)^2N)$
Mean m	$O(KN)$	$O(KN)$	$O(KN)$
Variance σ^2	$O(KN)$	$O(KN)$	$O(KN)$
Density f	$O((K+F)N)$	$O((K+F)N)$	$O((K+F)N)$
Fuzzy mean m_f	$O(F)$	$O(F)$	$O(F)$
Fuzzy variance σ	$O(F)$	$O(F)$	$O(F)$
Density g	$O((K+F)N)$	$O((K+F)N)$	$O((K+F)N)$

From the Table 9 the complexity level of EM is less than ICE and SEM complexity. EM doesn't execute the task of simulating the process X to estimate the parameters. ICE and SEM have the same level of complexity.

FHMC is more complex than classical HMC. FHMC takes account into fuzzy components, it calculates the mean and the variance of each fuzzy level. When, the number of level fuzzy or the size of data increase the calculus complexity increases. From the Mastser theorem [27], we can judge that EM, SEM and ICE have a linear complexity $O(N)$.

4 Conclusion

In this paper, we have carried out a comparative study between three algorithms estimators EM, SEM and ICE used for estimating the parameters of Fuzzy Hidden Markov Chain model applied to segment some fuzzy color and grey level images, we have compared these algorithms in term of quality of segmentation, running time, convergence and calculus complexity. The obtained values demonstrate that, FHMC provides the same results of segmentation under the three estimators used. The running time of EM is fast than those of SEM and ICE, the number of iterations to ensure convergence of ICE and SEM is less than EM. Approximately, These algorithms keep the same level of complexity. We have used a fuzzy classical model of Markov that considers two thematic classes. And it determinates the fuzzy level between them, as an open question that, we can address to the next computation is to generalize this model to be capable to segment more than two thematic classes and to determinate the different fuzzy levels between these classes independently.

References

1. Van Hadel, R.: Hidden Markov models, pp. 51–64. Accessed 28 July 2008
2. Zhang, Y., et al.: Segmentation of brain MR images through a hidden Markov field model and the expectation –maximization algorithm. In: IEEE Transactions on Medical Imaging, vol. 20, No. 1. Accessed 1 Jan 2001
3. Pieczynski, W.: Pairwise Markov chains. IEEE Trans. Pattern Anal. Mach. Intell. **25**(5) (2005)
4. Brunel, N., Pieczynski, W.: Signal restoration using hidden Markov chains with copulas. Signal Process. 2304–2315 (2005)
5. Pieczynski, W.: Multisensor triplet Markov chains and theory of evidence. Int. J. Approx. Reason. **45**(1), 1–16 (2007)
6. Boudaren, M.E.Y., An, L., Pieczynski, W.: Dempster-Shafer fusion of evidential pairwise Markov fields. Int. J. Approx. Reas. 13–29 (2016)
7. Boudaren, M.E.Y., Pieczynski, W.: Dempster–Shafer fusion of evidential pairwise Markov chains. IEEE Trans. Fuzzy Syst. (2016) https://doi.org/10.1109/TFUZZ.2016.2543750
8. Derrode, S., Pieczynski, W.: SAR image segmentation using generalized Pairwise Markov Chains. In: SPIE's International Symposium on Remote Sensing, September 22–27. Crete, Greece (2002)
9. Derrode, S., Pieczynski, W.: Segmentation non supervisée d'images par une chaine de Markov couple (2004)
10. Derrode, S., Pieczynski, W.: Signal and image segmentation using pairwise Markov Chains. IEEE Trans. Signal Process. **52**(9), 2477–2489 (2004). Sep
11. Lanchantin, P., Pieczynski, W.: Evidential Markov chains and trees with application to non-stationary processes segmentation, traitement du signal, vol. 22, No. 1 (2005)
12. Salzenstein, F., et al.: Non-stationary fuzzy Markov chain. Pattern Recognit. Lett. **8**, 2201–2208 (2007)
13. Bayes theorem: https://www.ucd.ie/msc/t4media/BayesTheorem.pdf
14. Zadeh, L.A.: Fuzzy sets as a basis for a theory of possibility. Fuzzy Sets Syst. J. **1**, 3–28 (1978)
15. Carincotte, C., Derrode, S., Bourennane, S.: Unsupervised change detection on SAR images using fuzzy hidden Markov chains. IEEE Trans. Geosci. Remote Sens. **44**(2) (2006)
16. Pieczynski, W.: EM and ICE in hidden and triplet Markov models, vol. 8–11 (2010)
17. Masson, P., Pieczynski, W.: SEM algorithm and unsupervised statistical segmentation of satellite images. Trans. Geosci. Remote Sens. **31**(3), 2201–2208 (1993)
18. Pieczynski, W.: Convergence of the iterative conditional estimation and application on the mixture proportion identification. In: IEEE Statistical Signal Workshop, SSP Madison, WI, USA, August 26–29 (2007)
19. Corner, M.L., Delp, E.J.: The EM/ MPM algorithm for segmentation of textured images. 1731–1744 (2000)
20. Devijver, P.: Baum's forward backward algorithm revisited. Pattern Recognit. Lett. **3**, 369–373 (1985)
21. Carincotte, C.: Unsupervised image segmentation based on a new fuzzy HMC model, ICASSP'04. Montreal, Canada (2004)
22. Lanchantin, P., Salzenstein, F.: Segmentation d'images multispectrales par arbre de Markov caché flou (2006)
23. Celeux, G., Diebolt, J.: Stochastic approximation type EM algorithm for the mixture (1991)
24. Biscarat, C., Celeux, G., Diebolt, J.: Stochastic versions of the EM algorithm (1985)
25. Tatiraju, S., et al.: Image segmentation using K-means clustering EM and normalized cuts (2008)
26. Al-Najjar, Y., Chen Soong, D.: August 2012. Comparison of Image Quality Assessment: PSNR, HVS, SSIM, UIQI. Int. J. Sci. Eng. Res. **3**(8) (2012)
27. Roura, S.: Improved master theorems for divide-and-conquer recurrences. J. ACM **48**(2), 170–205 (2001)

Solving the Intuitionistic Fuzzy Fractional Partial Differential Equations

Ali El Mfadel, Said Melliani, Mhamed Elomari, and Lalla Saadia Chadli

Abstract The present paper is devoted to discuss the existence and uniqueness of the solution of a fractional boundary problem. The main tools employed is fixed point of Schauder. we give an example in order to illustrate this situation.

1 Section Heading

Fractional differential equations(DEs) have been attracting more and more attention,since they are capable of modeling many different processes in physics, chemistry and engineering (see in [3, 4, 7]). Some different forms of fractional operators were introduced to study fractional DEs, such as the Grunwald-Letnikov, Riemann-Liouville and Caputo fractional derivatives, etc. Of these, the Caputo fractional derivative emerges as the best candidate to model many real-life processes due to its great advantages; namely that it allows traditional initial and boundary conditions to be included in the formulation of the problem [5]. In addition, its derivative at a constant value is zero. This lets the considered models adopt the theory of linear viscoelasticity and allows the utilization of physically interpretable initial conditions of other applied problems. In many real-word tasks, when we transform the behavior of a special phenomenon into a deterministic initial or boundary value problem of fractional DEs, it is worthwhile exploring the transformation of uncertainties in initial conditions, boundary conditions or the forcing functions of the model inputs into

A. El Mfadel · S. Melliani (✉) · M. Elomari · L. S. Chadli
Sultan Moulay Slimane University, 523 Beni Mellal, Morocco
e-mail: s.melliani@usms.ma

A. El Mfadel
e-mail: elmfadelali@gmail.com

M. Elomari
e-mail: m.elomari@usms.ma

L. S. Chadli
e-mail: sa.chadli@yahoo.fr

S. Melliani and O. Castillo (eds.), *Recent Advances in Intuitionistic Fuzzy Logic Systems and Mathematics*, Studies in Fuzziness and Soft Computing 395,
https://doi.org/10.1007/978-3-030-53929-0_7

model outputs. The uncertainty of the model input may come from many sources: the errors and approximations in input data measurement, parameter values of the model, model structure and model numerical solution algorithms,etc. In such situations, fractional DEs with uncertainty type of fuzziness are common notions if the underlying structure is vague or ambiguous. The analysis of the relative impacts of input variable uncertainties to the output variable leads us to study the qualitative behavior of the solutions of equations.

Along with fractional calculus techniques for crisp DEs, there have been significant developments in partial differential equations (PDEs) and partial differential inclusions. Some recent contributions can be seen in the monographs of Abbas et al. [1], Debnath and Bhatta [3], Moaddy et al. [7], Muslih et al. [8]. However when we fuzzify these models to adopt real-world problems containing uncertainties, we find that there has been no paper developed on this subject for intuitionistic fuzzy fractional PDEs up to now.

In this paper, we consider the fractional problem for a intuitionistic fuzzy hyperbolic involving the Caputo fractional gH-derivative in the form.

$$\begin{cases} D^q u(x,y) = f(x,y,u(x,y)); \ (x,y) \in J = [0,a] \times [0,b] \\ u(x,0) = g_1(x), \quad x \in [0,a] \\ u(0,y) = g_2(y), \quad y \in [0,b] \end{cases} \tag{1}$$

where $q = (q_1, q_2) \in [0, 1] \times [0, 1]$ is the fractional order of Caputo gH-Derivative operator D^q defined later in.

$g_1(x)$, $g_2(y)$ are two intuitionistic fuzzy functions.

The remainder of this paper is organized as follows. In Sect. 2,we set out some necessary preliminaries.In Sect. 3,we introduce the concepts of the intuitionistic fuzzy Riemann-Liouville integral and Caputo gH-derivatives for intuitionistic fuzzy-valued two-variable functions.In Sect. 4, we state the main problem for fractional intuitionistic fuzzy hyperbolic equations and prove the existence and uniqueness of The weak solution of this problem by using the Banach fixed point theorem. Finally, some conclusions and future work are discussed in Sect. 5.

2 Preliminaries

We denote by

$$\mathrm{IF}(\mathbb{R}) = \{\langle u, v\rangle : \mathbb{R} \longrightarrow [0,1]^2 \quad , \quad 0 \le u(x) + v(x) \le 1\}$$

Definition 1 ([4]) An element $\langle u, v\rangle \in \mathrm{IF}(\mathbb{R})$ is called an intuitionistic fuzzy nombber if it satisfy the following conditions:

1. $\langle u, v\rangle$ is normal, ie,there exists $x_0, x_1 \in \mathbb{R}$ such that $u(x_0) = 1$ et $u(x_1) = 1$.
2. u is fuzzy convex an v is fuzzy concave.

3. u is upper semi-continuous et v is lower semi-continuous.
4. $supp\langle u, v\rangle = \overline{\{x \in \mathbb{R} : v(x) < 1\}}$ is bounded.

we denote by IF^1 the collection of all intuitionistic fuzzy numbers. First,we define $0_{\mathrm{IF}} \in \mathrm{IF}^1$ by

$$0_{\mathrm{IF}}(t) = \begin{cases} (1, 0)\ if & t = 0 \\ (0, 1)\ if & t \neq 0 \end{cases}$$

For $\alpha \in [0, 1]$ and $\langle u, v\rangle \in \mathrm{IF}^1$ the upper and lower α-cuts of $\langle u, v\rangle$ are defined by

$$[\langle u, v\rangle]_\alpha = \{x \in \mathbb{R} : u(x) \geq \alpha\}$$

$$[\langle u, v\rangle]^\alpha = \{x \in \mathbb{R} : v(x) \leq 1 - \alpha\}$$

Definition 2 Let $\langle u_1, v_1\rangle$,$\langle u_2, v_2\rangle \in \mathrm{IF}^1$,$\lambda \in \mathbb{R}$ and $\alpha \in [0, 1]$,then

1. $(\langle u_1, v_1\rangle \oplus \langle u_2, v_2\rangle)(z) = (\sup_{z=x+y} min(u_1(x), u_2(y)), \inf_{z=x+y} max(u_1(x), u_2(y)))$
2. $\lambda\langle u_1, v_1\rangle = \langle \lambda u_1, \lambda v_1\rangle \quad if \quad \lambda \neq 0$
3. $\lambda\langle u_1, v_1\rangle = 0_{IF} \quad if \quad \lambda = 0$
4. $[\langle u_1, v_1\rangle \oplus \langle u_2, v_2\rangle]^\alpha = [\langle u_1, v_1\rangle]^\alpha + [\langle u_2, v_2\rangle]^\alpha$
5. $[\langle u_1, v_1\rangle \oplus \langle u_2, v_2\rangle]_\alpha = [\langle u_1, v_1\rangle]_\alpha + [\langle u_2, v_2\rangle]_\alpha$
6. $[\lambda\langle u_1, v_1\rangle]^\alpha = \lambda[\langle u_1, v_1\rangle]^\alpha$
7. $[\lambda\langle u_1, v_1\rangle]_\alpha = \lambda[\langle u_1, v_1\rangle]_\alpha$

Let $\langle u, v\rangle \in \mathrm{IF}^1$ and $\alpha \in [0, 1]$,then we define the following sets:

$$[\langle u, v\rangle]_l^+(\alpha) = \inf\{x \in \mathbb{R} : u(x) \geq \alpha\}$$

$$[\langle u, v\rangle]_r^+(\alpha) = \sup\{x \in \mathbb{R} : u(x) \geq \alpha\}$$

$$[\langle u, v\rangle]_l^-(\alpha) = \inf\{x \in \mathbb{R} : v(x) \leq 1 - \alpha\}$$

$$[\langle u, v\rangle]_r^-(\alpha) = \sup\{x \in \mathbb{R} : v(x) \leq 1 - \alpha\}$$

Remark 1 Let $\langle u, v\rangle \in \mathrm{IF}^1$ and $\alpha \in [0, 1]$,then we have:

$$[\langle u, v\rangle]^\alpha = [[\langle u, v\rangle]_l^-(\alpha), [\langle u, v\rangle]_r^-(\alpha)]$$

$$[\langle u, v\rangle]_\alpha = [[\langle u, v\rangle]_l^+(\alpha), [\langle u, v\rangle]_r^+(\alpha)]$$

Proposition 1 ([6]) Let $\alpha, \beta \in [0, 1]$ and $\langle u, v\rangle \in \mathrm{IF}^1$,then

1. $[\langle u, v\rangle]_\alpha \subset [\langle u, v\rangle]^\alpha$
2. $[\langle u, v\rangle]_\alpha et [\langle u, v\rangle]^\alpha$ are nonempty compact convex sets.
3. if $\alpha \leq \beta$ then $[\langle u, v\rangle]^\beta \subset [\langle u, v\rangle]^\alpha$ and $[\langle u, v\rangle]_\beta \subset [\langle u, v\rangle]_\alpha$
4. if $\alpha_n \nearrow \alpha$ then $[\langle u, v\rangle]^\alpha = \cap_n [\langle u, v\rangle]^{\alpha_n}$ and $[\langle u, v\rangle]_\alpha = \cap_n [\langle u, v\rangle]_{\alpha_n}$

Let $\alpha \in [0, 1]$ we put

$$M_\alpha = \{x \in \mathbb{R} : u(x) \geq \alpha\}$$

and

$$M^\alpha = \{x \in \mathbb{R} : v(x) \leq 1 - \alpha\}$$

Lemma 1 ([6]) Let $\{M^\alpha : \alpha \in [0, 1]\}$ and $\{M_\alpha : \alpha \in [0, 1]\}$ be two subset of $\mathbb{R}$ verify (1)-(4) of proposition1,if u and v are defined by

$$u(x) = \begin{cases} 0 & if \quad x \notin M_0 \\ sup\{\alpha \in [0, 1] : x \in M_\alpha\} & if \quad x \in M_0 \end{cases}$$

$$v(x) = \begin{cases} 1 & if \quad x \notin M^0 \\ 1 - sup\{\alpha \in [0, 1] : x \in M^\alpha\} & if \quad x \in M^0 \end{cases}$$

then $\langle u, v\rangle \in \mathrm{IF}^1$.

Lemma 2 ([6]) Let I be a dense subset in $[0, 1]$.
If $[\langle u, v\rangle]_\alpha = [\langle w, z\rangle]_\alpha$ and $[\langle u, v\rangle]^\alpha = [\langle w, z\rangle]^\alpha$,$\forall \alpha \in I$ then $\langle u, v\rangle = \langle w, z\rangle$.

Definition 3 ([2]) Let $\langle u_1, v_1\rangle$,$\langle u_2, v_2\rangle \in \mathrm{IF}^1$,if there exists $\langle w, z\rangle \in \mathrm{IF}^1$ such that, $\langle u_1, v_1\rangle = \langle u_2, v_2\rangle \oplus \langle w, z\rangle$ then $\langle w, z\rangle$ is called the Generalized Hukuhara difference of $\langle u_1, v_1\rangle$ and $\langle u_2, v_2\rangle$ denoted by $\langle u_1, v_1\rangle \ominus^{gH} \langle u_2, v_2\rangle$.

Definition 4 ([2]) Let $f : I \subset \mathbb{R}^2 \to \mathrm{IF}^1$ and $(x_0, y_0) \in I$.We say that f is generalized Hukuhara differentiable (gH-differentiable) with respect to x at at (x_0, y_0) if there exists $\frac{\partial f(x_0, y_0)}{\partial x} \in \mathrm{IF}^1$ such that:

$$\frac{\partial f(x_0, y_0)}{\partial x} = \lim_{h \to 0^+} \frac{f(x_0 + h, y) \ominus^{gH} f(x_0, y)}{h} = \lim_{h \to 0^-} \frac{f(x_0, y) \ominus^{gH} f(x_0 - h, y)}{h}$$

The gH-derivative of f with respect to y and higher order of fuzzy partial derivative of f at the point $(x_0, y_0) \in I$ are defined similarly

Notation.

- $L\left(J, IF^1\right)$ the space of all intuitionistic fuzzy-valued integrable functions on $I \subset \mathbb{R}^2$
- $C(I, \mathrm{IF}^1)$ the space of all intuitionistic fuzzy-valued continuous functions defined on $I \subset \mathbb{R}^2$
- $C^{i,j}(I, \mathrm{IF}^1)$ $(i, j = 0, 1, 2)$ the set of all intuitionistic fuzzy functions $f : I \subset \mathbb{R}^2 \to \mathrm{IF}^1$ which have partial $gH - derivatives$ up to order i with respect to x and up to order j with respect to y in I

Definition 5 ([2]) Assume that $f \in C^{1,0}(I, \mathrm{IF}^1)$,$[f(x, y)]^\alpha = \left[f_l^-(x, y)(\alpha), f_r^-\right.$ $\left.(x, y)(\alpha)\right]$,and
$[f(x, y)]_\alpha = \left[f_l^+(x, y)(\alpha), f_r^+(x, y; \alpha)\right]$ for all$\alpha \in [0, 1]$

We say that f is generalized Hukuhara differentiable (gH-differentiable) with respect to x at at (x_0, y_0) if,

$$\left[\frac{\partial f(x_0, y_0)}{\partial x}\right]^{\alpha} = \left[\frac{\partial f_l^-(x_0, y_0)(\alpha)}{\partial x}, \frac{\partial f_r^-(x_0, y_0)(\alpha)}{\partial x}\right]$$

and

$$\left[\frac{\partial f(x_0, y_0)}{\partial x}\right]_{\alpha} = \left[\frac{\partial f_l^+(x_0, y_0)(\alpha)}{\partial x}, \frac{\partial f_r^+(x_0, y_0)(\alpha)}{\partial x}\right]$$

Types of intuitionistic fuzzy gH-derivatives of f with respect to y and higher order of f at the point $(x_0, y_0) \in I$ are defined similarly

Let $d_\infty : \mathrm{IF}^1 \times \mathrm{IF}^1 \longrightarrow [0, +\infty]$ be a mapping defined by:

$$\begin{aligned} d_\infty(\langle u, v\rangle, \langle w, z\rangle) =& (\frac{1}{4} \sup_{0\leq\alpha\leq1} |[\langle u, v\rangle]_r^+(\alpha) - [\langle w, z\rangle]_r^+(\alpha)|^p d\alpha \\ &+ \frac{1}{4} \sup_{0\leq\alpha\leq1} |[\langle u, v\rangle]_l^+(\alpha) - [\langle w, z\rangle]_l^+(\alpha)|^p d\alpha \\ &+ \frac{1}{4} \sup_{0\leq\alpha\leq1} |[\langle u, v\rangle]_r^-(\alpha) - [\langle w, z\rangle]_r^-(\alpha)|^p d\alpha \\ &+ \frac{1}{4} \sup_{0\leq\alpha\leq1} |[\langle u, v\rangle]_l^-(\alpha) - [\langle w, z\rangle]_l^-(\alpha)|^p d\alpha)^{\frac{1}{p}} \end{aligned}$$

Then we have the following result.

Proposition 2 *(IF^1, d_∞) is a complet metric space.*

Definition 6 we consider the supremum metric D_∞ in the space $C(J, \mathrm{IF}^1)$ defined by

$$D_\infty(U, V) = \sup_{(x,y)\in J} d_\infty(U(x, y), V(x, y))$$

Remark 2 Since $(\mathrm{IF}^1, d_\infty)$ is a complet metric space,$(C(J, \mathrm{IF}^1), D_\infty)$ is also a complet metric space.

Definition 7 ([6]) $F : U \subset \mathbb{R}^2 \longrightarrow \mathrm{IF}^1$ is strongly measurable if $\forall\alpha \in [0, 1]$,the set-valued mappings $F_\alpha : U \longrightarrow \mathscr{P}_K(\mathbb{R})$ defined by $F_\alpha(t) = [F(t)]^\alpha$ and $F^\alpha : U \longrightarrow \mathscr{P}_K(\mathbb{R})$ defined by $F^\alpha(t) = [F(t)]_\alpha$ are Lebesgue measurable.

An intuitionistic fuzzy multivariable function $F : U \subset \mathbb{R}^2 \longrightarrow \mathrm{IF}^1$ is called integrable bounded if there exists an integrable functioin $H : U \subset \mathbb{R}^2 \longrightarrow [, +\infty[$ such that,

$$d_\infty(F(x, y), 0_{\mathrm{IF}}) < H(x, y); , \forall(x, y) \in U$$

A strongly measurable and integrable bounded intuitionistic fuzzy function is called integrable,and the intuitionistic fuzzy integral of the function $F : U \subset \mathbb{R}^2 \longrightarrow \mathrm{IF}^1$ is defined by the equation

$\left[\int_U F(u)du\right]^\alpha = \left[\int_U F_l^-(u;\alpha)du, \int_U F_r^-(u;\alpha)du\right], \forall\alpha \in [0,1]$
$\left[\int_U F(u)du\right]_\alpha = \left[\int_U F_l^+(u;\alpha)du, \int_U F_r^+(u;\alpha)du\right], \forall\alpha \in [0,1]$
we denote this integrale by $\int_U F(u)du$.

Definition 8 ([1]) Assum that $f \in C^{1,0}(I, \mathrm{IF}^1)$,such that $[f(x,y)]^\alpha = [f_l^-(x,y)(\alpha), f_r^-(x,y)(\alpha)]$ and $[f(x,y)]_\alpha = [f_l^+(x,y)(\alpha), f_r^+(x,y)(\alpha)]$ for all $\alpha \in [0,1]$ and $(x,y) \in I$.

Let $(x_0, y_0) \in I$,We say that f is intuitionistic fuzzy differentiable with respect to x at (x_0, y_0)if

$$\left[\frac{\partial f}{\partial x}(x_0, y_0)\right]^\alpha = \left[\frac{\partial f_l^-}{\partial x}(x_0, y_0)(\alpha), \frac{\partial f_r^-}{\partial x}(x_0, y_0)(\alpha)\right]$$

and

$$\left[\frac{\partial f}{\partial x}(x_0, y_0)\right]_\alpha = \left[\frac{\partial f_l^+}{\partial x}(x_0, y_0)(\alpha), \frac{\partial f_r^+}{\partial x}(x_0, y_0)(\alpha)\right]$$

the intuitionistic fuzzy gH-derivative of f with respect to y at the poin (x_0, y_0) is defined similarly.

Lemma 3 *Let $f \in C(\mathbb{R}^2, IF^1)$.*
If the function f is gH-differentiable with respect to y,then $\frac{\partial f(x,y)}{\partial y}$is integrable on $[b, y]$ and we have

$$\int_b^y \frac{\partial f(x,s)}{\partial s} ds = f(x,y) \ominus^{gH} f(x,b)$$

3 Caputo Differentiability of Intuitionistic Fuzzy Multivariable Functions

In this section,we develop the concept of the intuitionistic fuzzy Caputo derivatives expressed in [5] for one-variable functions,adapted for intuitionistic fuzzy valued multivariable functions.

Let $J = [0,a] \times [0,b], q = (q_1, q_2) \in [0,1] \times [0,1]$ and $f \in L^1(J, \mathbb{R})$ In [1],the authors presented the mixed Riemann-Liouville fractional integral notion of order q for real-valued functions $f(x,y)$ as follows.

$$I_0^q f(x,y) := \frac{1}{\Gamma(q_1)\Gamma(q_2)} \int_0^x \int_0^y (x-s)^{q_1-1}(y-t)^{q_2-1} f(x,y)dtds$$

provided that the expression on the right hand side is defined for almost every $(x,y) \in J$.

Let $F : J \longrightarrow \mathrm{IF}^1$ we have,

$$[F(x,y)]^\alpha = \left[F_l^-(x,y)(\alpha),\, F_r^-(x,y)(\alpha)\right]$$

$$[F(x,y)]_\alpha = \left[F_l^+(x,y)(\alpha),\, F_r^+(x,y)(\alpha)\right]$$

Since the family

$$\{[F_l^-(x,y)(\alpha), F_r^-(x,y)(\alpha)], [F_l^+(x,y)(\alpha), F_r^+(x,y)(\alpha)\}$$

builds an intuitionistic fuzzy element and the integral preserves the monotony then by lemma(1) we can define the integral of the function f as follows

Definition 9 ([1]) Let $F \in L\left(J, IF^1\right)$.the intuitionistic fuzzy fractional integral of order $q = (q_1, q_2) \in [0, 1]^2$ of F denoted by

$$I_0^q F(x,y) := \frac{1}{\Gamma(q_1)\Gamma(q_2)} \int_0^x \int_0^y (x-s)^{q_1-1}(y-t)^{q_2-1} F(s,t) t ds dt$$

is defined by

$$\left[I^q F(x,y)\right]^\alpha = \left[I^q F_l^-(x,y)(\alpha), I^q F_r^-(x,y)(\alpha)\right]$$

$$\left[I^q F(x,y)\right]_\alpha = \left[I^q F_l^+(x,y)(\alpha), I^q F_r^+(x,y)(\alpha)\right]$$

Lemma 4 *Let* $q = (q_1, q_2) \in [0,1] \times [0,1]$ *and* $[I^q F(x,y)]^\alpha = \left[I^q F_l^-(x,y)(\alpha),\right.$ $\left.I^q F_r^-(x,y)(\alpha)\right]$,
$[I^q F(x,y)]_\alpha = \left[I^q F_l^+(x,y)(\alpha), I^q F_r^+(x,y)(\alpha)\right]$ *for all* $(x,y) \in J$ *and* $\alpha \in [0,1]$. *If* $F_l^-, F_r^-, F_l^+, F_r^+ \in L(J, \mathbb{R})$,*then for each* $(x,y) \in J$ *, the family of closed intervals,*

$$G^\alpha := G^\alpha(x,y) = \left[I^q F_l^-(x,y)(\alpha), I^q F_r^-(x,y)(\alpha)\right]$$

$$G_\alpha := G_\alpha(x,y) = \left[I^q F_l^+(x,y)(\alpha), I^q F_r^+(x,y)(\alpha)\right]$$

defines an intuitionistic fuzzy number $\langle u, v\rangle \in IF^1$ *such that,* $[\langle u, v\rangle]^\alpha = G^\alpha(x,y)$ *and* $[\langle u, v\rangle]_\alpha = G_\alpha(x,y)$ *for all* $\alpha \in [0,1]$.

Proof Let $q = (q_1, q_2) \in [0,1] \times [0,1]$ and $(x,y) \in J)$.Set

$$G^\alpha = \frac{1}{\Gamma(q_1)\Gamma(q_2)} \left[\int_0^x \int_0^y (x-s)^{q_1-1}(y-t)^{q_2-1} F_l^-(s,t)(\alpha) ds dt, \int_0^x \int_0^y (x-s)^{q_1-1}(y-t)^{q_2-1} F_r^-(s,t)(\alpha) ds dt\right]$$

$$G_\alpha = \frac{1}{\Gamma(q_1)\Gamma(q_2)} \left[\int_0^x \int_0^y (x-s)^{q_1-1}(y-t)^{q_2-1} F_l^+(s,t)(\alpha) ds dt, \int_0^x \int_0^y (x-s)^{q_1-1}(y-t)^{q_2-1} F_r^+(s,t)(\alpha) ds dt\right]$$

By using lemma 1 we obtain the result.

Proposition 3 *Let $F, G \in L(J, IF^1)$ and $a \in IF^1$ then we have*

1. $I^q(aF)(x, y) = aI^q F(x, y)$.
2. $I^q(F \oplus G)(x, y) = I^q F(x, y) \oplus I^q G(x, y)$.
3. $I^q I^{q'} F(x, y) = I^{q+q'} F(x, y)$.

Definition 10 Let $J = [0, a] \times [0, b]$, $q = (q_1, q_2) \in [0, 1] \times [0, 1]$ and $u \in C^{2,2}(J, \mathrm{IF}^1)$.The intuitionistic fuzzy Caputo gH-derivatives of order q with respect to x, yof the function u are defined as follows :

$$^cD^q u(x, y) = I^{1-q}\left(\frac{\partial^2 u}{\partial x \partial y}(x, y)\right)$$

Where $1 - q = (1 - q_1, 1 - q_2) \in [0; 1]^2$.

Example 1 Consider the fuzzy function $u(x, y) = xyC$ where $C = (a_1; a_2; a_3; a_4; a_1'; a_2; a_3; a_4')$is a trapezoidal intuitionistic fuzzy number.

The gH-partial derivative of $u(x, y)$ with respect to x is calculated as follows

$$\frac{\partial u(x, y)}{\partial x} = \lim_{h \to 0} \frac{u(x + h, y) \ominus^{gH} f(x, y)}{h} = \lim_{h \to 0} \frac{(x + h)yC \ominus^{gH} xyC}{h} = yC$$

This implies that

$$\frac{\partial u(x, y)}{\partial x} = yC$$

Furthermore, using a similar argument, we also obtain

$$\frac{\partial^2 u(x, y)}{\partial y \partial x} = C$$

Since $[C]^\alpha = [a_2 - \alpha(a_2 - a_1'), a_3 + \alpha(a_4 - a_3)]$ and $[C]_\alpha = [a_1 + \alpha(a_2 - a_1), a_4 - \alpha(a_4' - a_3)]$ we have,

$$\left[{}^cD^q u(x, y)\right]^\alpha = \left[I^{1-q}\left(\frac{\partial^2 u}{\partial x \partial y}(x, y)\right)\right]^\alpha = \left[I^{q-1}C\right]^\alpha$$

$$\left[{}^cD^q u(x, y)\right]_\alpha = \left[I^{1-q}\left(\frac{\partial^2 u}{\partial x \partial y}(x, y)\right)\right]_\alpha = \left[I^{q-1}C\right]_\alpha$$

$$\left[I^{1-q}C\right]^\alpha = \frac{1}{\Gamma(q_1)\Gamma(q_2)} \int_0^x \int_0^y (x - s)^{q_1 - 1}(y - t)^{q_2 - 1} \left[a_2 - \alpha(a_2 - a_1'), a_3 + \alpha(a_4 - a_3)\right] ds dt$$

$$\left[I^{1-q}C\right]_\alpha = \frac{1}{\Gamma(q_1)\Gamma(q_2)} \int_0^x \int_0^y (x - s)^{q_1 - 1}(y - t)^{q_2 - 1} \left[a_1 + \alpha(a_2 - a_1), a_4 - \alpha(a_4' - a_3)\right] ds dt$$

$$[I^{1-q}C]^{\alpha} = \frac{x^{1-q_1}y^{1-q_2}}{\Gamma(1-q_1)\Gamma(1-q_2)\times(1-q_1)(1-q_2)}[a_2-\alpha(a_2-a_1'), a_3+\alpha(a_4-a_3)]$$

$$[I^{1-q}C]_{\alpha} = \frac{x^{1-q_1}y^{1-q_2}}{\Gamma(1-q_1)\Gamma(1-q_2)\times(1-q_1)(1-q_2)}[a_1+\alpha(a_2-a_1), a_4-\alpha(a_4'-a_3)]$$

$$[I^{1-q}C]^{\alpha} = \frac{x^{1-q_1}y^{1-q_2}}{\Gamma(2-q_1)\Gamma(2-q_2)}[C]^{\alpha}$$

$$[I^{1-q}C]_{\alpha} = \frac{x^{1-q_1}y^{1-q_2}}{\Gamma(2-q_1)\Gamma(2-q_2)}[C]_{\alpha}$$

Thus

$$[{}^cD^q u(x,y)] = \frac{x^{1-q_1}y^{1-q_2}}{\Gamma(2-q_1)\Gamma(2-q_2)}C$$

$${}^cD^q u(x,y) = \frac{x^{-q_1}y^{-q_2}}{\Gamma(2-q_1)\Gamma(2-q_2)}u(x,y)$$

4 Intuitionistic Fuzzy Fractional Partial Differential Equation

In this section we consider problem (1, 1) with $g_1 \in C([0,a], \mathrm{IF}^1)$ and $g_2 \in C([0,b], \mathrm{IF}^1)$ are given functions such that $g_1(0) \ominus g_2(y)$ exists for all $y \in [0,b]$ and $f \in C(J, \mathrm{IF}^1)$.

We denote $\varphi(x,y) = g_2(y) + [g_1(x) \ominus g_1(0)]$.

Lemma 5 *Let $u(.,.) \in C^{2,2}(J, IF^1)$ be an intuitionistic fuzzy-valued function satisfying the problem* (1.1) *Then $u(.,.)$ satisfies of the following integral equation*

$$u(x,y) = \varphi(x,y) + I^q f(x,y,u(x,y)), \quad \forall (x,y) \in J \tag{2}$$

Proof Let $u \in C^{2,2}(J, \mathrm{IF}^1)$ satisfy (1.1). Taking into account the definition of the fractional Caputo derivative${}^c D^q u(x,y)$ and from (1) we have

$$I^{1-q}\left(\frac{\partial^2 u}{\partial x \partial y}(x,y)\right) = f(x,y,u(x,y))$$

Then

$$I^q I^{1-q}\left(\frac{\partial^2 u}{\partial x \partial y}(x,y)\right) = I^q f(x,y,u(x,y))$$

That implies

$$I^1\left(\frac{\partial^2 u}{\partial x \partial y}(x,y)\right) = I^q f(x,y,u(x,y))$$

It follows

$$\int_0^x \int_0^y \frac{\partial^2 u(x,y)}{\partial x \partial y} dy dx = I^q f(x,y,u(x,y))$$

Since

$$\int_0^x \left(\int_0^y \frac{\partial}{\partial y}\left(\frac{\partial u}{\partial x}\right) dy\right) dx = \int_0^x \left(\frac{\partial u}{\partial x}(x,y) \ominus^{gH} \frac{\partial u}{\partial x}(x,0)\right) dx$$

Therefore

$$\int_0^x \left(\frac{\partial u}{\partial s}(s,t) \ominus^{gH} \frac{\partial u}{\partial s}(s,0)\right) ds = I^q f(x,y,u(x,y))$$

Or

$$\int_0^x \frac{\partial u}{\partial s}(s,t) ds = \int_0^x \frac{\partial u}{\partial s}(s,0) ds + I^q f(x,y,u(x,y))$$

It follows

$$u(x,y) \ominus^{gH} u(0,y) = u(x,0) \ominus^{gH} u(0,0) + I^q f(x,y,u(x,y))$$

$$u(x,y) = u(0,y) + u(x,0) \ominus^{gH} u(0,0) + I^q f(x,y,u(x,y))$$

That leads to the following equation.

$$u(x,y) = g_2(y) + g_1(x) \ominus^{gH} g_1(0) + I^q f(x,y,u(x,y))$$

Finally we obtain

$$u(x,y) = \varphi(x,y) + I^q f(x,y,u(x,y))$$

Definition 11 A function $u \in C(J, \mathrm{IF}^1)$ is called a weak solution of the problem (1.1) iif it satisfies the fractional integral equation (4.1) for all $(x, y) \in J$.

Theorem 1 *Let $f \in C(J, IF^1)$ satisfy the Lipschitz condition, i.e. there exists a positive real number K such that*

$$D_\infty(f(x,y,u), f(x,y,v)) < kD_\infty(u(,v); \quad \forall (u,v) \in C(J, IF^1) \times C(J, IF^1)$$

If $\frac{ka^{q_1}b^{q_2}}{\Gamma(q_1+1)\Gamma(q_2+1)} < 1$, then the problem (1.1) has a unique weak intuitionistic fuzzy solution defined on J.

Proof The proof is based on the application of Banach fixed point theorem,we define the operator $T : C(J, \mathrm{IF}^1) \longrightarrow C(J, \mathrm{IF}^1)$ by

$$Tu(x, y) := \varphi(x, y) + I^q f(x, y, u(x, y))$$

$$\begin{aligned}
D_\infty(Tu(x, y), Tv(x, y)) \leq & \frac{1}{\Gamma(q_1)\Gamma(q_2)} \int_0^x \int_0^y (x-s)^{q_1-1}(y-t)^{q_2-1} D_\infty(f(s,t,u), f(s,t,v))dsdt \\
\leq & \frac{1}{\Gamma(q_1)\Gamma(q_2)} \int_0^x \int_0^y (x-s)^{q_1-1}(y-t)^{q_2-1} k D_\infty(u, v)dsdt \\
\leq & \frac{k}{\Gamma(q_1)\Gamma(q_2)} D_\infty(u, v) \int_0^x \int_0^y (x-s)^{q_1-1}(y-t)^{q_2-1} dsdt \\
\leq & \frac{ka^{q_1}b^{q_2}}{\Gamma(q_1+1)\Gamma(q_2+1)} D_\infty(u, v)
\end{aligned}$$

Therefore,we obtain:

$$D_\infty(Tu, Tv) \leq \frac{ka^{q_1}b^{q_2}}{\Gamma(q_1+1)\Gamma(q_2+1)} D_\infty(u, v)$$

Example 2 Consider the following intuitionistic fuzzy fractional partial hyperbolic differential equation

$$\begin{cases} D^{(\frac{1}{2}, \frac{1}{2})}u(x, y) = xyu(x, y), & (x, y) \in J = [0, \frac{1}{2}] \times [0, \frac{1}{2}] \\ u(x, 0) = xC, & C \in \mathrm{IF}^1, x \in [0, \frac{1}{2}] \\ u(0, y) = yC, & C \in \mathrm{IF}^1, y \in [0, \frac{1}{2}] \end{cases} \tag{3}$$

where $C = (a_1; a_2; a_3; a_4; a_1'; a_2; a_3; a_4')$ is a trapezoidal intuitionistic fuzzy number. Here $f(x, y, u(x, y)) = xyu(x, y)$,therefore $k = 1$.

Since $q = (\frac{1}{2}, \frac{1}{2})$,$a = b = \frac{1}{2}$, $\Gamma(\frac{3}{2})^2 = 0.79$ and $\frac{1 \times a^{\frac{1}{2}} b^{\frac{1}{2}}}{\Gamma(\frac{3}{2})^2} = \frac{1}{1.58} < 1$ then the problem (3)has a unique weak intuitionistic fuzzy solution.

If $u(x, y) = C$ we can see that $\varphi(x, y) = (x + y)C$ and $I^q f(x, y, u(x, y)) = \frac{9\pi}{16} x^{\frac{3}{2}} y^{\frac{3}{2}} C$ Then $u(x, y) = (x + y)C + \frac{9\pi}{16} x^{\frac{3}{2}} y^{\frac{3}{2}} C$.

5 Conclusions

In this paper, we study intuitionistic fuzzy fractional PDEs under Caputo gH-differentiability. It is important to think about the value of embedding our results within fractional calculus for fuzzy-valued multivariable functions in the sense of gH-differentiability. By using the Banach fixed point theorem, we have proved some new results on the existence and uniqueness of an intuitionistic fuzzy solutions for the local boundary valued problems of fractional partial hyperbolic differential equations. Studying the existence and stability of global intuitionistic fuzzy solutions of the boundary valued problems for fractional PDEs is the next step that will be considered.

References

1. Abbas, S., Benchohra, M., N'Guerekata, G.M.: Topics in Fractional Differential Equations. Springer, New York (2012)
2. Bede B., Gal S.G.: Generalizations of the differentiability of fuzzy-number-valued functions with applications to fuzzy differential equations. Fuzzy Sets Syst. **151**, 581–99 (2005)
3. Debnath, L., Bhatta, D.: Solutions to few linear fractional inhomogeneous partial differential equations in fluid mechanics. Fract. Calc. Appl. Anal. **7**, 153–162 (2004)
4. Delbosco, D., Rodino, L.: Existence and uniqueness for a nonlinear fractional differential equation. J. Math. Anal. Appl. **204**, 609–625 (1996)
5. Kilbas A.A.A., Srivastava, H.M., Trujillo J.J.: Theory and applications of fractional differential Equations, North-Holland Mathematical studies, vol. 204, Ed van Mill. Elsevier, Amsterdam (2006)
6. Melliani, S., Elomari, M., Chadli, L.S., Ettoussi, R.: Intuitionistic fuzzy metric space. Notes Intuit. Fuzzy Sets **21**(1), 43–53 (2015)
7. Moaddy, K., Momani, S., Hashim, I.: The non-standard finite difference scheme for linear fractional PDEs in fluid mechanics. Comput. Math. Appl. **61**, 1209–1216 (2011)
8. Muslih, S., Agrawal, O.P.: Riesz fractional derivatives and fractional dimensional space. Int. J. Theor. Phys. **49**, 270–275 (2010)

Using Machine Learning with PySpark and MLib for Solving a Binary Classification Problem: Case of Searching for Exotic Particles

Mourad Azhari, Abdallah Abarda, Badia Ettaki, Jamal Zerouaoui, and Mohamed Dakkon

Abstract Searching for exotic particles in high-energy represents a major challenge for physicists. In this paper, we propose to solve the binary classification problem in the area of exotic particles using the Apache Spark environment with the Mlib library. Then, We compare the performance of four methods: Logistic Regression (LR), Decision Tree (DT), Random Forest (RF), and Gradient Boosted Tree (GBT). In this work, we use "SUSY" dataset, collected from UCI machine learning repository, for the experimentation phase.

Keywords Exotic particles · Spark · Pyspark · Machine learning (ML) · Logistic regression (LR) · Decision tree (DT) · Random forest (RF) · Gradient boosted tree (GBT) · AUC · Accuracy, and computation time

---

M. Azhari (✉) · J. Zerouaoui
Laboratory of Engineering Sciences and Modeling, Faculty of Sciences, Ibn Tofail University, Campus Universitaire, BP 133 Kenitra, Morocco
e-mail: azharimourad@yahoo.fr

J. Zerouaoui
e-mail: j_zerouaoui@yahoo.fr

A. Abarda
Laboratoire de Modélisation Mathématiques et de Calculs Economiques, FSJES, Université Hassan 1er, Settat, Morocco
e-mail: abardabdallah@gmail.com

B. Ettaki
Laboratory of Research in Computer Science, Data Sciences and Knowledge Engineering, Department of Data, Content and knowledge Engineering, School of Information Sciences Rabat , Rabat, Morocco
e-mail: ettakibadia@yahoo.fr

M. Dakkon
Département de Statistique et Informatique de Gestion, Université Abdelmalek Essaadi, Tétouan, Morocco
e-mail: m.dakkoun@gmail.com

S. Melliani and O. Castillo (eds.), *Recent Advances in Intuitionistic Fuzzy Logic Systems and Mathematics*, Studies in Fuzziness and Soft Computing 395,
https://doi.org/10.1007/978-3-030-53929-0_8

1 Introduction

Searching for exotic particles in high-energy is a major challenge for the High-energy Physics area. Machine learning methods are a powerful tool that can improve the performance of exotic particles discovery. Those methods are proposed to solve the problem of separation of signal from background events in the case of the "SUSY" dataset. Several studies related to exotic particle search have been proposed in the particle physics domain. In this context, ATLAS detector physicists, considered as a main experiment at the Large Hadron Collider (LHC) at the European Organization for Nuclear Research(CERN), studied the predictions of the Standard Model. With this in mind, Higgs Boson Machine Learning Challenge has been tenuous in 2014 [1]. Many algorithms were tested, among these classifiers, we cite the basic algorithms such as Decision Tree, Naive Bayes, K-Nearest Neighbors, etc., advanced algorithms: KMeans Clustering, Support Vector Machine, etc., and ensemble methods such as Bagging, Random Forest and Boosting 84% is the accuracy score obtained with Gradient Boosting Classifier and 76% achieved by Support vector machines with linear kernels. Chen and Tong [2] suggested a regularized version of the Gradient Boosting method with an efficient implementation [2]. Baldi et al. [3] have implemented the deep networks classifier on Higgs and Susy datasets [3–5]. In multivariate statistical analysis (MVA), Alves [6] proposed Stacking machine learning classifiers to identify Higgs bosons at the LHC, this approach exceeds the Boosted Decision trees and deep neural network applicable to particle physics [6]. This paper proposes to solve the exotic particles classification problem with four machine learning (ML) methods with Pyspark framework: Logistic Regression (LR), Decision Tree (DT), Random Forest (RF) and Gradient Boosted Tree (GBT). Hence, we compare the accuracy and AUC metrics of those ML methods using five million instances of the "SUSY" dataset. The paper is organised as follows. The first section gives a brief related work of exotic particles discovery, the second section presents the spark framework. The third section proposes machine learning methods to detect 'SUSY' particle. The paper is organized as follows. The first section will give brief related works of exotic particle discovery, the second section will present the spark framework. The third section will propose machine learning methods to detect the "SUSY" particle. The fourth section will describe 'SUSY' dataset. The last section will summarise our experimental results and their subsequent analysis.

2 Spark Environement

Apache Spark is a powerful tool of Big Data and an open-source distributed cluster-computing framework. It includes a common Machine Learning (ML) library (MLlib) designed to ML classifiers. Spark runs well in memory and supports many programming languages (java, scala, python, SQL, R) and works well with python. Apache Spark is an open-source distributed cluster-computing framework. It contains a Machine Learning (ML) library (MLlib) reserved for ML classifiers. Spark works well in memory with many languages (java, scala, python, SQL, R) and handles successfully Big data as such "SUSY" dataset with the python and Pyspark programs [7, 8]. Apache Spark has a rapid processing, dynamic, In-Memory Computation speed quick, and reusable [9].

3 Proposed Methods

This paper aims to applied different algorithms in order to solve the classification problems of exotic particle search: Logistic Regression (LR), Decision Tree (DT), Random Forest (RF) and Gradient Boosted Tree (GBT).

3.1 Logistic Regression (LR)

Logistic Regression (LR) is a classification method with powerful predictors. In the context of binary discrimination, LR aims to predict or explain a categorical response with two modalities (positive, negative) depending on predictors. It focuses only on the ratio of the probabilities. In practice, we use the spark.ml logistic regression module to predict a binary outcome [10].

3.2 Decision Tree (DT) Method

Decision Tree is a supervised learning machine classifier used to solving regression or classification problems [11, 12]. A Decision Tree is a tree where each node designs an attribute, each branch represents a decision rule and each leaf represents an outcome. We can build a decision tree for classification problems using many algorithms as ID3 and CART [13, 14].

3.3 Random Forest (RF)

The Random Forest is a learning method that consists of multiple decision trees in order to build the final decision using the majority voting [15–18]. The principal benefit of RF is that it reduces the risk of over-fitting, offers a high level of accuracy and runs efficiently in large databases [19].

3.4 Gradient Boosted Tree (GBT)

The Boosting method consists of converting weak learners into strong learners in order to minimize a loss function. The principal idea of Boosting is to join a new classifier to the ensemble sequentially. Gradient Boosted tree (GBT) is an ensemble method that uses an ensemble of decision trees to predict a target label.The spark.ml implementation supports GBTs for binary classification [20].

4 Distribution of "SUSY" Dataset

In this work, we use the "SUSY" dataset (Supersymmetric particles), downloaded from UCI site. We use 18 features as independent variables and a dependent variable with two classes (signal and background events) [21]:

- Eight Low-level input features: $'lepton1\_pT'$, $'lepton1\_eta'$, $'lepton1\_phi'$, $'lepton2\_pT'$, $'lepton2\_eta'$, $'lepton2\_phi'$, $'missing\_energy\_magnitude'$, $'missing\_energy\_phi'$.
- Ten high-level input features: $'MET\_rel'$, $'axial\_MET'$, $'M\_R'$, $'M\_TR\_2'$, $'R'$, $'MT2'$, $'S\_R'$, $'M\_Delta\_R'$, $'dPhi\_r\_b'$, $'cos\_theta\_r1'$.

The dependent variable is the class label (1 for signal, 0 for background). It consists of 54% (2 712 173 examples) of the signal and 46% of background events (2 287 827 examples). We conclude that the distribution of the SUSY dataset is nearly balanced (Table 1).

4.1 Split Data

In the aim to get an unbiased estimation of the performances of the algorithms, SUSY dataset is divided into a training sample (70%) and test sample (30%) (see Table 1).

Table 1 Susy dataset description: instances, features, classlabel and split data

	Description	
Instances	5 000 000	
Class label (signal, background)	Class label (s = 1, b = 0)	(1; 0)
Signal	2 712 173,00	54%
Background	2 287 827,00	46%
Training set	3 500 000,00	**70%**
Test set	1 500 000,00	30%

4.2 Evaluation Functions

We present important metrics that are evaluated for comparison [22].

4.2.1 Accuracy

Accuracy is the proportion of total examples classified correctly:

$$Accuracy = \frac{TP + TN}{TP + TN + FP + FN} = 1 - ErrorRate \qquad (1)$$

Where:

- TP: The true positive is the number of instances that are SUSY particles and are being classified as Higgs particles;
- TN: The true negative is the number of instances which are no-SUSY particles and being classified as no-SUSY particles;
- FN: The false negative is the number of true SUSY particles that are wrongly being classified as no- SUSY particles;
- FP: The false positive is the number of no-SUSY particles that are wrongly being classified as SUSY particles.

4.2.2 AUC-ROC

Receiver Operating Characteristic (ROC) Curve allows comparing various supervised learning classifiers. It is especially useful for cases of skewed class distribution [23]. Area Under Curve (AUC) is an area equivalent to the probability that the algorithm will place a randomly selected positive example higher than a randomly selected negative example [24].

4.2.3 Computation Time (CT)

Computation time estimates running time required to perform a computational process. All Algorithms were trained using machine with: intel (R) cores (TM) i7,7500U CPU@ 2.7 GHZ 2.9 GHZ, 8 Go memory (RAM) and processor x64.

5 Experimental Results and Analysis

5.1 Experimental Results

Tables 2, 3 and 4 present respectively the performance of low-level input features, high-level input features, and all input features (low and high-level input features).

5.2 Analysis and Discussion

In this part, we compare different models on susy dataset in terms of AUC, Accuracy and computation time metrics.

Table 2 Performance of low-level input features

Classifier	Accuracy	AUC	CT (s)
Logistic Regression (LR)	0.768	0.8312	89.869
Random Forest (RF)	0.775	0.7697	518.11
Decision Tree (DT)	0.7588	0.7474	219.173
Gradient Boosted Tree (GBT)	0.7816	0.7746	2358.773

Table 3 Performance of high-level input features

Classifier	Accuracy	AUC	CT (s)
Logistic Regression (LR)	0.7702	0.8336	254.702
Random Forest (RF)	0.7621	0.7528	467.497
Decision Tree (DT)	0.7485	0.7374	249.622
Gradient Boosted Tree (GBT)	0.7827	0.7748	4907.049

Table 4 Performance of alll input features: low-nevel and high-nevel

Classifier	Accuracy	AUC	CT (s)
Logistic Regression (LR)	0.7884	0.8576	2148.734
Random Forest (RF)	0.774	0.7658	564.944
Decision Tree (DT)	0.7546	0.7416	627.392
Gradient Boosted Tree (GBT)	0.793	0.7862	5468.769

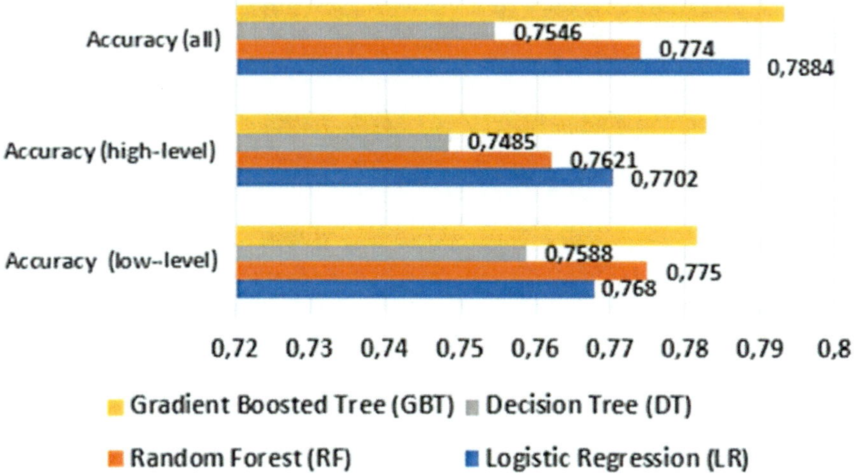

Fig. 1 Accuracy of low-level, high-level, and complete input features

In the context of low-level energy, the accuracy metric varies between 76% for the DT algorithm and 78% for the GBT model (See Fig. 1). However, the LR method achieved a higher AUC score with 83% better than GBT and RF with 77%, that better than DT with(75%). On the other hand, the Fig. 3 shows that RL classifier runs quickly (89 s), whereas the GBT method trained very slowly in 2358 s.

In the context of high-level energy, similarly, the accuracy metric varies between 75% for the DT algorithm and 78% for the GBT model (See Fig. 2). However, the LR method achieved a higher AUC score with 83% better than GBT with 77% and RF with 75%, that better than DT with(74%). On the other hand, the Fig. 3 shows that DT classifier, LR, and RF algorithm run quickly, whereas the GBT method trained very slowly in 4907 s.

In the context of the complete input features (low-level energy and high-level energy), the accuracy metric is improved. The score varies between 75% for the DT algorithm and 79% for the GBT model. However, the LR method achieved a higher AUC score with 86% better than GBT with 79% and RF with 76%, that better than DT with 74%. On the other hand, the Fig. 3 shows that DT classifier and RF algorithm run quickly, whereas the GBT method trained very slowly in 5468 s.

Those results are closes to the score of Higgs challenge 2014 (83%) [1] and these results are very close to those obtained using the deep learning method [3–5].

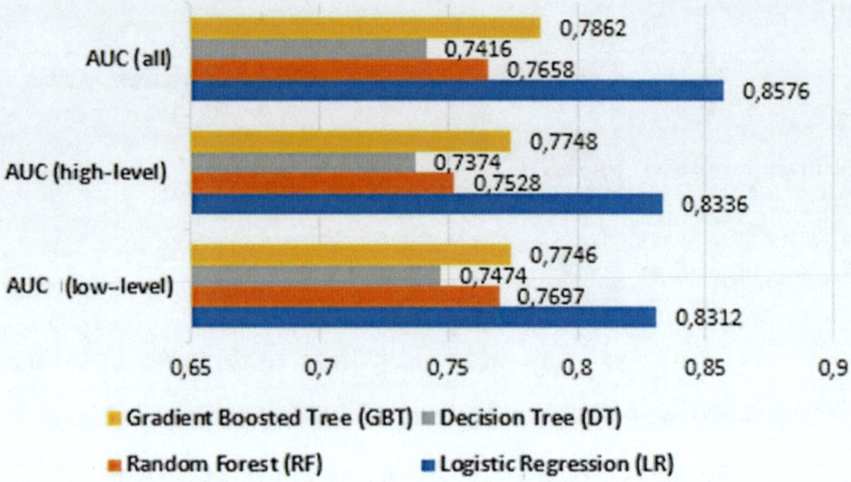

Fig. 2 AUC of low-level , high-level and complete input features

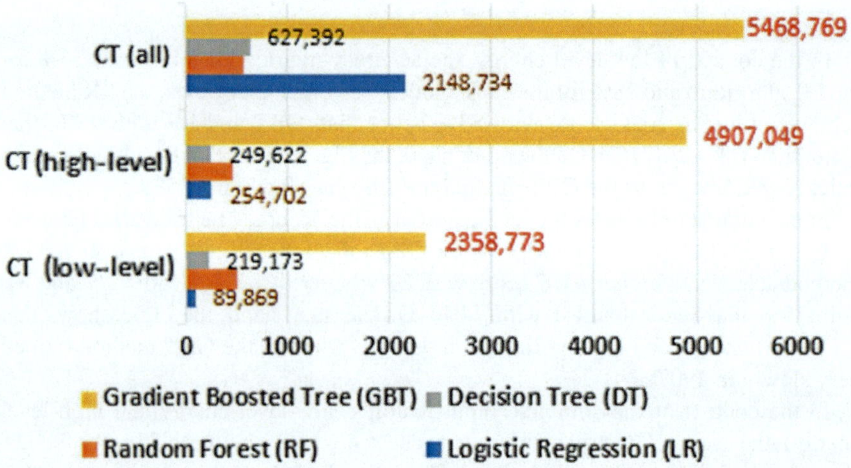

Fig. 3 CT at seconds of low-level , high-level and all input features

6 Conclusion

In the area of Searching for exotic particles, we conclude that:In the area of Searching for exotic particles, we conclude that:

In the context of low-level energy, The Gradient Boosted Tree (GBT) classifier runs well with accuracy achieved 78% While LR works efficiently with AUC (83%).

In the context of the high-level, we remark a similar conclusion, the performance of the Gradient Boosted Tree (GBT) classifier is good with accuracy 78%. Whereas, Logistic Regression outperforms the AUC score of(83%).

In the context of the complete input features, The Gradient Boosted Tree (GBT) classifier achieves a high accuracy score (79%), while the LR exceeds others classifiers with 86%. Concerning the computation time, the GBT method works very slowly, particularly in the case of the complete input features.

References

1. Adam-Bourdarios, C., Cowan, G., Germain-Renaud, C., Guyon, I., Kégl, B., Rousseau, D.: The higgs machine learning challenge. J. Phys.: Conf. Ser. **634**, 072015. https://doi.org/10.1088/1742-6596/664/7/072015
2. Chen, T., He, T.: Higgs Boson Discovery with Boosted Trees. In: JMLR: Workshop and Conference Proceedings, vol. 42, pp. 69–80 (2015)
3. Baldi, P., Cranmer, K., Faucett T., Sadowski P., Whiteson, D.: Parameterized machine learning for high-energy physics. Eur. Phys. J. **76**: 235–241 (2016). https://doi.org/10.1140/epjc/s10052-016-4099-4
4. Sadowski, P.J., Whiteson, D., Baldi, P.: Searching for higgs boson decay modes with deep learning. In: Advances in Neural Information Processing Systems (NIPS), vol. 27 (2014)
5. Sadowski, P., Collado, J., Whiteson, D., Baldi, P.: Deep learning, dark knowledge, and dark matter. In: JMLR: Workshop and Conference Proceedings, vol. 42, pp. 81–97 (2015)
6. Alves, A.: Stacking machine learning classifiers to identify higgs bosons at the LHC. J. Instrum. **12**, T05005–T05005 (2017). https://doi.org/10.1088/1748-0221/12/05/T05005
7. Meng, X., Joseph, B., Burak, Y., Evan, S., Shivaram, V., Davies, L., Jeremy, F., et al.: MLlib: Machine Learning in Apache Spark. J. Mach. Learn. Res. Boston, MA **17**, 1–7 (2016)
8. Assefi, M., Behravesh, E., Liu, G., Tafti, A.P.: Big data machine learning using apache spark MLlib. 2017. In: IEEE International Conference on Big Data (Big Data), Boston, MA, pp. 3492–3498 (2017). https://doi.org/10.1109/BigData.2017.8258338
9. Armbrust, M. et al.: Spark SQL: relational data processing in spark. In: Proceedings of the 2015 ACM SIGMOD International Conference on Management of Data—SIGMOD '15, Melbourne, Victoria, Australi, pp. 1383–1394 (2015). https://doi.org/10.1145/2723372.2742797
10. Peng, H., Liang, D., Choi, C.: Evaluating parallel logistic regression models. In: IEEE International Conference on Big Data, Silicon Valley, CA, USA, pp. 119–126 (2013). https://doi.org/10.1109/BigData.2013.6691743
11. Loh, W.: Classification and regression trees. In: Wiley Interdisciplinary Reviews: Data Mining and Knowledge Discovery, vol. 1, pp. 14–23 (2011)
12. Azhari, M., Alaoui, A., Achraoui, Z., Ettaki, B., Zerouaoui, J.: Detection of pulsar candidates using bagging method. In: The International Workshop on Statistical Methods and Artificial Intelligence - IWSMAI 2020. Warsaw, Poland (2020)
13. Genuer, R., Poggi, J.: Arbres CART et forets al éatoires-Importance et sélection de variables, HAL Id: hal-01387654 (2017). https://hal.archives-ouvertes.fr/hal-01387654v2/document
14. Abarda, A., Bentaleb, Y., El Moudden, M., Dakkon, M., Azhari, M., Zerouaoui, J., Ettaki, B.: Solving the problem of latent class selection. In: Proceedings of the International Conference on Learning and Optimization,ACM Algorithms: Theory and Applications, vol. 15 (2018)
15. Breiman, L.: Random Forests. Mach. Learn. **45**, 5–32 (2001)
16. Azhari, M., Alaoui, A., Achraoui, Z., Ettaki, B., Zerouaoui, J.: Adaptation of the random forest method. Proceedings of the 4th International Conference on Smart City Applications - SCA '19 (2019). https://doi.org/10.1145/3368756.336900

17. Azhari, M., Alaoui, A., Abarda, A., Ettaki, B., Zerouaoui, J.: Using Ensemble Methods to Solve the Problem of Pulsar Search. In: Farhaoui Y, (eds) Big Data and Networks Technologies, BDNT 2019, Lecture Notes in Networks and Systems, vol. 81, pp. 183–189. Springer, Berlin (2020)
18. Azhari, M., Alaoui, A., Abarda A., Ettaki, B., Zerouaoui, J.: A comparison of random forest methods for solving the problem of pulsar search. In: The Fourth International Conference on Smart City Applications, Springer, Cham (2020)
19. Liaw, A., Wiener, M.: classification and regression by random forest. R News **2**, 18–22 (2002)
20. Ganjisaffar, Y., Caruana, R., Lopes, C.V.: Bagging gradient-boosted trees for high precision, low variance ranking models. In: Proceedings of the 34th international ACM SIGIR conference on Research and development in Information - SIGIR '11, Beijing, China (2011). https://doi.org/10.1145/2009916.2009932
21. Baldi, P., Sadowski, P., Whiteson, D.: Searching for Exotic Particles in High-energy Physics with Deep Learning. Nat. Commun. **5**, 4308 (2014). https://doi.org/10.1038/ncomms5308
22. Congalton, R.G.: A review of assessing the accuracy of classifications of remotely sensed data. Remote Sens. Environ. **37**, 35–46 (1991). https://doi.org/10.1016/0034-4257(91)90048-B
23. Kumar, R., Indrayan, A.: Receiver operating characteristic (ROC) curve for medical researchers. Indian Pediatr. **48**, 277–287 (2011)
24. Fawcett, T.: An introduction to ROC analysis. Patt. Recogn. **27**, 861–874 (2006)

On Some Results of Fuzzy Super-Connected Space

Mariam El Hassnaoui, Said Melliani, and Mohamed Oukessou

Abstract The purpose of this paper, is to define the fuzzy super-connected spaces, and fuzzy super-connected subset also we prove some characterization between a fuzzy super-connected subset and using the notion of fuzzy continuity. Our approach is based on the idea of fuzzy connected topological spaces and its properties.

1 Introduction

At present there is a great deal of activity in the area of fuzzy topological spaces, a topology on X can be regarded as a family of characteristic function with the usual set operations $\subset$, $\cup$, $\cap$, and complementation replaced by function operations $\leq$, $\vee$, $\wedge$ and $1 - \mu_A$, respectively. In this paper we will introduce the notion of fuzzy connectedness and give its characterization, and in [2] Azad observed that fuzzy connectedness is preserved under fuzzy continuity, here we give one more characterization of this connectedness we also define fuzzy connected subset of fuzzy topological space and study their properties and introduce here the notion of fuzzy super-connected space (fuzzy D-space) [5], and as a result we proved some proved properties of fuzzy super-connected spaces.

M. El Hassnaoui · S. Melliani (✉) · M. Oukessou
Laboratory of Applied Mathematics and Scientific Competing, Department of Mathematics, Faculty of Sciences and Technics, Sultan Moulay Slimane University, Beni-Mellal, Morocco
e-mail: saidmelliani@gmail.com

M. El Hassnaoui
e-mail: mariam.hassnaoui@gmail.com

M. Oukessou
e-mail: ouk_mohamed@yahoo.fr

S. Melliani and O. Castillo (eds.), *Recent Advances in Intuitionistic Fuzzy Logic Systems and Mathematics*, Studies in Fuzziness and Soft Computing 395,
https://doi.org/10.1007/978-3-030-53929-0_9

2 Preliminaries

2.1 Fuzzy sets

Definition 1 Let X be a set. A fuzzy set in X is a function from X into [0, 1], the closed unit interval. We shall denote the characteristic function of a subset A of X by μ_A.

The union $\vee_{\alpha\in A}\lambda_\alpha$, and intersection $\wedge_{\alpha\in A}\lambda_\alpha$, of fuzzy sets λ'_α are the mappings defined respectively as follows: For $x \in X$,

$$(\vee_{\alpha\in A}\lambda_\alpha)(x) = \sup\{\lambda_\alpha(x)|\alpha \in A\}$$

and

$$(\wedge_{\alpha\in A}\lambda_\alpha)(x) = \inf\{\lambda_\alpha(x)|\alpha \in A\}$$

respectively.

Definition 2 For any two fuzzy sets λ and δ, $\lambda \leq \delta$ (or $\delta \geq \lambda$) iff for each $x \in X$, $\lambda(x) \leq \delta(x)$. The complement λ' of a fuzzy set λ in X is $1 - \lambda$ defined by

$$\lambda'(x) = (1-\lambda)(x) = 1 - \lambda(x) \text{ for each } x \in X;$$

0 and 1 stand for $\mu_\varnothing$ and μ_X, respectively.

Definition 3 Let $F : X \longrightarrow Y$ be a mapping from X to Y. If λ is a fuzzy set in X and δ in Y then $F(\lambda)$ and $F^{-1}(\lambda)$ are defined as follows: For $y \in Y$

$$F(\lambda)(y) = \begin{cases} \sup\limits_{y\in F^{-1}(y)} \lambda(y) & \text{if } F^{-1}(y) \neq \varnothing \\ 0 & \text{otherwise,} \end{cases}$$

and for $x \in X$, $F^{-1}(\delta)(x) = \delta(F(x))$.

2.2 Fuzzy Topology

Definition 4 A fuzzy topology on X is a family $\tau(X)$ of fuzzy sets which satisfy following conditions

- 0 and 1 belong to $\tau(X)$.
- Any union of members of $\tau(X)$ is in $\tau(X)$.
- Any finite intersection of members of $\tau(X)$ is in $\tau(X)$.

Remark 1 X together with $\tau(X)$ is called a fuzzy topological space. Briefly we shall call it fts. Members of $\tau(X)$ are called fuzzy open sets and their complements fuzzy closed sets in the fts. X. For a fuzzy set λ in X, the closure $\bar{\lambda}$ and interior λ^i of λ are defined as

$$\bar{\lambda} = \wedge\{\delta : \delta \geq \lambda,\ \delta' \in \tau(x)\},$$

and

$$\lambda^i = \vee\{\delta : \delta \leq \lambda,\ \delta \in \tau(x)\}.$$

Definition 5 A fuzzy open set λ is called a fuzzy regular open set if $\bar{\lambda}^i = \lambda$ and a fuzzy closed set δ is called a fuzzy regular closed set if $\bar{\delta^i} = \delta$.

Remark 2 In [1] it is proved that (1) Complement of a fuzzy regular open set is a fuzzy regular closet set and vice versa. (2) Closure of a fuzzy open set is a fuzzy regular closed set and interior of a fuzzy closed set is a fuzzy regular open set.

Definition 6
- A fuzzy set λ is called fuzzy semi-open [1] if there is a fuzzy open set δ such that $\delta \leq \lambda \leq \bar{\delta}$. It is proved [1] that λ is fuzzy semi-open iff $\lambda \leq \bar{\lambda^i}$. The Complement of a fuzzy semi-open set is called a fuzzy semi-closed set. Clearly, δ is fuzzy semi-closed iff $\bar{\delta}^i \leq \delta$ iff there exist a fuzzy closed set k such that $k^i \leq \delta \leq k$.
- A function F from an fts. X to an fts. Y is called fuzzy continuous (F-continuous) [7] if for each fuzzy open set E. in Y, $F^{-1}(\lambda)$ is a fuzzy open set in X.
- Warren [7] defined fuzzy subspace A of an fts $(X, \tau(X))$ as follows:
  If $A \in X$, then the family $\{T_A = \lambda/A : \lambda \in \tau(X)\}$ is a fuzzy topology on A, where i/A is the restriction of 1. to A. Then (A, T_A) is called the fuzzy subspace of the fts X with underlying set A. It is easy to see that a fuzzy set a in A is fuzzy closed in A iff there exists a fuzzy closed set h in X such that $a = b/A$.

Wong [8] defined basis and subbasis for a given fuzzy topology analogously to these concepts in the general topology. Warren [7] established among others the following results:

Theorem 1 *For any fuzzy set δ in an fts,*

$$1 - \bar{\delta} = (1 - \delta)^i \text{ and } 1 - \delta^i = \overline{(1 - \delta)}$$

Theorem 2 *Let (A, T_A) be a fuzzy subspace of an fts $(X, \tau(X))$ and let a be a fuzzy set in A. Further let b be the fuzzy set in X defined as $b(x) = a(x)$ if $x \in A$ and $b(x) = 0$ if $x \in X - A$. Then $\bar{a} = \bar{b}/A$, where $\bar{a}$ is the closure of a with respect to T, and $\bar{b}$ is the closure of b with respect to $\tau(X)$.*

2.3 Product Fuzzy Topology

Definition 7 Let $\{X_\alpha\}_{\alpha \in A}$, be a family of non-empty sets. Let $X = \prod_{\alpha \in A} X_\alpha$, be the usual product of X_α and let P_α, be the projection from X to X_α. Further assume that each X_α, is an fts with a fuzzy topology τ_α. Fuzzy topology generated by $\rho = \{P_\alpha^{-1}(b_\alpha) : b_\alpha \in \tau_\alpha, \alpha \in A\}$ as subbasis, is called the product fuzzy topology in X. Clearly if λ is a basic element in the product topology, then for $x = (x_\alpha)_{\alpha \in A} \in X$, there exist $\alpha_1, \alpha_2, ..., \alpha_n \in A$ such that

$$\lambda(x) = \min\{(b_{\alpha_i}) : i = 1, 2, ..., n\}.$$

Now we prove the following elementary results as we shall use them in the sequel.

Theorem 3 *If $A \subset X$ such that μ_A is fuzzy open (closed) in fts X and λ_A, is a fuzzy open (closed) set in the fuzzy subspace A of X, then the fuzzy set λ is fuzzy open (closed) in X where $\lambda(x) = \lambda_A(x)$ if $x \in A$ and $\lambda(x) = 0$ if $x \in X - A$.*

Proof $\lambda_A = \delta/A$ for some fuzzy open (closed) set δ in X. So $\lambda = \delta \wedge \mu_A$. Therefore λ is fuzzy open (closed) as δ is so.

Corollary 1 *If $B \subset A \subset X$ and μ_A is fuzzy open (closed) in fts X and μ_B/A is fuzzy open (closed) in the fuzzy subspace A of X, then μ_B is fuzzy open (closed) in X.*

Proof As μ_B/A is fuzzy open (closed) in A there is a fuzzy open (closed) set δ in X such that $\mu_B/A = \delta/A$. Now $\mu_B = \delta \wedge \mu_A$.

Theorem 4 *Let X be an fts and U, W, and Z be subsets of X such that $U \subset W \cap Z$. Further suppose that a is a fuzzy set in the fuzzy subspace U of X and b is defined as $b(x) = a(x)$ if $x \in U$ and $b(x) = 0$ if $x \in X - U$. If b/W is fuzzy open (closed) in W and b/Z is fuzzy open (closed) in Z, then $b/W \cup Z$ is fuzzy open (closed) in $W \cup Z$.*

Proof $b/W = \lambda/W$ and $b/Z = \delta/Z$ for some fuzzy open (closed) sets λ and δ in X, so $b/W \cup Z = (\lambda \wedge \delta)/W \cup Z$ which is fuzzy open (closed) in $W \cup Z$.

Corollary 2 *If μ_U/W is fuzzy open (closed) in W and μ_U/Z is fuzzy open (closed) in Z, then $\mu_U/W \cup Z$ is fuzzy open (closed) in $W \cup Z$.*

Proof This follows from Theorem 2.4.

3 Fuzzy Connectedness

Definition 8 An fts X is said to be fuzzy connected if it has no proper fuzzy clopen (closed and open) set. [A fuzzy set λ in X is proper if $\lambda \neq 0$ and $\lambda \neq 1$]

Now, we give a characterizations of fuzzy connected set.

Theorem 5 *An fts X is fuzzy connected iff it has no non-zero fuzzy open sets λ and δ such that $\lambda + \delta = 1$.*

Proof If such λ and δ exist, then λ is a proper clopen fuzzy set in X. If X is not fuzzy connected then it has a proper fuzzy clopen set λ. So $\delta = 1 - \lambda$ is a fuzzy open set such that $\delta \neq 0$ and $\lambda + \delta = 1$.

Corollary 3 *Fts X is fuzzy connected iff it has no non-zero fuzzy sets λ, and δ such that $\lambda + \delta = 1$, $\bar{\lambda} + \delta = \lambda + \bar{\delta} = 1$.*

Remark. The fuzzy product of fuzzy connected spaces may not be a fuzzy connected space. For example, let $X_i = [0, 11]$, $i \in I$. For some $j, k \in I$, let

$$\tau(X_j) = \{0, 1, \lambda\} \text{ and } \tau(X_k) = \{0, 1, \lambda'\},$$

where $\lambda(x) = \frac{1}{3}$ for $0 \leq x \leq 1$, and $\tau(X_i) = \{0, 1\}$ for each $i \in I$ and $i \neq j,\ i \neq k$. Then each X_i is fuzzy connected but $\prod_{i \in I} X_i$, is not so as $\tau(\prod_{i \in I} X_i)$ contains non-zero fuzzy open sets $P_j^{-1}(\lambda)$ and $P_k^{-1}(\lambda)$ such that for every $x \in \prod_{i \in I} X_i$

$$P_j^{-1}(\lambda) + P_k^{-1}(\lambda') = 1.$$

4 Fuzzy Connected Subsets in an FTS

Definition 9 If $A \subset X$, X is an fts, then A is said to be a fuzzy connected subset of X if A is a fuzzy connected space as a fuzzy subspace of X.
Remark It is easy to see that if $A \subset Y \subset X$, then A is a fuzzy connected subset of the fts X iff it is a fuzzy connected subset of the fuzzy subspace Y of X.

Theorem 6 *If X is an fts and A is a fuzzy connected subset of X, and λ and δ are non-zero fuzzy open sets in X satisfying $\lambda + \delta = 1$, then either $\lambda/A = 1$ or $\delta/A = l$.*

Proof Suppose there exists $x_0,\ y_0 \in A$ such that $\lambda(x_0) \neq 11$ and $\delta(y_0) \neq 1$. Then $\lambda + \delta = 1$ implies that $\lambda/A + \delta/A = 1$, where $\lambda/A \neq 0$ and $\delta/A \neq 0$. So by Theorem 3.1, A is not a fuzzy connected space.

Definition 10 Fuzzy sets λ and δ in an fts X are said to be separated from each other if $\bar{\lambda} + \delta \leq 1$ and $\lambda + \bar{\delta} \leq 1$.

Theorem 7 *Let $\{A_\alpha\}_{\alpha \in A}$ be a family of fuzzy connected subsets of X such that for each α and β in A and $\alpha \neq \beta$, μ_{A_α} and μ_{A_β} are not separated from each other. Then $\bigcup_{\alpha \in A} A_\alpha$ is a fuzzy connected subset of X.*

Proof suppose $Y = \cup_{\alpha \in A} A_\alpha$ is not a fuzzy connected subset of X. Then there exist non zero fuzzy open sets a and b in Y such that $a + b = 1$. Fix $\alpha_0 \in A$.Then $A_{\alpha 0}$ is fuzzy connected subset of Y as it is so in X. Therefore by Theorem 4.1 either

$$\mu_{A_{\alpha 0}}/A_{\alpha 0} = a/A_{\alpha 0} \text{ or } \mu_{A_{\alpha 0}}/A_{\alpha 0} = b/A_{\alpha 0}.$$

Without loss of generality assume that

$$\mu_{A_{\alpha 0}}/A_{\alpha 0} = a/A_{\alpha 0}, (i).$$

Define λ and δ as $\lambda(x) = a(x)$ if $x \in Y$ and $\lambda(x) = 0$ if $x \in X - Y$ $\delta(x) = b(x)$ if $x \in Y$ and $\delta(x) = 0$ if $x \in X - Y$.

By Theorem 2

$$\bar{a} = \bar{\lambda}/Y \text{ and } \bar{b} = \bar{\delta}/Y, (ii).$$

So (i) implies that $\mu_{A_{\alpha 0}} \leq \lambda$. Therefore

$$\overline{\mu_{A_{\alpha 0}}} \leq \bar{\lambda}, (iii).$$

Let $\alpha \in A - \{\alpha_0\}$. Since A_α is a fuzzy connected subset of Y either $\mu_{A_{\alpha 0}}/A_{\alpha 0} = a/A_{\alpha 0}$ or $\mu_{A_{\alpha 0}}/A_{\alpha 0} = b/A_{\alpha 0}$. We show that $\mu_{A_{\alpha 0}}/A_{\alpha 0} \neq b/A_{\alpha 0}$.

Suppose that $\mu_{A_\alpha}/A_\alpha = b/A_\alpha$. Therefore $\mu_{A_\alpha} \leq \delta$. Hence

$$\overline{\mu_{A_\alpha}} \leq \bar{\delta}(iv)$$

Since $\bar{a} + b = a + \bar{b} = 1, \bar{\lambda} + \delta \leq 1$ (by (ii) and definition of λ and δ). So (iii) and (iv) imply that

$$\bar{\mu}_{A_\alpha 0} + \mu_{A_\alpha} \leq 1 \text{ and } \mu_{A_{\alpha 0}} + \bar{\mu}_{A_\alpha} \leq 1.$$

This gives a contradiction as $\mu_{A_{\alpha_\alpha}}$ and μ_{A_α} are not separated from each other.

So $\mu_{A_\alpha}/A_\alpha \neq b/A_\alpha$. Hence $\mu_{A_\alpha}/A_\alpha = a/A_\alpha$ for each $\alpha \in A$. Which in turn implies that $\mu_Y = a$. But $a + b = 1$. So $b(x) = 0$ for every $x \in Y$. But $b \neq 0$. So our supposition that Y is not a fuzzy connected subset of X is false.

Corollary 4 *If $\{A_\alpha\}_{\alpha \in A}$ is a family of fuzzy connected subsets of an fts X and $\bigcap_{\alpha \in A} A_\alpha \in \varnothing$, then $\bigcup_{\alpha \in A} A_\alpha$ is a fuzzy connected subset of X.*

Hint: For any α, $\beta \in A$, we have $A_\alpha \cap A_\beta \neq \varnothing$. So $\overline{\mu_{A_\alpha}} + \mu_{A_\beta} > 1$ and $\mu_{A_\alpha} + \overline{\mu_{A_\beta}} > 1$. Thus characteristic functions of each pair of members of the family are not separated from each other.

Corollary 5 *If $\{A_n : n = 1, 2, ...\}$ is a sequence of fuzzy connected subsets of an fts X such that μ_{A_n} and $\mu_{A_{n+1}}$ are not separated from each other for n $= 1, 2, ...,$ then $\bigcup_{n=1}^{\infty} A_n$ is a fuzzy connected subset of X.*

We omit the proof.

Theorem 8 *If A and B are subsets of an fts. X and* $\mu_A \leq \mu_B \leq \overline{\mu_A}$ *and A is a fuzzy connected subset of X, then B is also a fuzzy connected subset of X.*

5 Fuzzy Super Connectedness

Introduction Levine introduced the notion of a D-space to be a topological space in which every non-empty open set is dense. It is defined a super-connected space to be a topological space which has no proper regular open subset and have shown that a space is super-connected iff it is a D-space. So we define a fuzzy D-space to be a fuzzy topological space in which there is no proper fuzzy regular open set and we shall also call such a space to be a fuzzy super-connected space. Since a fuzzy clopen set is a fuzzy regular open set, fuzzy super-connectedness implies fuzzy connectedness but the following example shows that the converse is not true.
Example. Let $X = [0, 1]$. For each $x \in X$ let is define $\lambda(x) = \frac{1}{6}$ and $\mu(x) = \frac{2}{3}$. Let $\tau(X) = \{0, 1, \lambda, \mu\}$. Then clearly fts X is fuzzy connected but it is not fuzzy super-connected, since it has a proper fuzzy regular open set μ.

We also define fuzzy super-connected subsets of an fts and study their properties

6 Characterization of Fuzzy Connectedness

Theorem 9 *If X is an fts then the following statements are equivalent:*

1. *X is fuzzy super-connected.*
2. *Closure of every non-zero fuzzy open set in X is* 1.
3. *Interior of every fuzzy closed set (in X), different from* 1, *is zero.*
4. *X does not have non-zero fuzzy open sets* λ *and* μ *such that* $\lambda + \mu \leq 1$.
5. *X does not have non-zero fuzzy sets* λ *and* μ *satisfying* $\bar{\lambda} + \mu = \lambda + \bar{\mu} = 1$.
6. *X does not have non-zero fuzzy closed sets* f *and* k *satisfying* $f^i + k = f + k^i = 1$.

Proof (1)$\Longrightarrow$ (2). If X has non-zero fuzzy open set λ such that $\bar{\lambda} \neq 1$, then $\bar{\lambda}^i$ is a proper fuzzy regular open set.

(2)$\Rightarrow$ (3). Let f be a fuzzy closed set in X different from 1. Now $f^i = 1 - \overline{1 - f} = 0$ as $1 - f$ is non-zero open set (Theorem 2).

(3)$\Rightarrow$(4). If X has non-zero fuzzy open set λ and μ such that $\lambda + \mu \leq 1$,then $\bar{\lambda} + \mu \leq 1$. So $\mu \neq 0$ implies $\bar{\lambda} \neq 1$. Since $\lambda \neq 0, \bar{\lambda}^i \neq 0$, which contradicts (3).

(4)$\Rightarrow$(1). if X has a proper fuzzy regular open set λ and $\mu = 1 - \bar{\lambda}$ are non-zero fuzzy open sets satisfying $\lambda + \mu \leq 1$.

(1)$\Leftrightarrow$ (5). if X is not fuzzy super-connected, then it has a proper fuzzy regular open set say λ. If we put $\mu = 1 - \bar{\lambda}$, then $\mu \neq 0$ and $\lambda + \bar{\mu} = 1$. Also by Theorem 2.0(a),

$\bar{\mu} = \overline{1-\lambda} = \overline{(1-\lambda)^i} = 1-\lambda$ as $1-\lambda$ is a fuzzy regular closed set. Therefore $\lambda + \bar{\mu} = 1$. So (5) is violated.

Conversely, if X has non-zero fuzzy open sets λ and μ such that $\bar{\lambda} + \mu = \lambda + \bar{\mu} = 1$, then

$$\bar{\lambda}^i = (1-\mu)^i = 1 - \bar{\mu} = \lambda$$

Since μ and $\bar{\lambda} + \mu = 1$, $\lambda \neq 1$. Also $\lambda \neq 0$ is given. Therefore λ is a proper fuzzy regular open set. Therefore X cannot be fuzzy super-connected.
(5)$\Leftrightarrow$ (6). (5)$\Rightarrow$ (6) follows if we take $f = 1 - \lambda$ and $k = 1 - \mu$. Reverse implication can be proved similarly.

Theorem 10 *An fts X is fuzzy super-connected iff it has no proper fuzzy open set which is also fuzzy semi-closed or equivalently iff it has no proper fuzzy closed set which is also fuzzy semi-open.*

Proof This follows immediately from the definition of fuzzy regular open sets, fuzzy semi-open sets, and fuzzy semi-closed sets, and fuzzy semi-closed sets.

Theorem 11 *If X and Y are fuzzy topological spaces and a function F from X onto Y is fuzzy continuous then X is fuzzy super-connected implies Y is fuzzy super-connected.*

Proof Deny. Then exists a fuzzy open set $\lambda \neq 0$ in Y such that $\lambda \neq 1$. F is fuzzy continuous implies

$$\overline{F^{-1}(\lambda)} \leq F^{-1}(\bar{\lambda})(v)$$

(see Theorem 4.2 of Warren [7]). Since $\lambda \neq 0$ and $\bar{\lambda} \neq 1$, there exist $y_1, y_2 \in Y$ such that $\lambda(y_1) \neq 0$ and $\bar{y_2} \neq 1$. Now F is onto. Therefore there exist $x_1, x_2 \in X$ such that

$$F(x_1) = y_1 \text{ and } F(x_2) = y_2.$$

So $F^{-1}(\lambda)(x_1) = \lambda(F(x_1) = \lambda(y_1) \neq 0$. similarly $\overline{F^{-1}(\bar{\lambda})}(x_2) \neq 0$. So by (v) $F^{-1}(\lambda)$ is a non zero fuzzy open set in X such that $\overline{F^{-1}\lambda} \neq 1$. This is a contradiction, as X is fuzzy super-connected.

Theorem 12 *A finite product of fuzzy super-connected spaces is fuzzy super-connected.*

Proof Let $(X, \tau(x)$ and $(Y, \tau(y))$ be fuzzy super-connected topological spaces. Suppose that $(X \times Y, \tau(X \times Y))$ is not fuzzy super-connected. Then there exist $\lambda, \mu \in \tau(X)$ and $\xi, \eta \in \tau(Y)$ such that $\lambda \times \xi \neq 0$. and

$$(\lambda \times \xi)(x, y) + (\mu \times \eta)(x, y) \leq 1,$$

for every $(x, y) \in X \times Y$, where $\lambda \times \xi$, $\mu \times \eta \in \tau(X \times Y)$, $\lambda \times \xi = P_X^{-1}(\lambda) \cap P_Y^{-1}(\xi)$ P_x is a projection map of $X \times Y$ onto X, etc. So $min\{\lambda(x), \xi(y)\} + min\{\mu(x), \eta(y)\} \leq 1$ for every $(x, y) \in X \times Y$ which implies that for any $(x, y) \in$

$X \times Y$ either (i) $\lambda(x) + \mu(x) \leq 1$ or (ii) $\lambda(x) + \mu(y) \leq 1$ Or (iii)$\eta(y) + \mu(x) \leq 1$ Or (iv)$\xi(y) + \eta(y) \leq 1$. Now $\lambda \wedge \mu \in \tau(X)$ and $\xi \wedge \eta \in \tau(Y)$.

As X and Y are fuzzy super-connected topological spaces, if $\lambda \wedge \mu \neq 0, \xi \wedge \eta \neq 0$,then there exist $x_1 \in X$ and $y_1 \in Y$ such that $(\lambda \wedge \mu)(x_1) > \frac{1}{2}$ and $(\xi \wedge \eta)(y_1) > \frac{1}{2}$. So $\lambda(x_1) > \frac{1}{2}$, $\mu(x_1) > \frac{1}{2}$, $\xi(y_1) > \frac{1}{2}$, and $\eta(y_1) > \frac{1}{2}$. Therefore if $x = x_1$ and $y = y_1$, then none of the above possibilities will be true.

If $\lambda \wedge \mu = 0$, then for each $x \in X$ either $\lambda(x) = 0$ or $\mu(x) = 0$. So for every $x \in X$, $\lambda(x) + \mu(x) \leq 1$. Note that λ, $\mu \neq 0$ as $\lambda \times \xi$, $\mu \times \eta \neq 0$ which implies that $(X, \tau(X))$ is not fuzzy super-connected. Similarly $\xi \wedge \eta = 0$will imply that $(Y, \tau(Y))$ is not fuzzy super-connected.

7 Fuzzy Super-Connected Subspace

Definition 11 A subset of an fts X is called a fuzzy super-connected subset of X if it is a fuzzy super-connected topological space as a fuzzy subspace of X.

Theorem 13 *If $A \subset Y \subset X$ then A is a fuzzy super-connected subset of X iff it is a fuzzy super-connected subset of the fuzzy subspace Y of X.*

Theorem 14 *Let A be a fuzzy super-connected subset of an fts X. if there exist fuzzy closed sets f and k in X such that*

$$f^i + k = f + k^i = 1$$

then $f/A = 1$ or $k/A = 1$.

Proof If $f(x_0) \neq 1$ and $k(y_0) \neq 1$ for x_0, $y_0 \in A$ then $f^i(y_0) + k(y_0) = 1$ and $f(x_0) + k^i(x_0) = 1$ imply that $f^i(y_0) \neq 0$ and $k^i(x_0) \neq 0$. Thus f^i/A and k^i/A are non-zero fuzzy open sets A such that $f^i/A + k^i/A \leq 1$, which contradicts the fact that A is a fuzzy super-connected subset of X.

Theorem 15 *Let X be an fts and $A \subset Y$ be a fuzzy super-connected subset of X such that μ_A is a fuzzy open set in X. If λ is a fuzzy regular open set in X, then either $\mu_A \leq \lambda$ or $\mu_A \leq 1 - \lambda$.*

Proof If $\lambda = 0$ or 1 then the result holds. Suppose that $\lambda \neq 0$ and $\lambda \neq 1$. Let $f = \bar{\lambda}$ and $k = 1 - \lambda$. Then f and k are such that $f^i + k = f + k^i = 1$. By the previous theorem $\mu_A \leq f$ or $\lambda_A \leq k$. So $\mu_A \leq \underline{f^i}$ or $\mu_A \leq k^i$, as μ_A is fuzzy open. Therefore $\mu_A \leq \bar{\lambda}^i = \lambda$; or $\mu_A \leq (1-\lambda)^i \leq \overline{(1-\lambda)}^i = 1 - \lambda$.

Theorem 16 *Let $\{O_\alpha\}_{\alpha \in A}$ be a family of subsets of an fts X such that each μ_{O_α} is fuzzy open. If $\bigcap_{\alpha \in A} O_\alpha \neq \varnothing$ and each O_α is a fuzzy super-connected subset of X, then $\bigcup_{\alpha \in A} O_\alpha$ is also a fuzzy super-connected subset of X.*

Proof Let $Y = \cup_{\alpha \in A} O_\alpha$ and suppose that Y is not a fuzzy super-connected subset of X .Then there exists a proper fuzzy regular open set λ_Y in the fuzzy subspace Y of X.
Each μ_{O_α} is fuzzy open in X. So each μ_{O_α}/Y is fuzzy open in Y. also each O_α is a fuzzy super-connected subset of the subspace Y as it is so in X. Therefore by previous result for each $\alpha \in A$ either $\mu_{O_\alpha}/Y \leq \lambda_Y$ or $\mu_{O_\alpha}/Y \leq 1 - \lambda_Y$. suppose $x_0 \in \cap_{\alpha \in A} O_\alpha$, Then either $\lambda_Y(x_0) = 1$ or $\lambda_Y(x_0) = 0$. If $\lambda_Y(x_0) = 1$, then $\mu_{O_\alpha}/Y \leq \lambda_Y$ for every $\alpha \in A$. Hence $\mu_Y/Y = \bigvee_{\alpha \in A}(\mu_{O_\alpha}/Y) \leq \lambda_Y$. But $\lambda_Y \leq \mu_Y/Y$; so $\lambda_Y = 1$, which is prohibited since $\lambda_Y \neq 1$. By similar argument, if $\lambda_Y(x_0) = 0$ then we shall get $\lambda_Y = 0$, which is also a contradiction.

Theorem 17 *If A and B are fuzzy super-connected subsets of an fts X and μ^i_B/A or $\mu^i_A \neq 0$, then $A \cup B$ is a fuzzy super-connected subset of X.*

Proof Suppose that $Y = A \cup B$ is not a fuzzy super-connected subset of X. Then there exist fuzzy open set λ and δ such that $\lambda/Y \neq 0, \delta/Y \neq 0$, and $\lambda/Y + \delta/Y \leq 1$. Since A is a fuzzy super-connected subset of X either $\lambda/A = 0$ or $\delta/A = 0$. Without loss of generality assume that $\delta/A = 0$. In that case since B is also fuzzy super-connected, We have (i) $\lambda/A \neq 0$, (ii)$\delta/B \neq 0$, (iii) $\delta/A = 0$, and (iv) $\lambda/B = 0$. Therefore

$$(v)\lambda/A + \mu^i_B/A \leq 1(\text{ because } \lambda/B = 0)$$

If $\mu^i_B/A \neq 0$ then (i) and (v) imply that A is not a fuzzy super-connected subset of X.

Similarly if $\mu^i_A/B \neq 0$, then (ii) and $\delta/B + \mu^i_A/B \leq 1$ imply that B is not a fuzzy super-connected subset of X. We thus get a contradiction.

Theorem 18 *If $\{A_\alpha\}_{\alpha \in A}$ is a family of fuzzy super-connected subset of an fts X such that*

$$\left[\bigwedge_{\alpha \in A} \mu_{A_\alpha}\right]^i \neq 0 \textit{ then } \cup_{\alpha \in A} A_\alpha$$

is a fuzzy super-connected subset of X.

Proof Suppose $Y = \cup_{\alpha \in A} A_\alpha$ is not a fuzzy super-connected subset of X. Then there exist fuzzy open sets λ and δ in X such that

$$\lambda/Y \neq 0(i)$$
$$\delta/Y \neq 0(ii) \text{ and } \lambda/Y + \delta/Y \leq 1.$$

Equations (i) and (ii) imply that exist β and γ in A such that $\lambda/A_\beta \neq 0$ and $\gamma/A_\gamma \neq 0$. Case (i). If $\beta = \gamma$, then A_β will not be a fuzzy super-connected subset of X, which is prohibited.

Case (ii). If $\beta \neq \gamma$, then

$$0 \neq \left[\bigwedge_{\alpha \in A} \mu_{A_\alpha}\right]^i \leq \bigwedge_{\alpha \in A} \mu'^{i}_{A_\alpha}$$

implies that $\mu'^{i}_{A_\beta} \wedge \mu'^{i}_{A_\gamma} \neq 0$. So $\mu'^{i}_{A_\beta}/A_\gamma \neq 0$. Hence by the previous theorem $A_\beta \cup A_\gamma$ is a fuzzy super-connected subset of X. On the contrary it can be seen that $\lambda/A_\beta \cup A_\gamma \neq 0$, $\delta/A_\beta \cup A_\gamma \neq 0$, and $\lambda/A_\beta \cup A_\gamma + \delta/A_\beta \cup A_\gamma \leq 1$. So $A_\beta \cup A\gamma$ is not a fuzzy super-connected subset of X.

Theorem 19 *Suppose an fts X is fuzzy super-connected and C is a fuzzy super-connected subset of X. Further suppose that $X - C$ contains a set V such that $\mu_v/X - C$ is a fuzzy open set in the fuzzy subspace $X - C$ of X then $C \cup V$ fuzzy super-connected subset of X.*

Proof Suppose $Y = C \cup V$ is not a fuzzy super-connected subset of X.

Then there exist fuzzy open sets λ and δ in A such that $\lambda/Y \neq 0$, $\delta/Y \neq 0$, and $\lambda/Y + \delta/Y \leq 1$. As C is a fuzzy super-connected subset of X, either $\lambda/C = 0$ or $\delta/C = 0$.

Without loss of generality assume that $\lambda/C = 0$. Therefore $\lambda/V \neq 0$ if we define a fuzzy set λ_V in X as $\lambda_v(c) = \lambda(x)$ if $X \in V$, $\lambda_v(x) = 0$ if $x \in X - V$, then λ_V is open in X as $\lambda_V = \lambda \wedge \mu_V$. So $\bar{\lambda_V}$ is a fuzzy regular closed set in X.

Now we show that $\bar{\lambda_V}$ is a proper fuzzy set in X. $\lambda/Y + \delta/Y \leq 1$ implies $\lambda_V + \delta \leq 1$. So $\bar{\lambda_V} + \delta \neq 1$ Therefore $\bar{\lambda_V} \neq 1$ as $\delta \neq 0$. Also if $\bar{\lambda_V} = 0$, then $\lambda_V = 0$, so $\lambda/V = 0$, but by (1), $\lambda/V \neq 0$. Thus X is not a fuzzy super-connected space, which is a contradiction.

8 Main Results

Theorem 20 *If A and B are subsets of an fts X and $\mu_A \leq \mu_B \leq \mu_{\bar{A}}$ and A is a fuzzy super connected subset of X then so is B. And as a conclusion we get that $\bar{A}$ is also fuzzy super connected subset.*

Proof If we suppose that B is not super connected then there exist fuzzy open set λ witch is also semi -closed such that: $\lambda/B \neq 0$ $\delta/B \neq 0$ and $\lambda/B + \delta/B = 1$ (i).

We first show that $\lambda/A \neq 0$, if $\lambda/A = 0 \Longrightarrow \lambda + \mu_A \leq 1 \Longrightarrow \lambda + \mu_{\bar{A}} \leq 1$ ($\mu_A \leq \lambda \leq \mu_{\bar{A}}$) $\Longrightarrow \lambda + \mu_B \leq 1$ ($\mu_B \leq \mu_{\bar{A}}$).

This in turn implies $\lambda/B = 0$ witch a contradiction as $\lambda/B \neq 0$ therefore $\lambda/A \neq 0$ similarly we can show that $\delta/A \neq 0$, now (i) and $\mu_A \leq \mu_B$ imply $\lambda/A + \delta/A = 1$. So A is not fuzzy super connected ,which is a contradiction. in particularly we have $\bar{A}$ is fuzzy super connected ($\mu_B \leq \mu_{\bar{A}}$).

Theorem 21 *Let* $I = [0, 1]$*: I is fuzzy super connected and*

$$f : I \longrightarrow I$$

a continuous mapping. Then we have the following assertion.

1. $F : I \longrightarrow I \times I$ *defined by* $F(x) = (x, f(x))$ *is continuous.*
2. $F(I)$ *is fuzzy super connected in* $I \times I$.

Proof • Since $F^{-1}(]x, y[\times]z, t[) =]x, y[\bigwedge f^{-1}(]z, t[)$ is an open set of I.
• Using the Theorem 6.4.

Acknowledgements The authors would like to thankful to the referee for his invaluable suggestions, which put the article in its present shape.

References

1. Azad, K.K.: On fuzzy semi-continuity, fuzzy almost continuity and fuzzy weakly continuity. J. Math. Anal. Appl. **82**, 14–32 (1981)
2. K. K. AZAD, Fuzzy connectedness, unpublished
3. Chang, Chin-Liang: Fuzzy topological spaces. J. Math. Anal. Appl. **24**, 182–190 (1968)
4. U.V. FATTEH AND D. S. BASSAN, A note on D-spaces, Bull. Calcutta Math. Soc. 75 (6)
5. Levine, Norman: Dense topologies. Amer. Math. Monthly **75**, 847–852 (1968)
6. Levine, Norman: Strongly connected sets in a topology. Amer. Math. Monthly **72**(10), 1098–1101 (1965)
7. Warren, R.H.: Neighborhoods, bases and continuity in fuzzy topological spaces. Rocky Mountain J. Math. **8**, 459–470 (1978)
8. Wong, Kun-Chun: Fuzzy topology: Product and Quotient theorems. J. Math. Anal. Appl. **45**, 512–521 (1974)
9. Lotfi Aliasker Zadeh: Fuzzy sets. Inform. and Control **8**, 338–353 (1965)

Regional Optimal Control Problem of a Heat Equation with Bilinear Bounded Boundary Controls

Zerrik El Hassan and EL Kabouss Abella

Abstract In this paper we discuss a regional optimal control problem with bilinear boundary controls for heat equation. We demonstrate the existence of a control minimizing a quadratic functional, and we give a characterization using an optimality system. Also we give the condition that ensures the uniqueness of the optimal control. The obtained results lead to an algorithm that we illustrate by simulations.

Keywords Heat equation · Boundary control · Bilinear control · Regional optimal control

1 Introduction

Bilinear systems are a special class of non-linear systems that are linear in input and linear in state. The attention on bilinear systems has been focused, both for its applicative interest and intrinsic simplicity. In fact, bi-linear systems can be used to represent a wide range of physical, chemical and biological systems (see [6]).

A wide literature is devoted to the optimal control of the controllability of such systems. For internal case, in [2] authors determined a spatio-temporal optimal control for a Kirchhoff equation, in [4] they showed the case of only time dependent optimal control, and in [3] they studied the case of a spacial optimal control. In case of boundary bilinear controls, in [5] author considered the problem of controlling the solution of the heat equation with the convective boundary condition taking the heat transfer coefficient as the bilinear control, they established existence and uniqueness of the optimal control, while in [7] author considered a bilinear boundary optimal control problem for a Kirchhoff plate equation.

Z. El Hassan (✉) · E. K. Abella
MACS laboratory, Faculty of Sciences, Univerisity of Moulay Ismail, Meknes, Morocco
e-mail: zerrik3@yahoo.fr

E. K. Abella
e-mail: elkabouss.abella@gmail.com

S. Melliani and O. Castillo (eds.), *Recent Advances in Intuitionistic Fuzzy Logic Systems and Mathematics*, Studies in Fuzziness and Soft Computing 395,
https://doi.org/10.1007/978-3-030-53929-0_10

The principal reason for considering the regional concept is that it is close to real applications, for example, the physical problem which concerns a tunnel furnace where one has to maintain a prescribed temperature only in a subregion of the furnace and there exist systems which are controllable in some subregion but not controllable in the whole domain, and controlling a system regionally is cheaper than controlling it in the whole domain. For bilinear systems, In [13] and [9] authors characterized a regional optimal control of distributed bilinear systems by minimizing a quadratic cost functional using an optimality system, and in [14] by the resolution of the Riccati equation. In [10] authors studied a regional bilinear optimal control of a system governed by a fourth-order parabolic operator with bounded and unbounded controls, minimizing a functional cost, then in [12] they studied this problem for a class of infinite bilinear system with unbounded control operator in both cases unbounded and bounded controls, while in [11] they considered a regional bilinear optimal control problem of a wave equation.

In this work we deal with a regional optimal control problem heat equation with bilinear boundary controls, we show existence and characterization of an optimal control.

More precisely, let Ω be an open bounded set of $\mathbb{R}^n$ $(n \geq 2)$, with a regular boundary $\partial\Omega$.

For $T > 0$, we denote by $Q = \Omega \times]0, T[$, $\Gamma = \partial\Omega \times]0, T[$, and we consider a system described by

$$\begin{cases} \dfrac{\partial z}{\partial t}(x,t) = \Delta z(x,t) & Q, \\ \dfrac{\partial z}{\partial \upsilon_A}(x,t) = u(x,t)Bz(x,t) & \Gamma, \\ z(x,0) = z_0(x) & \Omega, \end{cases} \tag{1}$$

where $\dfrac{\partial}{\partial \upsilon_A}$ is the directional derivative of the co-normal vector υ_A, and $u \in U = \{u \in L^\infty(\Gamma)/0 \leq u \leq M\}$ (m and M are non-negative constants) is the control where U is the set of admissible controls.

B is a linear continuous operator on $L^2(\Gamma)$ such that $< uBz, z >_{L^2(\partial\Omega)} \leq 0$.

For $z_0 \in L^2(\Omega)$ and $u \in U$ according to lemma 5.3 (p:373 [8]) system (1) has a unique weak solution $z \in L^2(0, T; H^1(\Omega))$ such that $\dfrac{\partial z}{\partial t} \in L^2(0, T; (H^1(\Omega))')$ which satisfies, for $\varphi \in H^1(\Omega)$,

$$\int_0^T \langle \frac{\partial z}{\partial t}, \varphi \rangle dt + \int_0^T < \nabla z, \nabla \varphi >_{L^2(\Omega)} \, dt = \int_\Gamma uBz\varphi \, dx \, dt.$$

where $\langle ., . \rangle$ denotes the duality between $H^1(\Omega)$ and $(H^1(\Omega))'$.

Let ω be an open subregion of Ω, $\chi_\omega : L^2(\Omega) \longrightarrow L^2(\omega)$ indicates the restriction function to $\omega \subset \Omega$.

$\chi_\omega^* : L^2(\omega) \longrightarrow L^2(\Omega)$ is the adjoint operator of χ_ω given by

$$\chi_\omega^* z(x,.) = \begin{cases} z(x) & if\ x \in \omega \\ 0 & if\ x \in \Omega \setminus \omega, \end{cases}$$

An optimal control problem my be stated as follows

$$\min_{u \in U} J(u), \tag{2}$$

where J is the cost functional given by

$$J(u) = \frac{1}{2}\int_0^T \|\chi_\omega z(x,t) - z_d(x)\|^2_{L^2(\omega)} dt + \frac{\beta}{2}\|u\|^2_{L^2(\Gamma)}. \tag{3}$$

β is a positive constant.

The paper is organized as follows, in Sect. 2, the existence of an optimal control solution of problem (2) is shown. In Sect. 3, we give a characterization of such a control and we discuss a condition of its uniqueness. In the last section, an example and simulations illustrate the obtained results.

2 Existence of an Boundary Optimal Control

In this section, we show the existence of solution of problem (2). Firstly, we prove a priori estimates which are necessary for the existence of an optimal control.

Lemma 1 *We consider the system:*

$$\begin{cases} \frac{\partial z}{\partial t}(x,t) = \Delta z(x,t) & Q, \\ \frac{\partial z}{\partial \upsilon_A}(x,t) = u(x,t)Bz(x,t) + f(x,t) & \Gamma, \\ z(x,0) = z_0(x) & \Omega, \end{cases} \tag{4}$$

where $z_0 \in L^2(\Omega)$ and $f \in L^2(\Gamma)$. Then the weak solution z of system (4) satisfies the following estimates.

$$\|z\|_{L^2(0,T;H^1(\Omega))} \le C_1\|f\|_{L^2(0,T;L^2(\partial\Omega))} + C_2\|z_0\|_{L^2(\Omega)} \tag{5}$$

$$\|\frac{\partial z}{\partial t}\|_{L^2(0,T;(H^1(\Omega))')} \le C_3\|f\|_{L^2(0,T;L^2(\partial\Omega))} + C_4\|z\|_{L^2(0,T;H^1(\Omega))} \tag{6}$$

Proof Multiplying system (4) by z and integrating over $(0, T)$ we have

$$\int_0^T \langle \frac{\partial z}{\partial t}, z\rangle dt + \int_0^T < \nabla z, \nabla z >_{L^2(\Omega)} dt = \int_\Gamma (uBz + f)z \, dxdt.$$

Since B is a bounded operator, we have

$$\frac{1}{2}\int_0^T \frac{d}{dt}\|z(t)\|^2_{L^2(\Omega)}dt + \int_0^T \|\nabla z(t)\|_{L^2(\Omega)}dt \leq \int_0^T \|f\|_{L^2(\partial\Omega)}\|z\|_{L^2(\partial\Omega)}dt$$

Hence, for $\alpha > 0$ we have

$$\frac{1}{2}\|z(T)\|^2_{L^2(\Omega)} + \int_0^T \|\nabla z(t)\|_{L^2(\Omega)}dt \leq \int_0^T \frac{\alpha^2}{2}\|f\|^2_{L^2(\partial\Omega)} + \frac{1}{2\alpha^2}\|z\|^2_{L^2(\partial\Omega)}dt + \frac{1}{2}\|z_0\|^2_{L^2(\Omega)}$$

where C depends on B.

For $n \geq 2$, we have

$$\forall z \in H^1(\Omega), \exists C_0 > 0, \|z\|_{L^2(\partial\Omega)} \leq C_0\|z\|_{H^1(\Omega)} (\text{see [1]}),$$

we obtain

$$\int_0^T \|z\|_{H^1(\Omega)}dt \leq \int_0^T \frac{\alpha^2}{2}\|f\|^2_{H^1(\Omega)} + \frac{C_0}{2\alpha^2}\|z\|^2_{H^1(\Omega)}dt + \frac{1}{2}\|z_0\|^2_{L^2(\Omega)}$$

Choose $\alpha > \sqrt{\dfrac{C_0}{2}}$, we deduce that

$$\|z\|_{L^2(0,T;H^1(\Omega))} \leq C_1\|f\|_{L^2(0,T;L^2(\partial\Omega))} + C_2\|z_0\|_{L^2(\Omega)}.$$

For $\varphi \in H^1(\Omega)$, we have

$$\int_0^T |\langle \frac{\partial z}{\partial t}, \varphi\rangle + < \nabla z, \nabla \varphi >_{L^2(\Omega)} |dt = \int_\Gamma |(uBz + f)\varphi| \, dx\, dt.$$

It follows

$$\int_0^T |\langle \frac{\partial z}{\partial t}, \varphi\rangle| dt \leq C \int_0^T (\|z\|_{H^1(\Omega)} + \|f\|_{L^2(\partial\Omega)})\|\varphi\|_{H^1(\Omega)}dt$$

Therefore using (5), we obtain

$$\|\frac{\partial z}{\partial t}\|_{L^2(0,T;(H^1(\Omega))')} \leq C_3\|f\|_{L^2(0,T;L^2(\partial\Omega))} + C_4\|z\|_{L^2(0,T;H^1(\Omega))}dt$$

Theorem 1 *There exists an optimal control $u \in U$, solution of problem (2).*

Proof The set $\{J(u)|u \in U\}$ is non-empty and is bounded from below by 0.
Let $(u_k)_{k\in\mathbb{N}}$ be a minimizing sequence in U.

$$\lim_{k\to\infty} J(u_k) = \inf_{h\in U} J(h).$$

Then $(J(u_k))_{k\in\mathbb{N}}$ is bounded.

Since $\|u_k\|_{L^2(\Gamma)} \leq \frac{2}{\beta} J(u_k)$ then $(u_k)_{k\in\mathbb{N}}$ is bounded.

Then there exists a sup-sequences still denoted $(u_k)_{k\in\mathbb{N}}$ that converges weakly to a limit $u^* \in L^2(\Gamma)$.

Let z_k and z_{u*} solutions of system (1) associated to u_k and u^* respectively.

Using the inequalities (5) and (6) there exists subsequence with the following convergence proprieties.

$$\begin{aligned} z_k &\rightharpoonup z_{u^*} \quad \text{in } L^2(0,T;H^1(\Omega)), \\ \frac{\partial z_k}{\partial t} &\rightharpoonup \frac{\partial z_{u^*}}{\partial t} \quad \text{in } L^2(0,T;(H^1(\Omega))'), \\ u_k &\rightharpoonup u^* \quad \text{in } L^2(\Gamma). \end{aligned}$$

Using the compact injection of $H^1(\Omega)$ into $H^{1/2+\varepsilon}(\Omega)$ (where $0 < \varepsilon < \frac{1}{2}$), we obtain

$$z_k \longrightarrow z_{u^*} \text{ in } L^2(0,T;H^{1/2+\varepsilon}(\Omega)).$$

The mapping $\varphi : H^{1/2+\varepsilon}(\Omega) \longrightarrow \varphi_{|\Gamma} \in L^2(\Gamma)$ is continuous then $z_k \longrightarrow z_{u^*}$ in $L^2(\Gamma)$ Since B is a linear operator then $u_k B z_k \longrightarrow u^* B z_{u^*}$ in $L^2(\Gamma)$ as $k \longrightarrow \infty$, we deduce that z_{u^*} solves system (1) with control u^*.

Since U is convex, $u^* \in U$.

Since J is lower semi-continuous with respect to weak convergence and χ_ω is a linear operator, we obtain

$$J(u^*) \leq \lim_{k\to\infty} \inf J(u_k),$$

leading to $J(u^*) = \inf\limits_{u\in U} J(u_k)$.

3 Characterization of an Optimal Control

In this section we derive a characterization of an optimal control solution of problem (2).

Firstly, we examine the differentiability of the mapping $u \longrightarrow z_u$ with respect to u, given by the following lemma.

Lemma 2 *The mapping*

$$u \in U \to z_u \in L^2(0, T; H^1(\Omega))$$

is differentiable in the sense:

$$\frac{z_{u+\varepsilon h} - z_u}{\varepsilon} \rightharpoonup \psi \quad \textit{weakly in } L^2(0, T; H^1(\Omega)),$$

as $\varepsilon \to 0$, *for any* $u, u + \varepsilon h \in U$, *where* ψ *is solution of*

$$\begin{cases} \dfrac{\partial \psi}{\partial t}(x, t) = \Delta \psi(x, t) & Q, \\ \dfrac{\partial \psi}{\partial \upsilon_A}(x, t) = u(x, t) B \psi(x, t) + h(x, t) B z(x, t) & \Gamma, \\ \psi(x, 0) = 0 & \Omega. \end{cases} \tag{7}$$

Proof Let consider $z_{u+\varepsilon h}$ solution of system (1) with control $u + \varepsilon h$, then $\varphi = \dfrac{z_{u+\varepsilon h} - z_u}{\varepsilon}$ is solution of the following system

$$\begin{cases} \dfrac{\partial \varphi}{\partial t}(x, t) = \Delta \varphi(x, t) & Q, \\ \dfrac{\partial \varphi}{\partial \upsilon_A}(x, t) = u(x, t) B \varphi(x, t) + h(x, t) B z(x, t) & \Gamma, \\ \varphi(x, 0) = 0 & \Omega, \end{cases}$$

Using estimates (5) and (6) we obtain

$$\begin{aligned} \left\| \frac{z_{u+\varepsilon h} - z_u}{\varepsilon} \right\|_{L^2(0,T;H^1(\Omega))} &\le C_1 \|hBz\|_{L^2(0,T;L^2(\partial\Omega))} \\ &\le C_2 \|z_0\|_{L^2(\Omega)} \end{aligned}$$

and

$$\begin{aligned} \left\| \frac{\partial}{\partial t} \left(\frac{z_{u+\varepsilon h} - z_u}{\varepsilon} \right) \right\|_{L^2(0,T;(H^1(\Omega))')} &\le \|hBz\|_{L^2(0,T;L^2(\partial\Omega))} \\ &\le C_3 \|z_0\|_{L^2(\Omega)} \end{aligned}$$

where C_2 and C_3 are independent of ε.

Thus as $\varepsilon \longrightarrow 0$ we have

$$\frac{z_{u+\varepsilon h} - z_u}{\varepsilon} \rightharpoonup \psi \quad \text{weakly in } L^2(0, T; H^1(\Omega)),$$

Similar to the proof of Theorem 1, we deduce that ψ is the weak solution of system (7).

Proposition 1 *Let $u^* \in U$ be an optimal control and let $h \in L^2(\Gamma)$ such that $u^* + \varepsilon h \in U$, for $\varepsilon > 0$, z_{u^*} is the solution of system (1) corresponding to u^*, and p solution of the following system*

$$\begin{cases} \dfrac{\partial p}{\partial t}(x,t) = \Delta p(x,t) + \chi_\omega^* z_d(x) - \chi_\omega^* \chi_\omega z_{u^*}(x,t) & Q \\ \dfrac{\partial p}{\partial \upsilon_A}(x,t) = B^* u^*(x,t) p(x,t) & \Gamma, \\ p(x,T) = 0 & \Omega. \end{cases} \tag{8}$$

Then an optimal control is given by

$$u^*(x,t) = \max\Big(m, \min\Big(-\frac{1}{\beta} B z_{|\Gamma}(x,t) p_{|\Gamma}(x,t), M\Big)\Big)$$

Proof Let u^* be an optimal control and z_{u^*} the corresponding solution to u^*, $h \in L^2(\Gamma)$ such that $u^* + \varepsilon h \in U$ for $\varepsilon > 0$. We calculate the directional derivative of the cost functional $J(u)$ with respect to u in the direction of h

$$\begin{aligned} dJ(u).h &= \lim_{\varepsilon \to 0^+} \frac{J(u^* + \varepsilon h) - J(u^*)}{\varepsilon} \\ &= \int_0^T \frac{1}{2\varepsilon}(\|\chi_\omega z_\varepsilon - z_d\|^2_{L^2(\omega)} - \|\chi_\omega z_{u^*} - z_d\|^2_{L^2(\omega)})dt + \frac{\beta}{2}\int_\Gamma (2hu^* + \varepsilon h^2)\,dxdt \\ &= \lim_{\varepsilon \to 0^+} \int_0^T \frac{1}{2\varepsilon} < \chi_\omega z_\varepsilon - z_d, \chi_\omega z_\varepsilon - z_{u^*} >_{L^2(\omega)} dt + \frac{\beta}{2}\int_\Gamma (2hu^* + \varepsilon h^2)\,dxdt \\ &= \lim_{\varepsilon \to 0^+} \int_Q \left(\frac{z_\varepsilon - z_{u^*}}{\varepsilon}\right)\left(\frac{\chi_\omega^* \chi_\omega z_\varepsilon + \chi_\omega^* \chi_\omega z_{u^*} - 2\chi_\omega^* z_d}{2}\right) dQ + \frac{\beta}{2}\int_\Gamma (2hu^* + \varepsilon h^2)\,dxdt \\ &= \int_Q \psi(\chi_\omega^* \chi_\omega z_{u^*} - \chi_\omega^* z_d) dQ + \beta \int_\Gamma u^* h\,dxdt\,. \end{aligned}$$

where ψ is solution of system (7) corresponding to u^*.

Using (8) we have

$$\begin{aligned} dJ(u).h &= (-\int_0^T < \psi, \frac{\partial p}{\partial t} > - < \psi, Ap >)dt + \beta \int_\Gamma u^* h dxdt. \\ &= \int_0^T (< \frac{\partial \psi}{\partial t}, p > + < \nabla \psi, \nabla p > - < hB\psi, p >_{L^2(\partial\Omega)} dt + \beta \int_\Gamma u^* h dxdt. \end{aligned}$$

Using system (7) we obtain

$$dJ(u).h = \int_{\Gamma} hBzp\, d\Gamma + \beta \int_{\Gamma} hu^*\, d\Gamma.$$

Since J achieves its minimum at u^*, we have

$$0 \leq \int_{\Gamma} hBzp\, d\Gamma + \beta \int_{\Gamma} hu^*\, d\Gamma.$$

Taking $h = \max\Big(m, \min\Big(-\frac{1}{\beta} Bz_{|\Gamma}(x,t)p_{|\Gamma}(x,t), M\Big)\Big) - u^*$, we show that $h(u^* + \frac{1}{\beta} Bz_{|\Gamma} p_{|\Gamma})$ is negative and then

$$\Big(\max\Big(m, \min\Big(-\frac{1}{\beta} Bz_{|\Gamma}(x,t)p_{|\Gamma}(x,t), M\Big)\Big) - u^*\Big)\Big(u^* + \frac{1}{\beta} Bz_{|\Gamma} p_{|\Gamma}\Big) = 0.$$

If $M \leq -\frac{1}{\beta} Bz_{|\Gamma} p_{|\Gamma}$ we have $(M - u^*)(u^* + \frac{1}{\beta} Bz_{|\Gamma} p_{|\Gamma}) = 0$, thus $u^* = M$.

If $m \leq -\frac{1}{\beta} Bz_{|\Gamma} p_{|\Gamma} \leq M$ we have $(-\frac{1}{\beta} Bz_{|\Gamma} p_{|\Gamma} - u^*)(u^* + \frac{1}{\beta} Bz_{|\Gamma} p_{|\Gamma}) = 0$.

Therefore $u^* = -\frac{1}{\beta} Bz_{|\Gamma} p_{|\Gamma}$.

Now, if $m \geq -\frac{1}{\beta} Bz_{|\Gamma} p_{|\Gamma}$, we have $(m - u^*)(u^* + \frac{1}{\beta} Bz_{|\Gamma} p_{|\Gamma}) = 0$ and then $u^* = m$.

We conclude that

$$u^*(x,t) = \max\Big(m, \min\Big(-\frac{1}{\beta} Bz_{|\Gamma}(x,t)p_{|\Gamma}(x,t), M\Big)\Big).$$

Corollary 1 *If we consider $U = \{u \in L^\infty(0,T), m \leq u \leq M\}$,an optimal control is given by*

$$u^*(t) = \max\Big(m, \min\Big(-\frac{1}{\beta} \int_{\partial\Omega} Bz_{|\Gamma}(x,t)p_{|\Gamma}(x,t)dx, M\Big)\Big). \tag{9}$$

Proposition 2 *Assume that z and p are bounded on $\overline{\Omega}$, for β large enough, the solution of problem (2) is unique.*

Proof For $i = 1, 2$, let z_i and p_i be solutions of system (1) and (7) respectively, then using (5) and (6) we have

$$\|z_i\|_{L^2(0,T;H^1(\Omega))} \leq C_2 \|z_0\|_{L^2(\Omega)}$$

and

$$\|p_i\|_{L^2(0,T;H^1(\Omega))} \leq C_1\|z_i - z_d\|_{L^2(\Omega)} + C_2\|z_0\|_{L^2(\Omega)}.$$

Since $z_1 - z_2$ is the solution of the following system

$$\begin{cases} \dfrac{\partial(z_1 - z_2)}{\partial t}(x,t) = \Delta(z_1 - z_2)(x,t) & Q \\ \dfrac{\partial(z_1 - z_2)}{\partial \upsilon_A}(x,t) = u_1(x,t)B(z_1 - z_2)(x,t) + (u_1 - u_2)(x,t)Bz_2(x;t) & \Gamma, \\ (z_1 - z_2)(x,0) = 0 & \Omega, \end{cases} \tag{10}$$

then using (5), we have

$$\|z_1 - z_2\|_{L^2(0,T;H^1(\Omega))} \leq C\|z_0\|_{L^2(\Omega)}\|u_1 - u_2\|_{L^\infty(\Gamma)}$$

Similarly for $(p_1 - p_2)$ we have

$$\begin{aligned} \|p_1 - p_2\|_{L^2(0,T;H^1(\Omega))} &\leq \|B^*(u_1 - u_2)p_2\|_{L^2(\Gamma)} \\ &\leq C\|u_1 - u_2\|_{L^\infty(\Gamma)}\|p_2\|_{L^2(0,T;H^1(\Omega))} \\ &\leq C\|z_2 - z_d\|_{L^2(\Omega)}\|u_1 - u_2\|_{L^\infty(\Gamma)}. \end{aligned}$$

Then

$$\|u_1 - u_2\|_{L^\infty(\Gamma)} \leq \frac{1}{\beta}(\|p_1 B(z_1 - z_2)\|_{L^\infty(\Gamma)} + \frac{1}{\beta}\|Bz_2(p_1 - p_2)\|_{L^\infty(\Gamma)}),$$

which gives

$$\begin{aligned} \|u_1 - u_2\|_{L^\infty(\Gamma)} &\leq \frac{1}{\beta}(\|p_1\|_{L^\infty(\Gamma)}\|B(z_1 - z_2)\|_{L^2(\Gamma)} + \|Bz_2\|_{L^2(\Gamma)}\|(p_1 - p_2)\|_{L^\infty(\Gamma)}) \\ &\leq \frac{C}{\beta}(\|(z_1 - z_2)\|_{L^2(0,T;H^1(\Omega))} + \|(p_1 - p_2)\|_{L^2(0,T;H^1(\Omega))}) \\ &\leq \frac{C}{\beta}(\|(z_1 - z_2)\|_{L^2(0,T;H^1(\Omega))} + \|z_2 - z_d\|_{L^2(\Omega)}\|u_1 - u_2\|_{L^\infty(\Gamma)}). \end{aligned}$$

It follows

$$\|u_1 - u_2\| \leq \frac{C}{\beta}(\|(u_1 - u_2)\|_{L^\infty(\Gamma)}(2\|z_0\|_{L^2(\Omega)} + \|z_d\|_{L^2(\Omega)})). \tag{11}$$

So for β large enough such that $(2C\|z_0\|_{L^2(\Omega)} + C\|z_d\|_{L^2(\Omega)}) < \beta$.

Then the solution of problem(2) is unique.

4 Algorithm and Simulations

An optimal control solution of problem (2) can be calculated by the following formula:

$$u_{k+1}(x,t) = \max\Big(m, \min\Big(-\frac{1}{\beta}Bz_{k|\Gamma}(x,t)p_{k|\Gamma}(x,t), M\Big)\Big) \tag{12}$$

where z_k is the solution of system (1) associated with u_k and p_k is the one of the adjoint Eq. (8). This allows to consider the following algorithm:
Algorithm

Step 1: initials data.

- $*$ An initial state z_0, a time T, a control $u_1 = 0$, a desired output z_d.
- $*$ A threshold accuracy $\varepsilon > 0$, and $k = 1$.

Step 2:

- $*$ Solving Eq. (3) gives z_k.
- $*$ Solving Eq. (8) gives p_k.
- $*$ Calculate u_{k+1} by formula (3).

Step 3 : while $\|u_{k+1} - u_k\|_{L^2(\Gamma)} > \varepsilon$, k=k+1 go to step 2.

Example
On $\Omega =]0,1[\times]0,1[$, we consider the following system

$$\begin{cases} \dfrac{\partial z}{\partial t}(x,y,t) = 0.01\Delta z(x,y,t) & Q \\ \dfrac{\partial z}{\partial \upsilon}(x,y,t) = -u(t)\mathbb{1}_{\Gamma_1} z(x,y,t) & \Gamma, \\ z(x,y,0) = z_0(x,y) & \Omega, \end{cases} \tag{13}$$

where $z_0(x,y) = x(y-x)$, $u \in U = \{u \in L^2(0,T)/0 \leq u \leq 0.04\}$, and $\mathbb{1}_{\Gamma_1}$ denotes the characteristic function to $\Gamma_1 = \{1\}\times]0,1[$.

Let $\omega =]0,1[\times]0,0,25[$, We consider a desired output $z_d(x) = 0 \in L^2(\omega)$. Applying control (9), Fig. 1 shows that the final state on ω is close to the desired one on ω with error $\|\chi_\omega z(x,y,T) - z_d(x,y)\|^2_{L^2(\omega)} = 8.26.10^{-5}$, and a cost $J(u^*) = 1.22.10^{-3}$.

An optimal control is presented by Fig. 2.

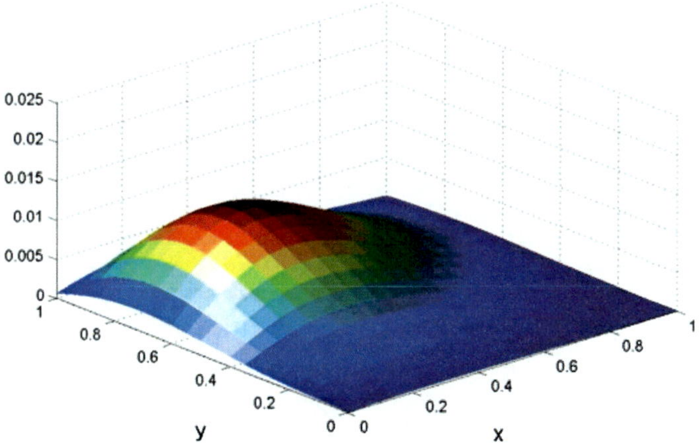

Fig. 1 Final state

Fig. 2 The evolution of an optimal control

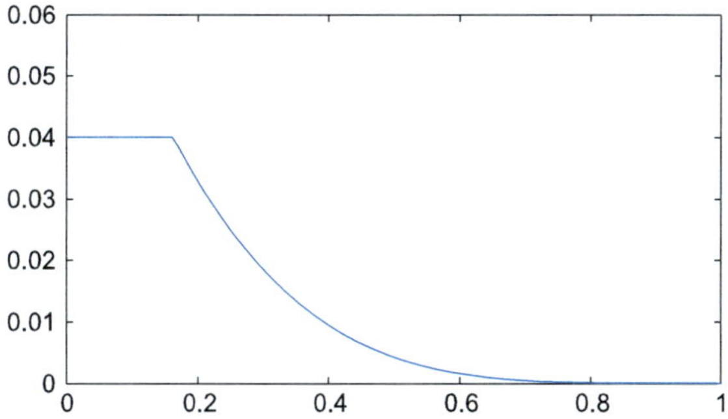

5 Conclusion

In this paper, we discussed a regional optimal control problem for a heat equation with bilinear boundary controls. This approach allows to study many open questions, for instance, a regional optimal control of a wave equation with bilinear boundary controls. This is under consideration.

Acknowledgements The work has been carried out with a grant from Hassan II Academy of Sciences and Technology.

References

1. Adams, R.A., Fournier, J.J.: Sobolev Spaces, vol. 140. Academic, New York (2003)
2. Bradley, M., Lenhart, S.: Bilinear optimal control of a kirchhoff plate. Syst. Control Lett. **22**(1), 27–38 (1994)

3. Bradley, M.E., Lenhart, S.: Bilinear spatial control of the velocity term in a kirchhoff plate equation. Electron. J. Differ. Equ. (EJDE)[electronic only] 2001 Paper–No (2001)
4. Bradley, M.E., Lenhart, S., Yong, J.: Bilinear optimal control of the velocity term in a kirchhoff plate equation. J. Math. Anal. Appl. **238**(2), 451–467 (1999)
5. Lenhart, S., Wilson, D.: Optimal control of a heat transfer problem with convective boundary condition. J. Optim. Theory Appl. **79**(3), 581–597 (1993)
6. Mohler, R.R.: Bilinear Control Processes: With Applications to Engineering, Ecology and Medicine. Academic, New York (1973)
7. Park, J.Y.: Bilinear boundary optimal control of the velocity terms in a kirchhoff plate equation. Trends Math. **9**, 41–44 (2006)
8. Tröltzsch, F.: Optimal Control of Partial Differential Equations: Theory, Methods, and Applications, vol. 112. American Mathematical Society, Providence (2010)
9. Zerrik, E.: Regional quadratic control problem for distributed bilinear systems with bounded controls. Int. J. Control **87**(11), 2348–2353 (2014)
10. Zerrik, E., ElKabouss, A.: Regional optimal control of a class of bilinear systems. IMA J. Math. Control Inf. **34**(4), 1157–1175 (2016)
11. Zerrik, E., ElKabouss, A.: Regional optimal control of a bilinear wave equation. Int. J. Control 1–10 (2017)
12. Zerrik, E., ElKabouss, A.: Regional optimal control problem of a class of infinite-dimensional bi-linear systems. Int. J. Control **90**(7), 1495–1504 (2017)
13. Zerrik, E., Sidi, M.O.: An output controllability of bilinear distributed system. Int. Rev. Autom. Control **3**(5), 466–473 (2010)
14. Ztot, K., Zerrik, E., Bourray, H.: Regional control problem for distributed bilinear systems: approach and simulations. Int. J. Appl. Math. Comput. Sci. **21**(3), 499–508 (2011)

Fuzzy Equations for Mixed Convection in a Rectangular Cavity

Atimad Harir, Hassan El Harfi, Said Melliani, and Lalla Saadia Chadli

Abstract In this paper, the fuzzy equations of mixed convection heat transfer has been introduced. The fuzzy boundary conditions for temperature is considered. We study the existence and uniqueness of fuzzy solutions, under some conditions. The results are presented and proved in terms of velocity, steram function and temperature profiles.

1 Introduction

Physical models often have some uncertainty in their parameters and estimates are usually based on statistical methods and experimental data. Since Zadeh [16] introduced the concept of fuzzy sets, there has been a great deal of research in this area, including studies of fuzzy partial differential equations (PDEs). Some studies considered application of PDEs with fuzzy parameters obtained through fuzzy rule-based systems [14]. Oberguggenberger described weak and fuzzy solutions for PDEs [15]. More recently Leite and Bassanezi used Zadeh's extension principle to determine a fuzzy solution for a PDE with initial fuzzy conditions [7].

A. Harir · S. Melliani (✉) · L. S. Chadli
Laboratory of Applied Mathematics and Scientific Computing, Sultan Moulay Slimane University, Beni Mellal, Morocco
e-mail: s.melliani@usms.ma

A. Harir
e-mail: atimad.harir@gmail.com

L. S. Chadli
e-mail: sa.chadli@yahoo.fr

H. El Harfi
Laboratory of Flows and Transfers Modelling (LAMET), Sultan Moulay Slimane University, Beni Mellal, Morocco
e-mail: elharfi.hassan@yahoo.fr

S. Melliani and O. Castillo (eds.), *Recent Advances in Intuitionistic Fuzzy Logic Systems and Mathematics*, Studies in Fuzziness and Soft Computing 395,
https://doi.org/10.1007/978-3-030-53929-0_11

The problem of mixed convection heat transfer of nanofluids in a lead-driven enclosure subject to Neumann fuzzy boundary conditions for temperature (i.e. fuzzy boundaries subject to heat fluxes) is not yet analyzed. So, in order to know more about the effect of the fuzzy boundary conditions kind on flow and heat transfer within nanofluids, the present paper focuses attention on such a problem within a two-dimensional shallow rectangular enclosure, filled with Cu-water nanofluids, whose short vertical sides are submitted to uniform heat fluxes while the long horizontal ones are maintained fuzzy with the top moving in the opposite direction to the heat flux. An fuzzy analytical one, based on the parallel flow approximation, is also proposed. The results are presented and proved, in terms of velocity, stream function and temperature profiles, and discussed for various values of the dimensionless parameters, controlling the problem, which are the Reynolds Re, and Richardson Ri, numbers, and the solid volume fraction of nanoparticles $\Phi = 0$.

The remainder of the paper is organized as follows Sect. 2 presents definitions and results from the basic theory of fuzzy sets. In Sect. 3 we present classic formulation of the mixed convection heat transfer of nanofluids in a lead-driven. In Sect. refch11sec:4 we present strategy for solving the full fuzzy differential equations(FDE) and we give figures of fuzzy analytical solutions.

2 Preliminaries

The basic defnition of fuzzy numbers is given in [11, 12].

Definition 1 A fuzzy number is a fuzzy set like $\mathrm{R}_{\mathcal{F}} = \Big\{\widetilde{u} \mid \widetilde{u} : \mathrm{R}^n \to [0, 1], \Big\}$ which satisfies:

1. $\widetilde{u}$ is upper semi-continuous on R^n.
2. $\widetilde{u} = 0$ outside some interval $[c, d]$.
3. There are real numbers a, b such that $c \leq a \leq b \leq d$ and
   - $\widetilde{u}$ is monotonic increasing on $[c, a]$,
   - $\widetilde{u}$ is monotonic decreasing on $[b, d]$,
   - $\widetilde{u} = 1, \quad a \leq x \leq b$.

The set of all these fuzzy numbers is denoted by $\mathrm{R}_{\mathcal{F}}$, in this paper we suppose $a = b$.

An equivalent parametric is also given in [12, 13] as follows.

Definition 2 A fuzzy number $\widetilde{u}$ in parametric form is a pair $(u_1(\alpha), u_2(\alpha))$ of functions $\widetilde{u}[\alpha] = [u_1(\alpha), u_2(\alpha)], \quad \alpha \in [0, 1]$, which satisfy the following requirements:

1. $u_1(\alpha)$ is a bounded monotonic increasing left continous function over $(0, 1]$, and right continuous at $\alpha = 0$,
2. $u_2(\alpha)$ is a bounded monotonic decreasing left continous function on $(0, 1]$, and right continuous at $\alpha = 0$,

3. $u_1(\alpha) \leq u_2(\alpha)$, for $0 \leq \alpha \leq 1$.

Remark 1 A crisp number ε_2 is simply represented by $u_1(1) = u_2(1) = \varepsilon_2$.

A popular fuzzy number is the triangular fuzzy number $\widetilde{u} = (\varepsilon_1, \varepsilon_2, \varepsilon_3)$, $(\varepsilon_1 < \varepsilon_2 < \varepsilon_3)$. With the membership function:

$$\widetilde{u}(t) = \begin{cases} 1 - \dfrac{(\varepsilon_2 - t)}{\varepsilon_2 - \varepsilon_1}, & \varepsilon_1 \leq t \leq \varepsilon_2 \\ 1 - \dfrac{(t - \varepsilon_2)}{\varepsilon_3 - \varepsilon_2}, & \varepsilon_2 \leq t \leq \varepsilon_3 \\ 0, & \mathit{otherwise.} \end{cases}$$

where $\varepsilon_2 \neq \varepsilon_1$, $\varepsilon_1 \neq \varepsilon_3$ and hence $[\widetilde{u}]^\alpha = [u_1(\alpha), u_2(\alpha)] = [\varepsilon_1 + \alpha(\varepsilon_2 - \varepsilon_1), \varepsilon_3 - \alpha(\varepsilon_3 - \varepsilon_2)]$

The addition and scalar multiplication of fuzzy numbers are defined by the exyension principle and can be equivalently represented as follows.

Definition 3 For arbitrary fuzzy numbers $[\widetilde{u}]^\alpha = \left[u_1(\alpha), u_2(\alpha)\right]$ and $[\widetilde{v}]^\alpha = \left[v_1(\alpha), v_2(\alpha)\right]$ we have algebraic operations as follows:

1. $[\widetilde{u} + \widetilde{v}]^\alpha = [u_1(\alpha) + v_1(\alpha), u_2(\alpha) + v_2(\alpha)]$
2.
$$[k\, u]^\alpha = k\, [u]^\alpha = \begin{cases} [k\, u_1(\alpha), k\, u_2(\alpha)]\; if\; k \geq 0 \\ [k\, u_2(\alpha), k\, u_1(\alpha)]\; if\; k < 0, \end{cases} \tag{1}$$

Definition 4 Let I be a real interval. A mapping $\widetilde{u} : I \to \mathrm{R}_{\mathscr{F}}$ is called a fuzzy process. We denote $[\widetilde{u}(y)]^\alpha = [u_1(y, \alpha), u_2(y, \alpha)]$ $y \in I$ $\alpha \in [0, 1]$.

The fuzzy integral $\int_a^b \widetilde{u}(y)dy$ is defined by

$$\left[\int_a^b \widetilde{u}(y)dy\right]^\alpha = \left[\int_a^b u_1(y, \alpha)dy, \int_a^b u_2(y, \alpha)dy\right] \tag{2}$$

3 Mixed Convection Heat Transfer for Nanofluids in a Lid-Driven Shallow Rectangular Cavity Uniformly Heated

The studied configuration is sketched in Fig. 1 [5, 6]. It is a shallow rectangular enclosure of height H' and length L', filled with Cu-water nanofluids. The long horizontal walls are adiabatic, while the vertical short ones are submitted to a uniform density of heat flux q'. All these boundaries are rigid, impermeable and motionless apart from the top one which moves in its own plane from right to left at uniform velocity.

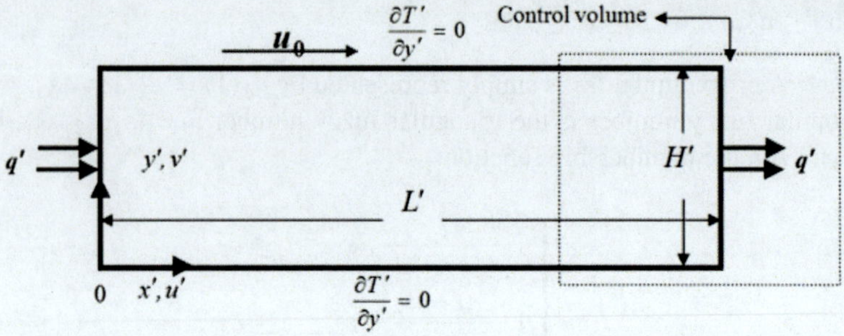

Fig. 1 Sketch of the cavity and co-ordinates system

The equations describing the conservation of mass (3), momentum (4)–(5) and energy (6), written in terms of velocity components $(u^{'}, v^{'})$, pressure $(p^{'})$ and temperature $(T^{'})$, are:

$$\frac{\partial u^{'}}{\partial x^{'}} + \frac{\partial v^{'}}{\partial y^{'}} = 0 \tag{3}$$

$$\frac{\partial u^{'}}{\partial t^{'}} + u^{'}\frac{\partial u^{'}}{\partial x^{'}} + v^{'}\frac{\partial u^{'}}{\partial y^{'}} = -\frac{1}{\rho_{nf}}\frac{\partial p^{'}}{\partial x^{'}} + \frac{\mu_{nf}}{\rho_{nf}}\left(\frac{\partial^2 u^{'}}{\partial^2 x^{'}} + \frac{\partial^2 u^{'}}{\partial^2 y^{'}}\right) \tag{4}$$

$$\frac{\partial v^{'}}{\partial t^{'}} + u^{'}\frac{\partial v^{'}}{\partial x^{'}} + v^{'}\frac{\partial v^{'}}{\partial y^{'}} = -\frac{1}{\rho_{nf}}\frac{\partial p^{'}}{\partial y^{'}} + \frac{\mu_{nf}}{\rho_{nf}}\left(\frac{\partial^2 v^{'}}{\partial^2 x^{'}} + \frac{\partial^2 v^{'}}{\partial^2 y^{'}}\right) + \frac{1}{\rho_{nf}}(\rho\beta)_{nf}\, g\left(T^{'} - T_0^{'}\right) \tag{5}$$

$$\frac{\partial T^{'}}{\partial t^{'}} + \frac{\partial (u^{'}T^{'})}{\partial x^{'}} + \frac{\partial (v^{'}T^{'})}{\partial y^{'}} = \alpha_{nf}\nabla^2 T^{'} \tag{6}$$

To close the problem, the following appropriate boundary conditions are applied:

$$u^{'} = v^{'} = 0 \;\; and \;\; \frac{\partial T^{'}}{\partial x^{'}} + \frac{q^{'}}{k_{nf}} = 0 \;\; for\; x^{'} = 0 \; and \; x^{'} = L^{'} \tag{7}$$

$$u^{'} = v^{'} = 0 \;\; and \;\; \frac{\partial T^{'}}{\partial y^{'}} = 0 \; for \; y^{'} = 0 \tag{8}$$

$$u^{'} - U_0^{'} = v^{'} = 0 \;\; and \;\; \frac{\partial T^{'}}{\partial y^{'}} = 0 \; for \; y^{'} = H^{'} \tag{9}$$

where $\rho_{nf}, \mu_{nf}, (\rho\beta)_{nf}, \alpha_{nf}$ *and* k_{nf} are the constants see [5].

The dimensionless governing equations and the corresponding boundary conditions are

$$\frac{\partial u}{\partial x} + \frac{\partial v}{\partial y} = 0 \tag{10}$$

$$\frac{\partial u}{\partial t} + u\frac{\partial u}{\partial x} + v\frac{\partial u}{\partial y} = -\frac{\partial p}{\partial x} + \frac{1}{Re}\left(\frac{\partial^2 u}{\partial^2 x} + \frac{\partial^2 u}{\partial^2 y}\right) \tag{11}$$

$$\frac{\partial v}{\partial t} + u\frac{\partial v}{\partial x} + v\frac{\partial v}{\partial y} = -\frac{\partial p}{\partial y} + \frac{1}{Re}\left(\frac{\partial^2 v}{\partial^2 x} + \frac{\partial^2 v}{\partial^2 y}\right) + RiT \tag{12}$$

$$\frac{\partial T}{\partial t} + u\frac{\partial T}{\partial x} + v\frac{\partial T}{\partial y} = \frac{1}{Pe}\left(\frac{\partial^2 T}{\partial^2 x} + \frac{\partial^2 T}{\partial^2 y}\right) \tag{13}$$

$$u = v = \frac{\partial T}{\partial x} + 1 = 0 \quad and\, x = 0 \tag{14}$$

$$u = v = \frac{\partial T}{\partial y} = 0\, for\ y = 0 \tag{15}$$

$$u - 1 = v = \frac{\partial T}{\partial y} = 0\, for\ y = 1 \tag{16}$$

To analysis the flow structure, the stream function Ψ, related to the velocity components via

$$u = \frac{\partial \Psi}{\partial y}\ and\ u = -\frac{\partial \Psi}{\partial x}\ (with\Psi = 0\ on\ all\ boundaries) \tag{17}$$

is used.

The above equations let appears some dimensionless parameters that govern the problem, namely, the aspect ratio of the enclosure, A, the Peclet, Pe, Reynolds, Re, and Richardson, Ri, numbers. For the last four, the expressions in [5] are

$$A = \frac{L'}{H'},\ Pe = \frac{U_0' H'}{\alpha_f},\ Re = \frac{U_0' H'}{\nu_f}\ and\ Ri = \frac{g\beta_f q' H'^2}{k_f U_0'^2} \tag{18}$$

Note that

$$Pe = PrRe\ and\ Ri = \frac{Ra}{PeRe} \tag{19}$$

where

$$Gr = \frac{g\beta_f' q' H'^4}{\nu_f^2 k_f},\ Pr = \frac{\nu_f}{\alpha_f}\ Ra = PrGr \tag{20}$$

are the Grashof, Prandtl and Rayleigh numbers, respectively.

Fig. 2 Streamlines (left) and isotherms (right) for $A = 8$, $Re = 1$, $\Phi = 0$ and $Ri = 10^3$

4 Approximate Parallel Flow Fuzzy Analytical Solution

As can be seen from Fig. 2, displaying streamlines (left) and isotherms (right), the flow and temperature fields exhibit a parallel aspect and a linear stratification, respectively, in the most part of the cavity, for and various values of Re, Ri.

Accordingly, the following simplifications

$$u(x, y) = u(y),\ v(x, y) = 0,\ \Psi(x, y) = \Psi(y)\ and\ T(x, y) = C\big(x - A/2\big) + \theta(y) \tag{21}$$

where C is unknown constant temperature gradient in the x-direction, are possible, which leads to the ordinary non-dimensional governing equations:

$$\frac{d^3u}{d^3y} = ReRi\frac{\partial T}{\partial x} = ReRiC \tag{22}$$

$$\frac{1}{Pe}\frac{d^2\theta}{d^2y} = Cu \tag{23}$$

with

$$u = k_1\ for\ y = 1, u = k_2\ for\ y = 0, \frac{d\theta}{dy} = 0\ for\ y = 0\ and\ 1 \tag{24}$$

$$\int_0^1 u(y)dy = 0 \tag{25}$$

$$\int_0^1 \theta(y)dy = 0 \tag{26}$$

as boundary, return flow and mean temperature conditions, respectively. And here for $Re = 1$ and $Ri = 10^3$ we find according to the equation of Bejan [1], $C = -0.149$.

Assume (22) has a solution

$$u(y) = G(y, k_1, k_2) \tag{27}$$

for continuous G.

Suppose the constants k_1 and k_2 are imprecise in their values (in the physical case mobile walls). We will model this uncertainty by substitute triangular fuzzy numbers for k_1 and k_2 (i.e the movement of the horizontal walls is fuzzy). If we fuzzify (22), then we obtian the fuzzy equation. Using the extension principle we compute $\widetilde{p}$ from

p and the functions u, θ become $\widetilde{u}, \widetilde{\theta}$ where $\widetilde{u} : I_2 \to \mathrm{R}_{\mathscr{F}}, \widetilde{\theta} : I_2 \to \mathrm{R}_{\mathscr{F}}$ (resp). That is $\widetilde{u}(y), \widetilde{\theta}(y)$ are fuzzy numbers and $\widetilde{T}(x, y) = C\big(x - A/2\big) + \widetilde{\theta}(y)$.

$$\frac{d^3\widetilde{u}}{d^3y} = ReRi\frac{\partial \widetilde{T}}{\partial x} = ReRiC \tag{28}$$

$$\frac{1}{Pe}\frac{d^2\widetilde{\theta}}{d^2y} = C\widetilde{u} \tag{29}$$

$$\frac{d\widetilde{\Psi}}{dy} = \widetilde{u} \tag{30}$$

and the boundary conditions can be of the form

$$\widetilde{u} = \widetilde{K}_1 \, for \; y = 1, \; \widetilde{u} = \widetilde{K}_2 \, for \; y = 0, \; \frac{d\widetilde{\theta}}{dy} = 0 \, for \; y = 0 \; and \; 1 \tag{31}$$

$$\int_0^1 \widetilde{u}(y)dy = 0 \tag{32}$$

$$\int_0^1 \widetilde{\theta}(y)dy = 0 \tag{33}$$

The $\widetilde{K}_1, \widetilde{K}_2$, are the extension principle of k_1, k_2 (resp). We wish to solve the problem given in (28). Finally, we fuzzify G in (27). Let $\widetilde{Z}(y) = \widetilde{G}(y, \widetilde{K}_1, \widetilde{K}_2)$ where $\widetilde{Z}$ is computed using the extension principle and is a fuzzy solution.

4.1 Fuzzy Solution

We first present the fuzzy solution [2–4] for (28). They define for all x, y and α

$$[\widetilde{Z}(y)]^\alpha = [z_1(y, \alpha), z_2(y, \alpha)] \; and \; [\widetilde{p}(x, y)]^\alpha = [p_1(x, y, \alpha), p_2(x, y, \alpha)]$$

and

$$z_1(y, \alpha) = \min\Big\{G(y, k_1, k_2) \; : \; k_1 \in [\widetilde{K}_1]^\alpha, \; k_2 \in [\widetilde{K}_2]^\alpha\Big\} \tag{34}$$

$$z_2(y, \alpha) = \max\Big\{G(y, k_1, k_2) \; : \; k_1 \in [\widetilde{K}_1]^\alpha, \; k_2 \in [\widetilde{K}_2]^\alpha\Big\} \tag{35}$$

and

$$p_1(x, y, \alpha) = \min\Big\{p(x, y, k_1, k_2) \; : \; k_1 \in [\widetilde{K}_1]^\alpha, \; k_2 \in [\widetilde{K}_2]^\alpha\Big\}, \tag{36}$$

$$p_2(x, y, \alpha) = \max\left\{p(x, y, k_1, k_2)\ :\ k_1 \in [\widetilde{K}_1]^\alpha,\ k_2 \in [\widetilde{K}_2]^\alpha\right\} \tag{37}$$

we assume that the $z_i(y, \alpha)$ and $p_i(x, y, \alpha)$ have continous partials so that

$$\varphi(D_x, D_y)\big(z_i(y, \alpha), p_i(x, y, \alpha)\big)$$

is continous for all $(x, y) \in I_1 \times I_2$, *all* α, $i = 1, 2$ were the operator $\varphi(D_x, D_y)$ will be a polynomial, with constant coefficients in D_x and D_y where $D_x(D_y)$ stands for the "pattial" with respect to $x(y)$. Define

$$\Gamma(x, y, \alpha) = \Big[\varphi(D_x, D_y)\big(z_1(y, \alpha), p_1(x, y, \alpha)\big), \varphi(D_x, D_y)\big(z_2(y, \alpha), p_2(x, y, \alpha)\big)\Big] \tag{38}$$

Sufficient conditions for $\Gamma(x, y, \alpha)$ to define α-cut of a fuzzy number are [3, 4, 10]:

(i) $\varphi(D_x, D_y)\big(z_1(y, \alpha), p_1(x, y, \alpha)\big)$ is an increasing function of α for each $(x, y) \in I_1 \times I_2$
(ii) $\varphi(D_x, D_y)\big(z_2(y, \alpha), p_2(x, y, \alpha)\big)$ is a decreasing function of α for each $(x, y) \in I_1 \times I_2$
(iii) and for $(x, y) \in I_1 \times I_2$

$$\varphi(D_x, D_y)\big(z_1(y, 1), p_1(x, y, 1)\big) \le \varphi(D_x, D_y)\big(z_2(y, 1), p_2(x, y, 1)\big)$$

We had already assumed that the $\big(z_i(y, \alpha), p_i(x, y, \alpha)\big)$ had continuous partial derivatives, so $\varphi(D_x, D_y)\big(z_i(y, \alpha), p_i(x, y, \alpha)\big)$ is continuous on $(x, y) \in I_1 \times I_2$, *all* α, $i = 1, 2$.

Hence, if conditions (i)–(iii) above are hold, $\big(\widetilde{Z}(y), \widetilde{p}(x, y)\big)$ is differentiable.

Definition 5 For $\widetilde{Z}(y)$ to be a fuzzy solution of the Eq. (28) we need

(a) $\widetilde{Z}(y)$ differentiable,
(b) Equation (28) hold for $\widetilde{u}(y) = \widetilde{Z}(y)$,
(c) $\widetilde{Z}(y)$ satisfies the initial and boundary conditions. Since no exist specified any particular initial and boundary conditions, then only is checked if (28) hold.

We will only say that $\widetilde{Z}(y)$ is a fuzzy solution (without the initial and boundary conditions) if $\widetilde{Z}(y)$ is differentiable and $\frac{d^3\widetilde{Z}}{d^3y}(y) = ReRiC$ or the following equations must hold

$$\frac{d^3 z_1}{d^3 y}(y, \alpha) = ReRiC \tag{39}$$

$$\frac{d^3 z_1}{d^3 y}(y, \alpha) = ReRiC \tag{40}$$

for all $y \in I_2$ and $\alpha \in [0, 1]$.

Theorem 1 *Assume* $\widetilde{Z}(y)$ *is differentiable.*

(a) $\widetilde{Z}(y)$ *is a fuzzy solution if*

$$\frac{\partial G}{\partial k_i}\frac{\partial p}{\partial k_i} > 0 \quad y \in I_2 \ \textit{for}\ i = 1, 2 \tag{41}$$

(b) if

$$\frac{\partial G}{\partial k_i}\frac{\partial p}{\partial k_i} < 0 \quad y \in I_2 \ \textit{for}\ i = 1, 2 \tag{42}$$

then $\widetilde{Z}(y)$ *is not a fuzzy solution.*

Proof (a) The proof for $\frac{\partial G}{\partial k_1} > 0$, $\frac{\partial p}{\partial k_1} > 0$ and $\frac{\partial G}{\partial k_2} < 0$, $\frac{\partial p}{\partial k_2} < 0$. Then from (34)–(37) we have

$$z_1(y, \alpha) = G\big(y, k_{11}(\alpha), k_{22}(\alpha)\big), \tag{43}$$

$$z_2(y, \alpha) = G\big(y, k_{12}(\alpha), k_{21}(\alpha)\big) \tag{44}$$

$$p_1(x, y, \alpha) = p\big(x, y, k_{11}(\alpha), k_{22}(\alpha)\big), \tag{45}$$

$$p_2(x, y, \alpha) = p\big(x, y, k_{12}(\alpha), k_{21}(\alpha)\big) \tag{46}$$

for all $\alpha \in [0, 1]$ where $\left[\widetilde{K}_1\right]^\alpha = \left[k_{11}(\alpha), k_{12}(\alpha)\right]$ and $\left[\widetilde{K}_2\right]^\alpha = \left[k_{21}(\alpha), k_{22}(\alpha)\right]$. Now G solves the partial differential equations (11) and (12)

$$\frac{\partial p}{\partial x}(x, y, k_1, k_2) = \frac{1}{Re}\Big(\frac{\partial^2 G}{\partial^2 y}(y, k_1, k_2)\Big) \tag{47}$$

$$\frac{\partial p}{\partial y}(x, y, k_1, k_2) = RiT \tag{48}$$

for all $x, y \in I_1 \times I_2$, so

$$\frac{\partial p_1}{\partial x}(x, y, \alpha) = \frac{1}{Re}\Big(\frac{\partial^2 z_1}{\partial^2 y}(y, \alpha)\Big) \tag{49}$$

$$\frac{\partial p_2}{\partial x}(x, y, \alpha) = \frac{1}{Re}\Big(\frac{\partial^2 z_2}{\partial^2 y}(y, \alpha)\Big) \tag{50}$$

$$\frac{\partial p_1}{\partial y}(x, y, \alpha) = RiT \tag{51}$$

$$\frac{\partial p_2}{\partial y}(x, y, \alpha) = RiT \tag{52}$$

and (z_i, p_i), *for* $i = 1, 2$ are a continuous functions, having continuous partials with respect to both x and y then by partial derivative at y in (49), (50) and partial derivative at x in (51), (52). So

$$\frac{d^3 z_1}{d^3 y}(y, \alpha) = ReRiC$$

$$\frac{d^3 z_1}{d^3 y}(y, \alpha) = ReRiC$$

for all $y \in I_2$ and $\alpha \in [0, 1]$.

Hence (39) and (40) hold and $\widetilde{Z}(y)$ is a fuzzy solution.

(b) $\frac{\partial G}{\partial k_1} > 0$, $\frac{\partial p}{\partial k_1} < 0$ and $\frac{\partial G}{\partial k_2} < 0$, $\frac{\partial p}{\partial k_2} < 0$. Then (43) and (44) are still true but (45) and (46) became

$$p_1(t, x, \alpha) = p\big(x, y, k_{12}(\alpha), k_{22}(\alpha)\big),$$
$$p_2(t, x, \alpha) = p\big(x, y, k_{11}(\alpha), k_{21}(\alpha)\big)$$

for all α Hence (49) and (50) do not hold then (39) and (40) do not hold. So $\widetilde{Z}(y)$ is not a fuzzy solution.

Theorem 2 *Let* $y_0 \in I_2$ *and assume that* $\widetilde{u} : I_2 \to \mathrm{R}_{\mathcal{F}}$ *is continuous. Consider the boundary conditions problem (29). A mapping* $\widetilde{\theta} : I_2 \to \mathrm{R}_{\mathcal{F}}$ *is a solution to (29) if and only if* $\widetilde{\theta}$, $\frac{d\widetilde{\theta}}{dy}$ *are continuous and satisfy the integral equation*

$$\widetilde{\theta}(y) = \eta_1(y - y_0) + \eta_2 \int_{y_0}^{y} \Big(\int_{y_0}^{y} \widetilde{u}(s)ds\Big)ds + \eta_3 \tag{53}$$

Proof Since $\widetilde{u}$ is continuous by [8], it must be integrable. So, for

$$\frac{d^2\widetilde{\theta}}{d^2 y} = PeC\widetilde{u},\ y \in I_2,\ where\ \eta_2 = PeC$$

we have equivalently

$$\frac{d\widetilde{\theta}}{dy}(y) = \eta_2 \int_{y_0}^{y} \widetilde{u}(s)ds + \frac{d\widetilde{\theta}}{dy}(y_0)$$

see [9]. Since $\frac{d\widetilde{\theta}}{dy}(y_0) = \eta_1$, we have

$$\frac{d\widetilde{\theta}}{dy}(y) = \eta_2 \int_{y_0}^{y} \widetilde{u}(s)ds + \eta_1$$

Thus, by [9],

$$\widetilde{\theta}(y) = \eta_2 \int_{y_0}^{y} \Big(\int_{y_0}^{y} \widetilde{u}(s)ds + \eta_1 \Big) ds + \widetilde{\theta}(y_0)$$

equivalently see[8],

$$\widetilde{\theta}(y) = \eta_2 \int_{y_0}^{y} \Big(\int_{y_0}^{y} \widetilde{u}(s)ds \Big) ds + \int_{y_0}^{y} \eta_1 ds + \widetilde{\theta}(y_0)$$

equivalently

$$\widetilde{\theta}(y) = \eta_2 \int_{y_0}^{y} \Big(\int_{y_0}^{y} \widetilde{u}(s)ds \Big) ds + \eta_1 (y - y_0) + \eta_3$$

Remark 2 Let $[\widetilde{\theta}]^{\alpha} = [\theta_1(y, \alpha), \theta_2(y, \alpha)]$ and $[\widetilde{u}]^{\alpha} = [u_1(y, \alpha), u_2(y, \alpha)]$. Consider the Definition 4. So

$$\theta_1(y, \alpha) = \eta_1 (y - y_0) + \eta_2 \int_{y_0}^{y} \Big(\int_{y_0}^{y} u_1(s, \alpha)ds \Big) ds + \eta_3$$
$$\theta_2(y, \alpha) = \eta_1 (y - y_0) + \eta_2 \int_{y_0}^{y} \Big(\int_{y_0}^{y} u_2(s, \alpha)ds \Big) ds + \eta_3$$

Lemma 1 *Let $y_0 \in I_2$ and assume that $\widetilde{u} : I_2 \to \mathrm{R}_{\mathscr{F}}$ is continuous. Consider the boundary conditions problem (30). A mapping $\widetilde{\Psi} : I_2 \to \mathrm{R}_{\mathscr{F}}$ is a solution to (30) if and only if $\widetilde{\Psi}$ is continuous and satisfy the integral equation*

$$\widetilde{\Psi}(y) = \int_{y_0}^{y} \widetilde{u}(s)ds + \sigma_1 \tag{54}$$

Proof The proof is similar to Theorem 2.

Using such an approach, the solution of Eq. (28), satisfying Eqs. (31), and (32), is

$$u_1(y, \alpha) = \frac{1}{12} ReRiC\Big(2y^3 - 3y^2 + y\Big) + 3\big(2\alpha - 1\big)y^2 - 2(3\alpha - 2)y + \alpha - 1$$
$$u_2(y, \alpha) = \frac{1}{12} ReRiC\Big(2y^3 - 3y^2 + y\Big) + 3\big(3 - 2\alpha\big)y^2 - 2(4 - 3\alpha)y + 1 - \alpha$$

where $[\widetilde{K}_1]^{\alpha} = [\alpha, 2 - \alpha]$ and $[\widetilde{K}_2]^{\alpha} = [\alpha - 1, 1 - \alpha]$. Or the fuzzy solution of (28) is

$$\widetilde{u}(y) = \frac{1}{12} ReRiC\Big(2y^3 - 3y^2 + y\Big) + 3\big(\widetilde{K}_1 + \widetilde{K}_2\big)y^2 - 2\big(\widetilde{K}_1 + 2\widetilde{K}_2\big)y + \widetilde{K}_2 \tag{55}$$

and the solution of Eq. (29), satisfying Eqs. (31), and (33), is

$$\theta_1(y,\alpha) = \frac{1}{12}RaC^2\left(\frac{y^5}{10} - \frac{y^4}{4} + \frac{y^3}{6} - \frac{1}{120}\right) + PeC\left((2\alpha-1)\frac{y^4}{4} - (3\alpha-2)\frac{y^3}{3} + \left(\frac{9\alpha-7}{60}\right)\right)$$
$$+ (\alpha-1)\frac{y^2}{2} - \left(\frac{\alpha-1}{6}\right)$$

$$\theta_2(y,\alpha) = \frac{1}{12}RaC^2\left(\frac{y^5}{10} - \frac{y^4}{4} + \frac{y^3}{6} - \frac{1}{120}\right) + PeC\left((3-2\alpha)\frac{y^4}{4} - (4-3\alpha)\frac{y^3}{3} + \left(\frac{11-9\alpha}{60}\right)\right)$$
$$+ (1-\alpha)\frac{y^2}{2} - \left(\frac{1-\alpha}{6}\right)$$

or

$$\widetilde{\theta}(y) = \frac{1}{12}RaC^2\left(\frac{y^5}{10} - \frac{y^4}{4} + \frac{y^3}{6} - \frac{1}{120}\right) + PeC\left((\widetilde{K}_1+\widetilde{K}_2)\frac{y^4}{4} - (\widetilde{K}_1+2\widetilde{K}_2)\frac{y^3}{3} + \left(\frac{2\widetilde{K}_1+7\widetilde{K}_2}{60}\right)\right)$$
$$+ \widetilde{K}_2\frac{y^2}{2} - \frac{\widetilde{K}_2}{6}$$

The expression of the stream function Ψ, can be deduced by Lemma 1 of Eq. (30), taking into account of the corresponding boundary conditions and Eq. (55), which gives:

$$\Psi_1(y,\alpha) = \frac{1}{12}ReRiC\left(\frac{y^4}{2} - y^3 + \frac{y^2}{2}\right) + (2\alpha-1)y^3 - (3\alpha-2)y^2 + (\alpha-1)y$$

$$\Psi_2(y,\alpha) = \frac{1}{12}ReRiC\left(\frac{y^4}{2} - y^3 + \frac{y^2}{2}\right) + (3-2\alpha)y^3 - (4-3\alpha)y^2 + (1-\alpha)y$$

Or

$$\widetilde{\Psi} = \frac{1}{12}ReRiC\left(\frac{y^4}{2} - y^3 + \frac{y^2}{2}\right) + (\widetilde{K}_1+\widetilde{K}_2)y^3 - (\widetilde{K}_1+2\widetilde{K}_2)y^2 + \widetilde{K}_2 y$$

5 Conclusion

In this paper a fuzzy analytical study on mixed convection in a two-dimensional horizontal shallow enclosure, of aspect ratio $A = 8$, filled with a nanofluid, has been conducted in the case where both short vertical sides are submitted to uniform heat fluxes while the long horizontal ones are assumed adiabatic, with the top one uniformly moving in the same direction to heat flux.

The computations, which have been limited to Cu-water mixtures, with $Pr = 7$, have been carried out with governing parameters, Re, Ri and Φ, varying, respectively, in the ranges $Re = 1$, $Ri = 10^3$ and $\Phi = 0$. Fuzzy analytical solution is derived on

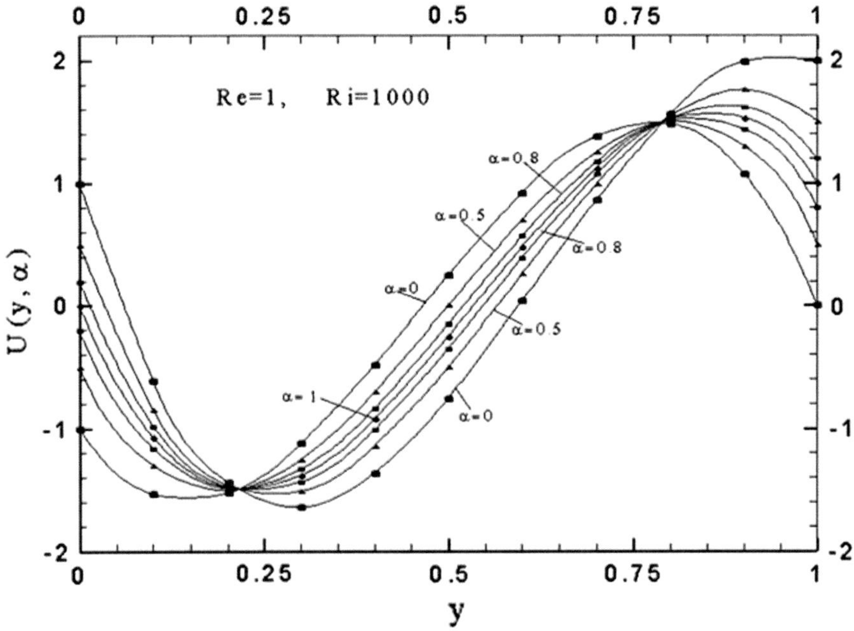

Fig. 3 The velocity profiles at mid-length of the cavity, along the vertical coordinate for $A = 8$ where $Re = 1,\ \ Ri = 10^3$, and various values of α

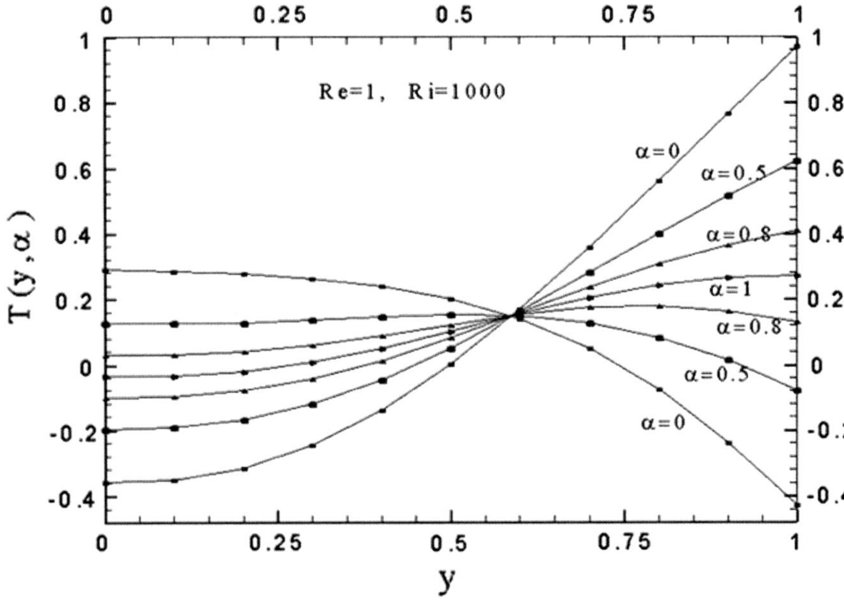

Fig. 4 The temperature profiles at mid-length of the cavity, along the vertical coordinate for $A = 8$ where $Re = 1,\ \ Ri = 10^3$, and various values of α

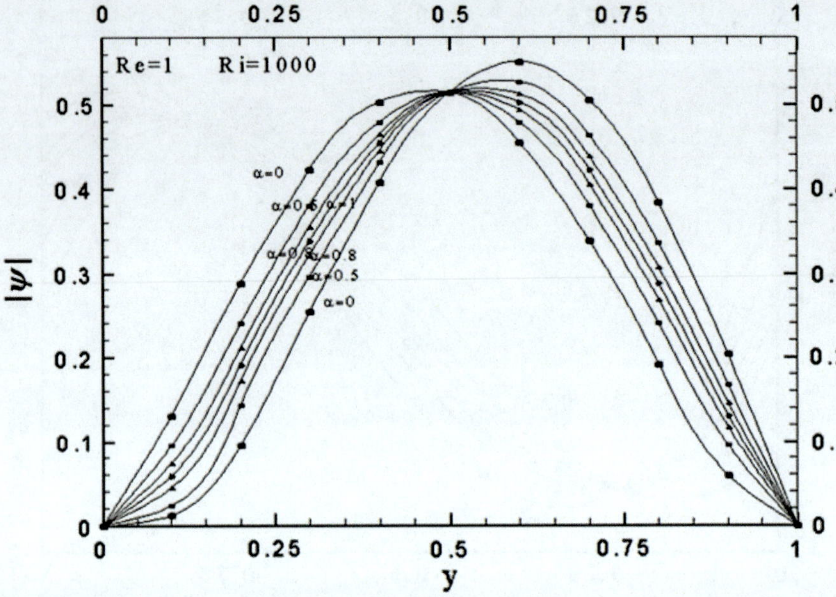

Fig. 5 The stream function profiles at mid-length of the cavity, along the vertical coordinate for $A = 8$ where $Re = 1$, $Ri = 10^3$, and various values of α

the basis of a parallel flow assumption in the core region of the enclosure. The main findings of such investigation can be summarized as follows:

1. We presented the strategy for solving fuzzy differential equations (FDE). These analytical solutions are based on α-cut of a fuzzy set. If these solutions define α-cuts of a fuzzy number, then the solutions of FDE, would exist.
2. Denotes that by this strategy the authors extended the results for the proposed models, have the higher accuracy and a validation mutually both the corresponding approaches if $\alpha = 1$ in the crisp problem [5].
3. In Fig. 3, we can explain the points mentioned $y = 0.21$ and $y = 0.79$, the velocity is independent of the boundary conditions for all α, which represents in physics the velocity becomes constant at these points.
4. For all $\alpha \in [0, 1]$ the temperature becomes independent of the boundary conditions for $y = 0.59$ as shown in Fig. 4, i.e it is an isothermal point. And the stream for $y = 0.49$ for all α there is only one point where all the streams have the same flow intensity as shown in Fig. 5.

References

1. Bejan, A.: The boundary layer regime in a porous layer with uniform heat flux from side. Int. Heat Mass Trans. **26**, 1339–1346 (1983)
2. Buckley, J.J., Qu, Y.: On using α-cuts to evaluate fuzzy equations. Fuzzy Sets Syst. **38**, 309–312 (1990)
3. Chadli, L.S., Harir, A., Melliani, S.: Solutions of fuzzy heat-like equations by variational iterative method. Ann. Fuzzy Math. Inf. **10**(1), 29–44 (2015)
4. Chadli, L.S., Harir, A., Melliani, S.: Solutions of fuzzy wave-like equations by variational iteration method. Int. Ann. Fuzzy Math. Inf. **8**(4), 527–547 (2014)
5. El Harfi, H., Naimi, M., Lamsaadi, M., Raji, A., Hasnaoui, M.: Mixed convection heat transfer for nanofluids in a lid-driven shallow rectangular cavity uniformly heated and cooled from the vertical sides: the cooperative case. J. Electron. Cool. Therm. Control **3**(3) (2013)
6. Lamsaadi, M., Naimi, M., Hasnaoui, M.: Natural convection heat transfer in shallow horizontal rectangular enclosures uniformly heated from the side and filled with non-Newtonian power law fluids. En. Con. Man. **47**, 2535–2551 (2006)
7. Leite, J., Bassanezi, R.C.: Sistemas dinamicos fuzzy aplicados a processos difusivos. Biomatemtica **20**, 157–166 (2010). (in Portuguese)
8. Kaleva, O.: Fuzzy differential equations. Fuzzy Sets Syst. **24**(3), 301–317 (1987)
9. Kaleva, O.: The Cauchy problem for fuzzy differential equations. Fuzzy Sets Syst. **35**, 389–396 (1990)
10. Getschel, R., Voxman, W.: Elementary fuzzy calculus. Fuzzy Sets Syst. **18**, 31–43 (1986)
11. Seikkala, S.: On the fuzzy initial value problem. Fuzzy Sets Syst. **24**, 319–330 (1987)
12. Diamond, P., Kloeden, P.E.: Metric Spaces of Fuzzy Sets: Theory and Applications. World Scienific, Singapore (1994)
13. Ma, M., Friedman, M., Kandel, A.: A new fuzzy arithmetic. Fuzzy Sets Syst. **108**, 83–90 (1999)
14. Jafelice, R.M., Almeida, C.G., Meyer, J.F., Vasconcelos, H.L.: Fuzzy parameter in a partial differential equation model for population dispersal of leaf-cuttingants. Nonlinear Anal. Real World Appl. **12**, 3397–3412 (2011)
15. Oberguggenberger, M.: Fuzzy and weak solutions to differential equations. In: Proceedings of the 10th International IPMU Conference, pp. 517–524 (2004)
16. Zadeh, L.A.: Fuzzy sets. Inf. Control **8**, 338–353

Likelihood and Decoding in a Partially Hidden Markov Model

Karima Elkimakh and Abdelaziz Nasroallah

Abstract A Hidden Markov Model (HMM) is composed of two processes, a Markov chain whose evolution is hidden and an emission process that is observable. The resolution of fundamental problems in HMM is based on the computation of the probability of an observed emission sequence using all possible corresponding hidden sequences. In this paper, we assume that the Markov chain is not completely hidden and we know one or two states at fixed times during the evolution interval time. The obtained HMM is then partially hidden. Principally we interest to the computation of the probability of an observed emission sequence given that the Markov chain goes through one or two fixed states at fixed times. Mainly we resolve the likelihood and decoding problems by developing formulas for forward and backward probabilities and the related Viterbi algorithm, assuming that we have some precise information at fixed times. Numerical examples are studied to show the smooth running of the proposed partially HMM.

Keywords Hidden Markov Model · Markov chain · Emission · Forward and Backward probabilities · Viterbi algorithm

1 Introduction

Hidden Markov Models (HMMs) are powerful mathematical tools generally used to study random phenomena or systems for which we can observe information induced by a hidden one. Such phenomena can be encountered: in DNA where an observed sequence of bases (A, C, G, T) is assumed to be generated from two hidden states (coding and non-coding region) [4], in speech recognition where we interest to determine which speech is present based on spoken information [8], in image processing

K. Elkimakh (✉) · A. Nasroallah
Faculty of Sciences Semlalia, Cadi Ayyad University, 2390 Marrakesh, Morocco
e-mail: karima.elkimakh@gmail.com

A. Nasroallah
e-mail: nasroallah@uca.ma

S. Melliani and O. Castillo (eds.), *Recent Advances in Intuitionistic Fuzzy Logic Systems and Mathematics*, Studies in Fuzziness and Soft Computing 395,
https://doi.org/10.1007/978-3-030-53929-0_12

where, for example, we are concerned to detect vehicles in a traffic scene using video features through the analyzing of images [1], in finance where it is important to find the future trends of the stock market given the daily stock data [6].

An HMM is a model where the hidden information is modeled by a Markov chain. In such a model, each state generates an emission that is visible. So generally in HMMs, we study the observed emissions (outputs) emitted by hidden Markov states. Sometimes we know some information about some hidden states like in the case of coding and non-coding regions of DNA, where sometimes, we suppose that the first region is non-coding. Also, some hidden information, such as the presence of the Markov chain in a fixed state at a specified time, can be discovered by an expert and therefore it can be exploited. Motivated by this idea, in this work, we propose to study the case of an HMM given that one or two states are known in fixed times. Such a model can be called partially HMM.

Other ideas of partial HMM can be found: in [2] where partial information on the latent Markov chain is given since this one reaches a fixed state and until it leaves this state, but not in a specified time, and in [7] where each hidden state is observed, with a fixed probability and might be observed with noise.

In our approach, we assume that we have precise information at fixed times. Mainly we study the likelihood and decoding problems given that one or two states are known at specified times. The generalization to more than two known states is without difficulties except that the formulas will be tedious.

The paper is organized as follows, in the second section we resolve the likelihood and decoding problem in two cases: In the first case we know just one state and in the second one we know two states. In the third section, we present a numerical example of the resolution of the likelihood and decoding problems using the proposed approach. A conclusion is given in the fourth section.

We adopt the following notations: $I\!P_\lambda(.)$, x_0^T, $X_{[x_0^T]}$ and $(i_0^T \setminus i_t)$ to signify $I\!P(.|\lambda)$, $x_0x_1 \dots x_T$, $X_0 = x_0, \dots, X_T = x_T$ and $i_0i_1 \dots i_{t-1}i_{t+1} \dots i_T$ respectively.

2 Partially Hidden Markov Model

Let $E = \{1, \dots, r\}$ and $F = \{1, \dots, m\}$ be two finite sets, where E is the set of hidden states and F is the set of observed emissions.

Let $X = (X_t)_{t\geq 0}$ be an homogeneous Markov chain on E, with initial distribution $\mu = (\mu(i))_{i\in E}$ and transition probability kernel $A = (A(k,i))_{1\leq k,i\leq r}$: $A(k,i) := I\!P(X_{t+1} = i|X_t = k)$.

Given $X = (X_t)_{t\geq 0}$ which will be the hidden process, the emission process $Y = (Y_t)_{t\geq 0}$ is such that, at time t, for $k \in E$, $b_k(j)$ is the probability of emission $j \in F$ observed at time t, given that X is in state k at time t: $b_k(j) := I\!P(Y_t = j|X_t = k)$.

The probability distribution matrix $B = (b_k(j))_{k\in E, j\in F}$ is called the emission probability.

As usual, we note the parameter set of the HMM (defined by X and Y) by $\lambda = (\mu, A, B)$.

We assume the following standard HMM hypothesis:

$$\begin{array}{c}\textbf{(H1)} : \text{Given} X_t, X_{t+1} \text{is independent of } Y_{[j_0^t]},\\ \textbf{(H2)} : \text{Given } X_t, Y_t \text{is independent of } (X_s, Y_s, s < t).\end{array}$$

In all the following calculations we use the above hypotheses.

For a non-negative integer T, the three basic problems of HMMs can be formulated as follows: *(P1)* ***Likelihood problem***: Given model λ and a sequence of observations j_0^T, what is the probability $I\!P_\lambda\left(Y_{[j_0^T]}\right)$?

(P2) ***Decoding problem***: Given model λ and a sequence of observations j_0^T, how do we find the state sequence i_0^T, that maximizes the probability

$I\!P_\lambda\left(X_{[i_0^T]}|Y_{[j_0^T]}\right)$?

(P3) ***Learning problem***: How do we adjust the model parameter set $\lambda = (\mu, A, B)$ so as to maximize $I\!P_\lambda\left(Y_{[j_0^T]}\right)$?

The solution of this problem attempts to determine an optimal model to best describe the observation sequence.

The resolution of these three fundamental problems relating to HMMs in the standard case can be found in [3, 5, 8].

Remark 1 In this work, we restrict ourselves to the study of problems $(P1)$ and $(P2)$.

Now to propose our approach, based on the knowledge of one or two states, we need some additional notations:

$\rho(t, i, j) := \frac{A^t(i,j)}{\sum\limits_{k=0}^{r} \mu(k)A^t(k,j)}$ which is the average proportion for the process X to go from i at time 0 to j at time t, and $\phi(t', t, i, k, j) := \frac{A^{t'-t}(i,j)}{A^{t'-(t-1)}(k,j)}$, where $A^n(i, j)$ is the entry (i, j) of matrix A^n.

2.1 HMM with One Known State at a Fixed Time

For the process X, we assume that all states are hidden except the state h which will be visited by the process X at the fixed time t'.

2.1.1 Brute Formula for Likelihood

The first problem $(P1)$ consists of computing the probability $I\!P_\lambda\left(Y_{[j_0^T]}\right)$ which can be written as follows:

$$I\!P_\lambda\left(Y_{[j_0^T]}\right) = \sum_h I\!P_\lambda(X_{t'} = h) I\!P_\lambda\left(Y_{[j_0^T]} | X_{t'} = h\right) \tag{1}$$

We will interest in computation of the probability $I\!P_\lambda\left(Y_{[j_0^T]} | X_{t'} = h\right)$, of the observed sequence j_0^T given that the process X will be in state $h \in E$ at time t'. It's easy to show that this probability is given by the following formula (See Appendix 1).

$$I\!P_\lambda\left(Y_{[j_0^T]} | X_{t'} = h\right) = \sum_{(i_0^T \backslash i_{t'}) \in E^T} \mu(i_0)\rho(t', i_0, h) b_{i_0}(j_0) \times \tag{2}$$
$$\left[\prod_{t=1}^{t'-1} A(i_{t-1}, i_t)\phi(t', t, i_t, i_{t-1}, h) b_{i_t}(j_t) \times b_h(j_{t'}) \prod_{t=t'+1}^{T} A(i_{t-1}, i_t) b_{i_t}(j_t)\right]$$

Since for large T, the computation of the probability (2) is expensive, there exists recursive formulas called Forward and Backward probabilities or algorithms generally used to reduce this computation cost.

As can be seen in [8], the computation cost of the brute formula (2) is $O(Tr^T)$ order, when it is of $O(r^2T)$ using Forward and Backward probabilities.

2.1.2 Forward Algorithm for Likelihood

Relative to our situation we define a Forward variable as the probability of the partial observation sequence j_0^t, $0 \le t \le T$ that ends to state i at time t, given that the process X is in state h at t' by:

$$\alpha_t^{t'}(i) := I\!P_\lambda\left(Y_{[j_0^t]}, X_t = i | X_{t'} = h\right) \tag{3}$$

This probability can be calculated inductively using the following steps:

Forward algorithm steps

Step 1. *Initialization:* give j_0^T and set, for $i = 1, \ldots, r$,

$$\alpha_0^{t'}(i) = \begin{cases} b_i(j_0)\mathbf{1}(h = i) & \text{if t'=0,} \\ \mu(i) b_i(j_0)\rho(t', i, h) & \text{if } t' > 0 \end{cases}$$

Step 2. *Recurrence: for* $i = 1, \ldots, r$ *and* $t = 0, \ldots, T-1$

$$\alpha_{t+1}^{t'}(i) = \begin{cases} b_i(j_{t+1}) \sum_{k=1}^{r} \alpha_t^{t'}(k) A(k,i) \phi(t', t+1, i, k, h) & \text{if} \quad t < t' - 1 \\ b_i(j_{t+1}) \mathbf{1}(h = i) \sum_{k=1}^{r} \alpha_t^{t'}(k) & \text{if} \quad t = t' - 1 \\ \alpha_t^{t'}(h) b_i(j_{t+1}) A(h, i) & \text{if} \quad t = t' \\ b_i(j_{t+1}) \sum_{k=1}^{r} \alpha_t^{t'}(k) A(k, i) & \text{if} \quad t > t' \end{cases}$$

Step 3. *Termination: the desired probability is then given by*

$$I\!P_\lambda \left(Y_{[j_0^T]} | X_{t'} = h \right) = \sum_{i=1}^{r} \alpha_T^{t'}(i) \tag{4}$$

Throughout the paper the notation $\mathbf{1}(i = h)$ is equal to one if $h = i$ and zero otherwise.

2.1.3 Backward Algorithm for Likelihood

Similarly to Forward variable we define the Backward variable as the probability of emitting the partial observation sequence j_{t+1}^T given that $X_t = i$, $X_{t'} = h$ and λ by:

$$\beta_t^{t'}(i) := I\!P_\lambda \left(Y_{[j_{t+1}^T]} | X_t = i, X_{t'} = h \right) \tag{5}$$

This probability can be calculated inductively using the following steps:

Backward algorithm steps

Step 1. *Initialization:* give j_0^T and set $\beta_T^{t'}(i) = 1$, for $i = 1, \ldots, r$.
Step 2. *Recurrence:* for $i = 1, \ldots, r$ and $t = T - 1, T - 2, \ldots, 0$,

$$\beta_t^{t'}(i) = \begin{cases} \sum_{k=1}^{r} \beta_{t+1}^{t'}(k) b_k(j_{t+1}) A(i, k) & \text{if} \quad t' < t < T \\ \sum_{k=1}^{r} \beta_{t+1}^{t'}(k) b_k(j_{t+1}) A(i, k) \mathbf{1}(h = i) & \text{if} \quad t = t' \\ \beta_{t+1}^{t'}(h) b_h(j_{t+1}) & \text{if} \quad t = t' - 1 \\ \sum_{k=1}^{r} \beta_{t+1}^{t'}(k) b_k(j_{t+1}) A(i, k) \phi(t', t+1, k, i, h) & \text{if} \quad 0 \le t < t' - 1 \end{cases}$$

Step 3. *Termination: the desired probability is given by*

$$\mathbb{P}_\lambda\left(Y_{[j_0^T]}|X_{t'}=h\right)=\begin{cases} b_h(j_0)\beta_0^{t'}(h) & \text{if } \ t'=0\\ \sum_{i=1}^{r}\mu(i)b_i(j_0)\beta_0^{t'}(i)\rho(t',i,h) & \text{if } \ t'>0\end{cases}$$

The recursive formulas in Forward and Backward algorithms are detailed in Appendix 1.

Remark 2 Forward and Backward probabilities can simultaneously be used to write the target probability for any fixed $t \in 0, \ldots, T$ in the form

$$\mathbb{P}_\lambda\left(Y_{[j_0^T]}|X_{t'}=h\right)=\sum_{i=1}^{r}\alpha_t^{t'}(i)\beta_t^{t'}(i) \tag{6}$$

2.1.4 Decoding Problem

We seek to find a sequence $(i_0^T)^*$ that maximizes the probability $\mathbb{P}_\lambda(Y_{[j_0^T]}, X_{[i_0^T]}|X_{t'}=h)$ over all possibilities $i_0^T \in E^{T+1}$, given that the process X is in state h at the fixed time $t' \in [0, T]$. For this aim, we define the quantity $\delta_t^{t'}(k)$ as the highest probability of producing observation sequence j_0^t when X is moving from i_0 to i_{t-1} and gets into hidden state k at time t given $X_{t'}=h$ and the model parameter λ.

$$\delta_t^{t'}(i):=\begin{cases}\mathbb{P}_\lambda(X_0=i, Y_0=j_0|X_{t'}=h) & \text{for } \ t=0\\ \max\limits_{i_0^{t-1}}\mathbb{P}_\lambda\left(X_{[i_0^{t-1}]}, X_t=i, Y_{[j_0^t]}|X_{t'}=h\right) & \text{for } \ t\geq 1\end{cases}$$

We use $\psi_t^{t'}(i)$ to keep a local optimal path i_0^t that maximizes $\delta_t^{t'}(i)$.
The following lemma gives a recursive formula to compute $\delta_t^{t'}(i)$.

Lemma 1 *For* $0 \leq t' \leq T$

$$\delta_t^{t'}(i)=\begin{cases}\mu(i)b_i(j_0)\rho(t',i,h) & \text{if } \ t=0\\ \max\limits_k\left(\delta_{t-1}^{t'}(k)A(k,i)\phi(t',t,i,k,h)\right)b_i(j_t) & \text{if } \ 0<t<t'\\ \max\limits_k\left(\delta_{t-1}^{t'}(k)\right)b_i(j_t)\mathbf{1}(h=i) & \text{if } \ t=t'\\ \delta_{t-1}^{t'}(h)A(h,i)b_i(j_t) & \text{if } \ t=t'+1\\ \max\limits_k\left(\delta_{t-1}^{t'}(k)A(k,i)\right)b_i(j_t) & \text{if } \ t>t'+1\end{cases}$$

$$\psi_t^{t'}(i)=\arg\max_{1\leq k\leq r}\begin{cases}\delta_{t-1}^{t'}(k)A(k,i)\phi(t',t,i,k,h) & \text{if } \ 0<t<t'\\ \delta_{t-1}^{t'}(k)\mathbf{1}(h=i) & \text{if } \ t=t'\\ h & \text{if } \ t=t'+1\\ \delta_{t-1}^{t'}(k)A(k,i) & \text{if } \ t>t'+1\end{cases}$$

Proof See Appendix 2.

Among all possible paths, the following Viterbi algorithm finds an optimal path that maximizes $I\!P_\lambda\left(X_{[i_0^T]}, Y_{[j_0^T]} \middle| X_{t'} = h\right)$:

Algorithm 1: The Viterbi algorithm

Input: $\lambda = (\mu, A, B)$

Initialization:

$\delta_0^{t'}(i) = \mu(i) b_i(j_0) \rho(t', i, h), \forall i \in \{1, \ldots, r\}$

$\psi_0^{t'}(i) = 0, \forall i \in \{1, \ldots, r\}$

Recursion:

for *i from 1 to r* **do**

    for *t from* 1 *to* T **do**

        compute $\delta_t^{t'}(i)$

        and find $\psi_t^{t'}(i)$

Termination:

$p^* = \max_{1 \le i \le r} [\delta_T^{t'}(i)]$

$i_T^* = \arg\max_{1 \le i \le r} [\delta_T^{t'}(i)]$

Backtracking:

for *t from* $T - 1$ *to* 0 **do**

    $i_t^* = \psi_{t+1}^{t'}(i_{t+1}^*)$

$I\!P_\lambda\left(Y_{[j_0^T]}, X_{[i_0^T]^*} \middle| X_{t'} = h\right) = p^*$

Return: $I\!P_\lambda\left(Y_{[j_0^T]}, X_{[i_0^T]^*} \middle| X_{t'} = h\right)$ and $(i_0^T)^*$

2.2 *HMM with Two Known States at Fixed Times*

Here, we consider that all states are hidden except two states h_1 and h_2 which will be visited by the process X at the fixed times t_1 and t_2 respectively. We assume that $t_1 < t_2$ but h_1 and h_2 can be the same.

2.2.1 Brute Formula for Likelihood

In this case the probability $I\!P_\lambda\left(Y_{[j_0^T]}\right)$ can be written as

$$I\!P_\lambda\left(Y_{[j_0^T]}\right) = \sum_{h_1, h_2} I\!P_\lambda(X_{t_1} = h_1, X_{t_2} = h_2) I\!P_\lambda\left(Y_{[j_0^T]} \middle| X_{t_1} = h_1, X_{t_2} = h_2\right) \quad (7)$$

We interest in computation of the probability $I\!P_\lambda\left(Y_{[j_0^T]} \middle| X_{t_1} = h_1, X_{t_2} = h_2\right)$, of the observed sequence j_0^T given that the process X will be in states h_1 and h_2 at times t_1 and t_2 respectively. As in the previous section (see Appendix 1), it's easy to show that this probability is given by the following formula:

$$\begin{aligned} I\!P_\lambda\left(Y_{[j_0^T]} \middle| X_{t_1} = h_1, X_{t_2} = h_2\right) = & \sum_{(i_0^T \setminus (i_{t_1}, i_{t_2})) \in E^{T-1}} \mu(i_0)\rho(t_1, i_0, h_1) b_{i_0}(j_0) \times \\ & \left[\prod_{t=1}^{t_1-1} A(i_{t-1}, i_t)\phi(t_1, t, i_t, i_{t-1}, h_1) b_{i_t}(j_t) b_{h_1}(j_{h_1}) \times \right. \\ & \prod_{t=t_1+1}^{t_2-1} A(i_{t-1}, i_t)\phi(t_2, t, i_t, i_{t-1}, h_2) b_{i_t}(j_t) b_{h_2}(j_{h_2}) \times \\ & \left. \prod_{t=t_2+1}^{T} A(i_{t-1}, i_t) b_{i_t}(j_t) \right], \end{aligned}$$

where $i_0^T \setminus (i_{t_1}, i_{t_2})$ means that states i_{t_1} and i_{t_2} are deleted from the sequence i_0^T.

2.2.2 Forward and Backward Algorithm for Likelihood

Similar to the case where one state is known, we define the Forward and backward probabilities as follows:

$$\alpha_t^{t_1,t_2}(i) := I\!P_\lambda\left(Y_{[j_0^t]}, X_t = i \middle| X_{t_1} = h_1, X_{t_2} = h_2\right) \tag{8}$$

and

$$\beta_t^{t_1,t_2}(i) := I\!P_\lambda\left(Y_{[j_{t+1}^T]} \middle| X_t = i, X_{t_1} = h_1, X_{t_2} = h_2\right) \tag{9}$$

Forward algorithm steps

Step 1. *Initialization:* give j_0^T, and set, for $i = 1, \ldots, r$

$$\alpha_0^{t_1,t_2}(i) = \begin{cases} b_i(j_0)\mathbf{1}(h_1 = i) & \text{if } \; t_1 = 0 \\ \mu(i) b_i(j_0)\rho(t_1, i, h_1) & \text{if } \; t_1 > 0 \end{cases}$$

Step 2. *Recurrence: for* $\quad i = 1, \ldots, r \quad$ *and* $\quad t = 0, \ldots, T-1$, compute

$$\alpha_{t+1}^{t_1,t_2}(i) = \begin{cases} b_i(j_{t+1}) \sum_{k=1}^{r} \alpha_t^{t_1,t_2}(k) A(k,i) \phi(t_1, t+1, i, k, h_1) & \text{if} \quad 0 < t < t_1 - 1 \\ b_i(j_{t+1}) \sum_{k=1}^{r} \alpha_t^{t_1,t_2}(k) \mathbf{1}(i = h_1) & \text{if} \quad t = t_1 - 1 \\ b_i(j_{t+1}) \alpha_t^{t_1,t_2}(h_1) A(h_1, i) \phi(t_2, t_1 + 1, i, h_1, h_2) & \text{if} \quad t = t_1 \\ b_i(j_{t+1}) \sum_{k=1}^{r} \alpha_t^{t_1,t_2}(k) A(k,i) \phi(t_2, t+1, i, k, h_2) & \text{if} \quad t_1 < t < t_2 - 1 \\ b_i(j_{t+1}) \sum_{k=1}^{r} \alpha_t^{t_1,t_2}(k) \mathbf{1}(i = h_2) & \text{if} \quad t = t_2 - 1 \\ \alpha_t^{t_1,t_2}(h_2) b_i(j_{t+1}) A(h_2, i) & \text{if} \quad t = t_2 \\ b_i(j_{t+1}) \sum_{k=1}^{r} \alpha_t^{t_1,t_2}(k) A(k,i) & \text{if} \quad t_2 < t < T \end{cases}$$

Step 3. *Termination:*

$$I\!P_\lambda \left(Y_{[j_0^T]} \middle| X_{t_1} = h_1, X_{t_2} = h_2 \right) = \sum_{i=1}^{r} \alpha_T^{t_1,t_2}(i) \tag{10}$$

Backward algorithm steps

Step 1. *Initialization:* give j_0^T and set $\beta_T^{t_1,t_2}(i) = 1$ for $i = 1, \ldots, r$.

Step 2. *Recurrence:* for $i = 1, \ldots, r$ and $t = T-1, T-2, \ldots, 0$,

$$\beta_t^{t_1,t_2}(i) = \begin{cases} \sum_{k=1}^{r} \beta_{t+1}^{t_1,t_2}(k) b_k(j_{t+1}) A(i,k) & \text{if} \quad t_2 < t < T \\ \sum_{k=1}^{r} \beta_{t+1}^{t_1,t_2}(k) b_k(j_{t+1}) A(i,k) \mathbf{1}(i = h_2) & \text{if} \quad t = t_2 \\ \beta_{t+1}^{t_1,t_2}(h_2) b_{h_2}(j_{t+1}), & \text{if} \quad t = t_2 - 1 \\ \sum_{k=1}^{r} \beta_{t+1}^{t_1,t_2}(k) b_k(j_{t+1}) A(i,k) \phi(t_2, t+1, k, i, h_2) & \text{if} \quad t_1 < t < t_2 - 1 \\ \sum_{k=1}^{r} \beta_{t+1}^{t_1,t_2}(k) b_k(j_{t+1}) A(i,k) \phi(t_2, t_1 + 1, k, h_1, h_2) \mathbf{1}(i = h_1) & \text{if} \quad t = t_1 \\ \beta_{t+1}^{t_1,t_2}(h_1) b_{h_1}(j_{t+1}) & \text{if} \quad t = t_1 - 1 \\ \sum_{k=1}^{r} \beta_{t+1}^{t_1,t_2}(k) b_k(j_{t+1}) A(i,k) \phi(t_1, t+1, k, i, h_1) & \text{if} \quad 0 \le t < t_1 - 1 \end{cases}$$

Step 3. *Termination:*

$$I\!P_\lambda \left(Y_{[j_0^T]} \middle| X_{t_1} = h_1, X_{t_2} = h_2 \right) = \begin{cases} b_{h_1}(j_0) \beta_0^{0,t_2}(h_1) & \text{if} \quad t_1 = 0 \\ \sum_{i=1}^{r} \mu(i) b_i(j_0) \beta_0^{t_1,t_2}(i) \rho(t_1, i, h_1) & \text{if} \quad t_1 > 0 \end{cases}$$

2.2.3 Decoding Problem

Let $\delta_t^{t_1,t_2}(k)$ be the highest probability of producing observation sequence j_0^t when X is moving from i_0 to i_{t-1} and gets into hidden state k at time t given that $X_{t_1} = h_1$, $X_{t_2} = h_2$ and the model parameter λ.

$$\delta_t^{t_1,t_2}(i) := \begin{cases} I\!P_\lambda\left(X_0 = i, Y_0 = j_0 \middle| X_{t_1} = h_1, X_{t_2} = h_2\right) & \text{for} \quad t = 0 \\ \max\limits_{i_0^{t-1}} I\!P_\lambda\left(X_{[i_0^{t-1}]}, X_t = i, Y_{[j_0^t]} \middle| X_{t_1} = h_1, X_{t_2} = h_2\right) & \text{for} \quad t \geq 1 \end{cases}$$

The following lemma gives a recursive formula to compute $\delta_t^{t_1,t_2}(i)$.

Lemma 2 *For* $\quad 0 \leq t_1 < t_2 \leq T$

$$\delta_t^{t_1,t_2}(i) = \begin{cases} \mu(i)b_i(j_0)\rho(t_1, i, h_1) & \text{if} \quad t = 0 \\ \max\limits_k \left(\delta_{t-1}^{t_1,t_2}(k)A(k,i)\phi(t_1, t, i, k, h_1)\right)b_i(j_t) & \text{if} \quad 0 < t < t_1 \\ \max\limits_k \left(\delta_{t-1}^{t_1,t_2}(k)\right)b_i(j_t)\mathbf{1}(h_1 = i) & \text{if} \quad t = t_1 \\ \delta_{t-1}^{t_1,t_2}(h_1)b_i(j_t)A(h_1,i)\phi(t_2, t_1+1, i, h_1, h_2) & \text{if} \quad t = t_1 + 1 \\ \max\limits_k \left(\delta_{t-1}^{t_1,t_2}(k)A(k,i)\phi(t_2, t, i, k, h_2)\right)b_i(j_t) & \text{if} \quad t_1 + 1 < t < t_2 \\ \max\limits_k \left(\delta_{t-1}^{t_1,t_2}(k)\right)b_i(j_t)\mathbf{1}(h_2 = i) & \text{if} \quad t = t_2 \\ \delta_{t-1}^{t_1,t_2}(h_2)b_i(j_t)A(h_2,i) & \text{if} \quad t = t_2 + 1 \\ \max\limits_k \left(\delta_{t-1}^{t_1,t_2}(k)A(k,i)\right)b_i(j_t) & \text{if} \quad t > t_2 + 1 \end{cases}$$

$$\psi_t^{t_1,t_2}(i) = \arg\max_{1 \leq k \leq r} \begin{cases} \delta_{t-1}^{t_1,t_2}(k)A(k,i)\phi(t_1, t, i, k, h_1) & \text{if} \quad 0 < t < t_1 \\ \delta_{t-1}^{t_1,t_2}(k)\mathbf{1}(h_1 = i) & \text{if} \quad t = t_1 \\ h1 & \text{if} \quad t = t_1 + 1 \\ \delta_{t-1}^{t_1,t_2}(k)A(k,i)\phi(t_2, t, i, k, h_2) & \text{if} \quad t_1 + 1 < t < t_2 \\ \delta_{t-1}^{t_1,t_2}(k)\mathbf{1}(h_2 = i) & \text{if} \quad t = t_2 \\ h2 & \text{if} \quad t = t_2 + 1 \\ \delta_{t-1}^{t_1,t_2}(k)A(k,i) & \text{if} \quad t > t_2 + 1 \end{cases}$$

Proof Similar to the proof of Lemma 1, given in Appendix 2.

The following Viterbi algorithm finds an optimal path that maximizes $I\!P_\lambda\left(X_{[i_0^T]}, Y_{[j_0^T]} \middle| X_{t_1} = h_1, X_{t_2} = h_2\right)$:

Algorithm 2: The Viterbi algorithm

Input: $\lambda = (\mu, A, B)$

Initialization:

$\delta_0^{t_1,t_2}(i) = \mu(i) b_i(j_0) \rho(t_1, i, h_1),\ \forall i \in \{1, \dots, r\}$

$\psi_0^{t_1,t_2}(i) = 0,\ \forall i \in \{1, \dots, r\}$

Recursion:

for *i from 1 to r* **do**

  for *t from* 1 *to* T **do**

    compute $\delta_t^{t_1,t_2}(i)$

    and find $\psi_t^{t_1,t_2}(i)$

Termination:

$p^* = \max_{1 \le i \le r} [\delta_T^{t_1,t_2}(i)]$

$i_T^* = \arg\max_{1 \le i \le r} [\delta_T^{t_1,t_2}(i)]$

Backtracking:

for *t from* $T - 1$ *to* 0 **do**

  $i_t^* = \psi_{t+1}^{t_1,t_2}(i_{t+1}^*)$

$I\!P_\lambda\left(Y_{[j_0^T]}, X_{[i_0^T]^*} \middle| X_{t_1} = h_1, X_{t_2} = h_2\right) = p^*$

Return: $I\!P_\lambda\left(Y_{[j_0^T]}, X_{[i_0^T]^*} \middle| X_{t_1} = h_1, X_{t_2} = h_2\right)$ and $(i_0^T)^*$

In the following we study a numerical example to show the usefulness of the proposed algorithms.

3 Numerical Example

Let $E = \{0, 1, 2\}$, $F = \{a, b, c\}$ and consider the following model λ:

$$\mu = \begin{pmatrix} 0.50 \\ 0.30 \\ 0.20 \end{pmatrix}, \quad A = \begin{pmatrix} 0.45 & 0.35 & 0.20 \\ 0.10 & 0.50 & 0.40 \\ 0.15 & 0.25 & 0.60 \end{pmatrix}, \quad B = \begin{pmatrix} 0.20 & 0.50 & 0.30 \\ 0.40 & 0.40 & 0.20 \\ 0.30 & 0.30 & 0.40 \end{pmatrix}$$

We resolve the likelihood and decoding problems for the two cases (one and two known states) for two different sequences.

Table 1 Likelihood results for the observed sequence j_0^9

	Brute formula	Forward	Backward
$IP_\lambda(Y_{[j_0^9]} \mid X_2 = 1)$	0.205093×10^{-4}	0.205093×10^{-4}	0.205093×10^{-4}
$IP_\lambda(Y_{[j_0^9]} \mid X_2 = 1, X_6 = 0)$	0.202061×10^{-4}	0.205093×10^{-4}	0.205093×10^{-4}

Table 2 Likelihood (column two) and Decoding (column three) of the first case results

Known states	$IP_\lambda Y_{[j_0^9]} \mid known\ states)$	Optimal path $(i_0^9)^*$	$IP_\lambda(X_{[i_0^9]^*}, Y_{[j_0^9]})$
$X_2 = 1$	0.127476×10^{-4}	0011112222	0.126338×10^{-6}
$X_5 = 2$	0.126098×10^{-4}	0011122222	0.937335×10^{-7}
$X_9 = 1$	0.201187×10^{-4}	0011111111	0.421762×10^{-7}

Table 3 Likelihood (column two) and Decoding (column three) of the second case results

Known states	$IP_\lambda Y_{[j_0^9]} \mid known\ states)$	Optimal path $(i_0^9)^*$	$IP_\lambda(X_{[i_0^9]^*}$
$X_1 = 0$ and $X_9 = 2$	0.154460×10^{-4}	0011112222	0.364464×10^{-6}
$X_2 = 2$ and $X_6 = 1$	0.151692×10^{-4}	2221111112	0.453037×10^{-7}
$X_4 = 2$ and $X_5 = 2$	0.163768×10^{-4}	0011222222	0.141566×10^{-6}

Table 4 Likelihood results for the observed sequence j_0^{12}

	Brute formula	Forward	Backward
$IP_\lambda Y_{[j_0^{12}]} \mid X_3 = 0)$	0.435823×10^{-6}	0.435823×10^{-6}	0.435823×10^{-6}
$IP_\lambda Y_{[j_0^{12}]} \mid X_3 = 0, X_8 = 2)$	0.416061×10^{-6}	0.416061×10^{-6}	0.416061×10^{-6}

- For the observed sequence $j_0^9 = cbaabbcaac$, see Tables 1, 2 and 3.
  The time CPU of the brute formula is of 12, where it's near to zero for the forward or backward probabilities.
- For the observed sequence $j_0^{12} = bccacbcbacabb$, see Tables 4, 5 and 6.
  The time CPU of the brute formula is between 429 and 474, where it's near to zero for the forward or backward probabilities.

Tables 1 and 4 summarise the results obtained for the likelihood of two different sequences assumed to be generated from the partial HMM λ. The same probabilities obtained for each sequence increase the confidence in the methods used as well as on the accuracy of the used formulas.

We notice that using the forward or backward probabilities reduces the computation cost compared to the brute formula.

In Tables 2, 3, 5 and 6 we find the desired state sequence $(i_0^T)^*$.

Table 5 Likelihood (column two) and Decoding (column three) of the first case results

Known states	$I\!P_\lambda Y_{[j_0^{12}]} \| known\ states)$	Optimal path $(i_0^{12})^*$	$I\!P_\lambda(X_{[i_0^12]^*}, Y_{[j_0^{12}]})$
$X_1 = 2$	0.919075×10^{-6}	0222222222222	0.119482×10^{-8}
$X_4 = 0$	0.871336×10^{-6}	0000012222222	0.294907×10^{-9}
$X_{12} = 1$	0.509785×10^{-6}	0222222222111	0.777226×10^{-9}

Table 6 Likelihood (column two) and Decoding (column three) of the second case results

Known states	$I\!P_\lambda Y_{[j_0^{12}]} \| known\ states)$	Optimal path $(i_0^{12})^*$	$I\!P_\lambda(X_{[i_0^12]^*}, Y_{[j_0^{12}]}$
$X_4 = 2$ and $X_8 = 1$	0.107437×10^{-5}	0222222112222	0.104599×10^{-8}
$X_1 = 0$ and $X_5 = 2$	0.864470×10^{-6}	0022222222222	0.181781×10^{-8}
$X_6 = 0$ and $X_7 = 1$	0.100130×10^{-5}	0222200112222	0.830192×10^{-10}

The proposed approach adapts well to the resolution of the likelihood and decoding problems of the HMM models. The obtained results are similar to those obtained in other numerical examples.

4 Conclusion

We have studied an example of a partially hidden Markov model. Among what is hidden, we suppose that one or two states of the Markov chain are known at fixed moments. Such information can be a result of expertise when observing the studied phenomenon during periods of time. For such a so-called partially hidden Markov model, we have elaborated formulas and algorithms used to calculate the probabilities involved in likelihood and decoding problems. We also tested the progress of the developed formulations through numerical investigation. The obtained results show the smooth running of the proposed partially HMM. The proposed approach can be generalized to the case when more than two hidden states are known.

Appendix 1: Likelihood Problem of the First Case

- **Brute Formula for Likelihood**:

$$\begin{aligned}
I\!P_\lambda\left(Y_{[j_0^T]}|X_{t'}=h\right) &= \sum_{(i_0^T\setminus i_t)\in E^T} I\!P_\lambda\left(Y_{[j_0^T]}|X_{[i_0^T]}\right) I\!P_\lambda\left(X_{[i_0^{t'-1}]}, X_{[i_{t'+1}^T]}|X_{t'}=h\right)\\
&= \sum_{(i_0^T\setminus i_t)\in E^T} b_{i_0}(j_0)\ldots b_{i_T}(j_T) I\!P_\lambda(X_0=i_0|X_{t'}=h)\times\\
&\quad I\!P_\lambda(X_1=i_1|X_{t'}=h, X_0=i_0)\ldots I\!P_\lambda(X_{t'-1}=i_{t'-1}|X_{t'}=h, X_{t'-2}=i_{t'-2})\times\\
&\quad I\!P_\lambda(X_{t'+1}=i_{t'+1}|X_{t'}=h) I\!P_\lambda(X_{t'+2}=i_{t'+2}|X_{t'+1}=i_{t'+1})\times\\
&\quad \ldots I\!P_\lambda(X_T=i_T|X_{T-1}=i_{T-1})\\
&= \sum_{(i_0^T\setminus i_t)\in E^T} \mu(i_0)\rho(t', i_0, h) b_{i_0}(j_0)\times\\
&\quad \left[\prod_{t=1}^{t'-1} A(i_{t-1}, i_t)\phi(t', t, i_t, i_{t-1}, h) b_{i_t}(j_t)\times b_h(j_{t'}) \prod_{t=t'+1}^{T} A(i_{t-1}, i_t) b_{i_t}(j_t)\right]
\end{aligned}$$

• **Likelihood by Forward Probabilities**:

$$\alpha_t^{t'}(i) := I\!P_\lambda\left(Y_{[j_0^t]}, X_t = i\middle|X_{t'}=h\right)$$

· For $t = t' = 0$

$$\alpha_0^0(i) = I\!P_\lambda(Y_0=j_0, X_0=i|X_0=h) = I\!P_\lambda(X_0=i|X_0=h) I\!P_\lambda(Y_0=j_0|X_0=i) = b_i(j_0)\mathbf{1}(h=i).$$

For $t = 0$ and $t \neq 0$

$$\begin{aligned}
\alpha_0^{t'}(i) &= I\!P_\lambda(Y_0=j_0, X_0=i|X_{t'}=h) = I\!P_\lambda(X_0=i|X_{t'}=h) I\!P_\lambda(Y_0=j_0|X_0=i)\\
&= b_i(j_0)\frac{I\!P_\lambda(X_0=i, X_{t'}=h)}{I\!P_\lambda(X_{t'}=h)} = b_i(j_0)\mu(i)\rho(t', i, h)
\end{aligned}$$

The recursion equation for $\alpha_t^{t'}(i)$ can be obtained such that: For $0 < t < T$

$$\begin{aligned}
\alpha_{t+1}^{t'}(i) &= I\!P_\lambda\left(Y_{[j_0^{t+1}]}, X_{t+1}=i\middle|X_{t'}=h\right)\\
&= \sum_{k=1}^{r} I\!P_\lambda\left(Y_{[j_0^{t+1}]}, X_{t+1}=i, X_t=k\middle|X_{t'}=h\right)\\
&= \sum_{k=1}^{r} I\!P_\lambda\left(Y_{[j_0^t]}, X_t=k\middle|X_{t'}=h\right) I\!P_\lambda\left(Y_{t+1}=j_{t+1}, X_{t+1}=i\middle|X_{t'}=h, X_t=k\right)\\
&= \sum_{k=1}^{r} I\!P_\lambda\left(Y_{[j_0^t]}, X_t=k\middle|X_{t'}=h\right) I\!P_\lambda\left(X_{t+1}=i\middle|X_{t'}=h, X_t=k\right) I\!P_\lambda\left(Y_{t+1}=j_{t+1}\middle|X_{t+1}=i\right)
\end{aligned}$$

By using the Markov property, we obtain:

$$\alpha_{t+1}^{t'}(i) = \begin{cases} \sum_{k=1}^{r} \alpha_t^{t'}(k) b_i(j_{t+1}) \frac{A(k,i)A^{t'-t-1}(i,h)}{A^{t'-t}(k,h)}, & t < t'-1 \\ \sum_{k=1}^{r} \alpha_t^{t'}(k) b_i(j_{t+1}) \mathbf{1}(h = i), & t = t'-1 \\ \alpha_t^{t'}(h) b_i(j_{t+1}) A(h,i), & t = t' \\ \sum_{k=1}^{r} \alpha_t^{t'}(k) b_i(j_{t+1}) A(k,i), & t > t' \end{cases}$$

Similarly to the Forward probabilities, we obtain the corresponding recursion for the Backward probabilities.

Appendix 2: Decoding Problem When of the First Case

Proof of Lemma 1

The probability $\delta_t^{t'}(i)$ is defined by:

$$\delta_t^{t'}(i) := \begin{cases} I\!P_\lambda(X_0 = i, Y_0 = j_0 | X_{t'} = h) & \text{for} \quad t = 0 \\ \max_{i_0^{t-1}} I\!P_\lambda\left(X_{[i_0^{t-1}]}, X_t = i, Y_{[j_0^t]} | X_{t'} = h\right) & \text{for} \quad t \geq 1 \end{cases}$$

For $t = 0$ and $0 < t' \leq T$

$$\begin{aligned} \delta_0^{t'}(i) &= I\!P_\lambda(X_0 = i, Y_0 = j_0 | X_{t'} = h) \\ &= I\!P_\lambda(X_0 = i, Y_0 = j_0, X_{t'} = h) / I\!P_\lambda(X_{t'} = h) \\ &= I\!P_\lambda(X_0 = i) I\!P_\lambda(Y_0 = j_0 | X_0 = i) I\!P_\lambda(X_{t'} = h | X_0 = i) / I\!P_\lambda(X_{t'} = h) \\ &= \mu(i) b_i(j_0) \rho(t', i, h) \end{aligned}$$

If $t' = 0$, we get

$$\begin{aligned} \delta_0^0(i) &= I\!P_\lambda(X_0 = i, Y_0 = j_0 | X_0 = h) \\ &= I\!P_\lambda(Y_0 = j_0 | X_0 = i) \frac{I\!P_\lambda(X_0 = i)}{I\!P_\lambda(X_0 = h)} \mathbf{1}(h = i) \\ &= b_i(j_0) \mathbf{1}(h = i). \end{aligned}$$

For all $0 < t < t'$

$$
\begin{aligned}
\delta_t^{t'}(i) &= \max_{i_0^{t-2},k} I\!P_\lambda\left(X_{[i_0^{t-2}]}, X_{t-1}=k, X_t=i, Y_{[j_0^t]} \middle| X_{t'}=h\right) \\
&= \max_k \max_{i_0^{t-2}} I\!P_\lambda\left(X_{[i_0^{t-2}]}, X_{t-1}=k, Y_{[j_0^{t-1}]} \middle| X_{t'}=h\right) \\
&\quad I\!P_\lambda\left(X_t=i, Y_t=j_t \middle| X_{t'}=h, X_{t-1}=k\right) \\
&= \max_k \delta_{t-1}^{t'}(k) b_i(j_t) \frac{I\!P_\lambda\left(X_t=i, X_{t'}=h \middle| X_{t-1}=k\right)}{I\!P_\lambda(X_{t'}=h|X_{t-1}=k)} \\
&= \max_k \left(\delta_{t-1}^{t'}(k) A(k,i)\phi(t',t,i,k,h)\right) b_i(j_t).
\end{aligned}
$$

For $t = t'$ we get

$$
\begin{aligned}
\delta_{t'}^{t'}(i) &= \max_{i_0^{t'-2},k} I\!P_\lambda\left(X_{[i_0^{t'-2}]}, X_{t'-1}=k, X_{t'}=i, Y_{[j_0^{t'}]} \middle| X_{t'}=h\right) \mathbf{1}(h=i) \\
&= \max_k \max_{i_0^{t'-2}} I\!P_\lambda\left(X_{[i_0^{t'-2}]}, X_{t'-1}=k, Y_{[j_0^{t'-1}]} \middle| X_{t'}=h\right) \\
&\quad I\!P_\lambda(Y_{t'}=j_{t'}, X_{t'}=i|X_{t'}=h, X_{t'-1}=k)) \\
&= \max_k \delta_{t'-1}^{t'}(k) \frac{I\!P_\lambda(Y_{t'}=j_{t'}, X_{t'}=i, X_{t'}=h|X_{t'-1}=k))}{I\!P_\lambda(X_{t'}=h|X_{t'-1}=k))} \\
&= \max_k \delta_{t'-1}^{t'}(k) \frac{A(k,i) b_i(j_{t'}) \mathbf{1}(h=i)}{A(k,h)} \\
&= \max_k \left(\delta_{t'-1}^{t'}(k)\right) b_i(j_{t'}) \mathbf{1}(h=i).
\end{aligned}
$$

For $t = t' + 1$

$$
\begin{aligned}
\delta_{t'+1}^{t'}(i) &= \max_{i_0^{t'-2},k} I\!P_\lambda\left(X_{[i_0^{t'-2}]}, X_{t'-1}=k, X_{t'+1}=i, Y_{[j_0^{t'+1}]} \middle| X_{t'}=h\right) \\
&= \max_{i_0^{t'-2},k} I\!P_\lambda\left(X_{[i_0^{t'-2}]}, X_{t'-1}=k, Y_{[j_0^{t'}]} \middle| X_{t'}=h\right) \\
&\quad I\!P_\lambda(X_{t'+1}=i, Y_{t'+1}=j_{t'+1}|X_{t'}=h)) \\
&= \delta_{t'}^{t'}(h) A(h,i) b_i(j_{t'+1}).
\end{aligned}
$$

For $t = t' + 2$, we get

$$\delta_{t'+2}^{t'}(i) = \max_{i_0^{t'-1},k} I\!P_\lambda \left(X_{[i_0^{t'-1}]}, X_{t'+1} = k, X_{t'+2} = i, Y_{[j_0^{t'+2}]} | X_{t'} = h \right)$$
$$= \max_k \max_{i_0^{t'-1}} I\!P_\lambda \left(X_{[i_0^{t'-1}]}, X_{t'+1} = k, Y_{[j_0^{t'+1}]} | X_{t'} = h \right)$$
$$I\!P_\lambda(X_{t'+2} = i, Y_{t'+2} = j_{t'+2} | X_{t'} = h, , X_{t'+1} = k))$$
$$= \max_k \left(\delta_{t'+1}^{t'}(k) A(k,i) \right) b_i(j_{t'+2}).$$

Finally, we get for all $t > t' + 1$,

$$\delta_t^{t'}(i) = \max_k \delta_{t-1}^{t'}(k) A(k,i) b_i(j_t)$$

The proofs for the second case (when two states are known) are similar to the first case (when one state is known).

References

1. Aas, K., Eikvil, L., Huseby, R.B.: Applications of hidden Markov chains in image analysis. Pattern Recognit. **32(4)**, 703–713 (1999)
2. Bordes, L., Vandekerkhove, P.: Statistical inference for partially hidden Markov models. Commun. Stat. Theory Methods. **34**(5), 1081–1104 (2005)
3. Cappé, O., Moulines, E., Rydén, T.: Inference in Hidden Markov Models. Springer (2005)
4. Churchill, G.: Hidden Markov chains and the analysis of genome structure. Comput. Chem. **16**(2), 107–115 (1992)
5. Nasroallah, A., Elkimakh, K.: HMM with emission process resulting from a special Combination of Independent Markovian Emissions. Monte Carlo Methods Appl. **23**(4), 287–306 (2017)
6. Gupta, A., Dhingra, B.: Stock market prediction using hidden Markov models. In: 2012 Students Conference on Engineering and Systems, pp. 1–4. IEEE (2012)
7. Ozkan, H., Akman, A., Kozat, S.S.: A novel and robust parameter training approach for HMMs under noisy and partial access to states. Signal Process. **94**, 490–497 (2014)
8. Rabiner, L.R.: A tutorial on Hidden Markov Models and selected applications in speech recognation. Proc. IEEE. **77**(2), 257–286 (1989)

Solution of First Order Linear Intuitionistic Fuzzy Differential Equations by the Variation of Constants Formula

Razika Ettoussi, Said Melliani, and Lalla Saadia Chadli

Abstract In this paper, we provide solution of first order linear intuitionistic fuzzy differential equations by using variation of constant formula and we present the general form of their solutions. Finally, we present some examples to illustrate this work.

1 Introduction

In 1965, Zadeh [10] first introduced the fuzzy set theory. Later many researchers have applied this theory to the well known results in the classical set theory.

The idea of intuitionistic fuzzy set was first published by Atanassov [1, 2] as a generalization of the notion of fuzzy set. The notions of differential and integral calculus for intuitionistic fuzzy-set-valued are given using Hukuhara difference in intuitionistic Fuzzy theory [4]. The authors of papers [7, 8] are discussed differential and partial differential equations under intuitionistic fuzzy environment respectively. The existence and uniqueness of the solution of intuitionistic fuzzy differential equations by using successive approximations method have been discussed in [4], while in [3] the theorem of the existence and uniqueness of the solution for differential equations with intuitionistic fuzzy data are proved by using the theorem of fixed

R. Ettoussi · S. Melliani (✉) · L. S. Chadli
Department of Mathematics, Laboratory of Applied Mathematics and Scientific Computing, Sultan Moulay Slimane University, PO Box 523, 23000 Beni Mellal, Morocco
e-mail: saidmelliani@gmail.com

R. Ettoussi
e-mail: razika.imi@gmail.com

L. S. Chadli
e-mail: sa.chadli@yahoo.fr

S. Melliani and O. Castillo (eds.), *Recent Advances in Intuitionistic Fuzzy Logic Systems and Mathematics*, Studies in Fuzziness and Soft Computing 395,
https://doi.org/10.1007/978-3-030-53929-0_13

point in the complete metric space, also the explicit formula of the solution are given by using the α-cuts method.

This paper is organized as follows: in Sect. 2 we give preliminaries which we will use throughout this work. In Sect. 3 we construct a procedure for solving first order linear intuitionistic fuzzy differential equations by using variation of constant formula and we present the general form of their solutions. In the last section, we present some examples for illustrate this work.

2 Preliminaries

Let us $I = [b, c] \subset \mathbb{R}$ be a compact interval.

Definition 1 We denote by

$$\mathrm{IF}_1 = \mathrm{IF}(\mathbb{R}) = \left\{ \langle u, v\rangle \; : \; \mathbb{R} \to [0,1]^2 \, , |\forall\, x \in \mathbb{R} | 0 \le u(x) + v(x) \le 1 \right\}$$

An element $\langle u, v\rangle$ of IF_1 is said an intuitionistic fuzzy number if it satisfies the following conditions

(i) $\langle u, v\rangle$ is normal i.e there exists $x_0, x_1 \in \mathbb{R}$ such that $u(x_0) = 1$ and $v(x_1) = 1$.
(ii) u is fuzzy convex and v is fuzzy concave.
(iii) u is upper semi-continuous and v is lower semi-continuous
(iv) $supp\, \langle u, v\rangle = \mathrm{cl}\{x \in \mathbb{R} \; : \, | \, v(x) < 1\}$ is bounded.

so we denote the collection of all intuitionistic fuzzy number by IF_1.

Definition 2 ([5]) An intuitionistic fuzzy number $\langle u, v\rangle$ in parametric form is a pair $\langle u, v\rangle = \left((\underline{\langle u, v\rangle}^+, \overline{\langle u, v\rangle^+}), (\underline{\langle u, v\rangle}^-, \overline{\langle u, v\rangle}^-) \right)$ of functions $\underline{\langle u, v\rangle}^-(\alpha)$, $\overline{\langle u, v\rangle}^-(\alpha)$, $\underline{\langle u, v\rangle}^+(\alpha)$ and $\overline{\langle u, v\rangle}^+(\alpha)$, which satisfies the following requirements:

1. $\underline{\langle u, v\rangle}^+(\alpha)$ is a bounded monotonic increasing continuous function,
2. $\overline{\langle u, v\rangle}^+(\alpha)$ is a bounded monotonic decreasing continuous function,
3. $\underline{\langle u, v\rangle}^-(\alpha)$ is a bounded monotonic increasing continuous function,
4. $\overline{\langle u, v\rangle}^-(\alpha)$ is a bounded monotonic decreasing continuous function,
5. $\underline{\langle u, v\rangle}^-(\alpha) \le \overline{\langle u, v\rangle}^-(\alpha)$ and $\underline{\langle u, v\rangle}^+(\alpha) \le \overline{\langle u, v\rangle}^+(\alpha)$, for all $0 \le \alpha \le 1$.

Example 2.1 A Triangular Intuitionistic Fuzzy Number (TIFN) $\langle u, v\rangle$ is an intuitionistic fuzzy set in $\mathbb{R}$ with the following membership function u and non-membership function v :

$$u(x) = \begin{cases} \dfrac{x - a_1}{a_2 - a_1} & \text{if } a_1 \le x \le a_2 \\ \dfrac{a_3 - x}{a_3 - a_2} & \text{if } a_2 \le x \le a_3, \\ 0 & \text{otherwise} \end{cases}$$

$$v(x) = \begin{cases} \dfrac{a_2 - x}{a_2 - a_1'} & \text{if } a_1' \le x \le a_2 \\ \dfrac{x - a_2}{a_3' - a_2} & \text{if } a_2 \le x \le a_3', \\ 1 & otherwise. \end{cases}$$

where $a_1' \le a_1 \le a_2 \le a_3 \le a_3'$

This TIFN is denoted by $\langle u, v\rangle = \langle a_1, a_2, a_3; a_1', a_2, a_3'\rangle$.

Its parametric form is

$$\underline{\langle u, v\rangle}^+(\alpha) = a_1 + \alpha(a_2 - a_1), \ \overline{\langle , v\rangle}^+(\alpha) = a_3 - \alpha(a_3 - a_2)$$

$$\underline{\langle u, v\rangle}^-(\alpha) = a_1' + \alpha(a_2 - a_1'), \ \overline{\langle u, v\rangle}^-(\alpha) = a_3' - \alpha(a_3' - a_2)$$

For $\alpha \in [0, 1]$ and $\langle u, v\rangle \in \mathrm{IF}_1$, the upper and lower α-cuts of $\langle u, v\rangle$ are defined by

$$[\langle u, v\rangle]^\alpha = \{x \in \mathbb{R} : v(x) \le 1 - \alpha\}$$

and

$$[\langle u, v\rangle]_\alpha = \{x \in \mathbb{R} : u(x) \ge \alpha\}$$

Remark 1 If $\langle u, v\rangle \in \mathrm{IF}_1$, so we can see $[\langle u, v\rangle]_\alpha$ as $[u]^\alpha$ and $[\langle u, v\rangle]^\alpha$ as $[1 - v]^\alpha$ in the fuzzy case.

We define $0_{\langle 1,0\rangle} \in \mathrm{IF}_1$ as

$$0_{\langle 1,0\rangle}(t) = \begin{cases} \langle 1, 0\rangle & t = 0 \\ \langle 0, 1\rangle & t \ne 0 \end{cases}$$

For $\langle u, v\rangle$, $\langle z, w\rangle \in \mathrm{IF}_1$ and $\lambda \in \mathbb{R}$, the addition and scaler-multiplication are defined as follows

$$\Big[\langle u, v\rangle \oplus \langle z, w\rangle\Big]^\alpha = \Big[\langle u, v\rangle\Big]^\alpha + \Big[\langle z, w\rangle\Big]^\alpha, \quad \Big[\lambda \langle z, w\rangle\Big]^\alpha = \lambda\Big[\langle z, w\rangle\Big]^\alpha$$

$$\Big[\langle u, v\rangle \oplus \langle z, w\rangle\Big]_\alpha = \Big[\langle u, v\rangle\Big]_\alpha + \Big[\langle z, w\rangle\Big]_\alpha, \quad \Big[\lambda \langle z, w\rangle\Big]_\alpha = \lambda\Big[\langle z, w\rangle\Big]_\alpha$$

Definition 3 Let $\langle u, v\rangle$ an element of IF_1 and $\alpha \in [0, 1]$, we define the following sets:

$$\Big[\langle u, v\rangle\Big]_l^+(\alpha) = \inf\{x \in \mathbb{R} \mid u(x) \ge \alpha\}, \quad \Big[\langle u, v\rangle\Big]_r^+(\alpha) = \sup\{x \in \mathbb{R} \mid u(x) \ge \alpha\}$$

$$\Big[\langle u, v\rangle\Big]_l^-(\alpha) = \inf\{x \in \mathbb{R} \mid v(x) \le 1 - \alpha\}, \quad \Big[\langle u, v\rangle\Big]_r^-(\alpha) = \sup\{x \in \mathbb{R} \mid v(x) \le 1 - \alpha\}$$

Remark 2

$$\Big[\langle u,v\rangle\Big]_\alpha = \Big[\Big[\langle u,v\rangle\Big]_l^+(\alpha), \Big[\langle u,v\rangle\Big]_r^+(\alpha)\Big]$$

$$\Big[\langle u,v\rangle\Big]^\alpha = \Big[\Big[\langle u,v\rangle\Big]_l^-(\alpha), \Big[\langle u,v\rangle\Big]_r^-(\alpha)\Big]$$

On the space IF_1 we will consider the following metric,

$$\begin{aligned} d_\infty\Big(\langle u,v\rangle, \langle z,w\rangle\Big) &= \frac{1}{4}\sup_{0<\alpha\le 1}\Big|\Big[\langle u,v\rangle\Big]_r^+(\alpha) - \Big[\langle z,w\rangle\Big]_r^+(\alpha)\Big| \\ &+ \frac{1}{4}\sup_{0<\alpha\le 1}\Big|\Big[\langle u,v\rangle\Big]_l^+(\alpha) - \Big[\langle z,w\rangle\Big]_l^+(\alpha)\Big| \\ &+ \frac{1}{4}\sup_{0<\alpha\le 1}\Big|\Big[\langle u,v\rangle\Big]_r^-(\alpha) - \Big[\langle z,w\rangle\Big]_r^-(\alpha)\Big| \\ &+ \frac{1}{4}\sup_{0<\alpha\le 1}\Big|\Big[\langle u,v\rangle\Big]_l^-(\alpha) - \Big[\langle z,w\rangle\Big]_l^-(\alpha)\Big| \end{aligned}$$

Theorem 1 ([6])
The metric space (IF_1, d_∞) is complete.

Proposition 1 *For all $\alpha, \beta \in [0,1]$ and $\langle u,v\rangle \in IF_1$*

(i) $\Big[\langle u,v\rangle\Big]_\alpha \subset \Big[\langle u,v\rangle\Big]^\alpha$

(ii) $\Big[\langle u,v\rangle\Big]_\alpha$ *and* $\Big[\langle u,v\rangle\Big]^\alpha$ *are nonempty compact convex sets in* $\mathbb{R}$

(iii) if $\alpha \le \beta$ *then* $\Big[\langle u,v\rangle\Big]_\beta \subset \Big[\langle u,v\rangle\Big]_\alpha$ *and* $\Big[\langle u,v\rangle\Big]^\beta \subset \Big[\langle u,v\rangle\Big]^\alpha$

(iv) If $\alpha_n \nearrow \alpha$ *then* $\Big[\langle u,v\rangle\Big]_\alpha = \bigcap_n \Big[\langle u,v\rangle\Big]_{\alpha_n}$ *and* $\Big[\langle u,v\rangle\Big]^\alpha = \bigcap_n \Big[\langle u,v\rangle\Big]^{\alpha_n}$

Let M any set and $\alpha \in [0,1]$ we denote by

$$M_\alpha = \{x \in \mathbb{R} : u(x) \ge \alpha\} \quad \text{and} \quad M^\alpha = \{x \in \mathbb{R} : v(x) \le 1-\alpha\}$$

Lemma 1 ([6]) *let* $\Big\{M_\alpha,\ \alpha \in [0,1]\Big\}$ *and* $\Big\{M^\alpha,\ \alpha \in [0,1]\Big\}$ *two families of subsets of* $\mathbb{R}$ *satisfies (i)–(iv) in Proposition 1, if u and v define by*

$$u(x) = \begin{cases} 0 & \textit{if } x \notin M_0 \\ \sup\{\alpha \in [0,1] : x \in M_\alpha\} & \textit{if } x \in M_0 \end{cases}$$

$$v(x) = \begin{cases} 1 & \textit{if } x \notin M^0 \\ 1 - \sup\{\alpha \in [0,1] : x \in M^\alpha\} & \textit{if } x \in M^0 \end{cases}$$

Then $\langle u,v\rangle \in IF_1$

Definition 4 ([3]) Let $F : I \to \mathrm{IF}_1$ be an intuitionistic fuzzy valued mapping and $t_0 \in I$. Then F is called intuitionistic fuzzy continuous in t_0 iff:

$$(\forall \varepsilon > 0)(\exists \delta > 0)\Big(\forall t \in I \text{such as} \mid t - t_0 \mid < \delta\Big) \Rightarrow d_\infty(F(t), F(t_0)) < \varepsilon.$$

Definition 5 ([3]) A mapping $F : I \to \mathrm{IF}_1$ is said to be differentiable at $t_0 \in (b, c)$ if there exist $F'(t_0) \in \mathrm{IF}_1$ such that limits:

$$\lim_{\Delta t \to 0^+} \frac{F(t_0 + \Delta t) \ominus F(t_0)}{\Delta t} \quad and \quad \lim_{\Delta t \to 0^+} \frac{F(t_0) \ominus F(t_0 - \Delta t)}{\Delta t}$$

exist and they are equal to $F'(t_0)$.

If $F : I \to \mathrm{IF}_1$ is differentiable at $t_0 \in I$, then we say that $F'(t_0)$ is the intuitionistic fuzzy derivative of $F(t)$ at the point t_0.

Here the limit is taken in the metric space $(\mathrm{IF}_1, d_\infty)$. At the end points of I we consider only the one-sided derivatives.

3 Solving First Order Linear Intuitionistic Fuzzy Differential Equations

In this section,we consider the following first order linear intuitionistic fuzzy differential equations

$$\begin{cases} y'(t) = a(t)y(t) + b(t) \\ y(0) = \Big(y_l^+(0, \alpha), y_r^+(0, \alpha), y_l^-(0, \alpha), y_r^-(0, \alpha)\Big) \end{cases} \tag{1}$$

where $a : I \to \mathbb{R}$ and $b : I \to \mathrm{IF}_1$
In the rest of the paper the interval $I = [0, c]$ where $c > 0$.

Definition 6 $y : I \to \mathrm{IF}_1$ is a solution of the initial value problem (1), if and only if it is continuous and satisfies the integral equation

$$y(t) = y(0) \oplus \int_0^t f(s, y(s))ds, \quad \text{for all } t \in I.$$

In the next, we will study problem (1) in three cases $a < 0$, $a > 0$ and $a \equiv 0$.

Case 1. $a < 0$: For finding solution we translate problem (1) into the following ODEs system:

$$\begin{cases} y_l'^{+}(t,\alpha) = a(t) y_r^{+}(t,\alpha) + b_l^{+}(t,\alpha), \\ y_r'^{+}(t,\alpha) = a(t) y_l^{+}(t,\alpha) + b_r^{+}(t,\alpha), \\ y_l'^{-}(t,\alpha) = a(t) y_r^{-}(t,\alpha) + b_l^{-}(t,\alpha), \\ y_r'^{-}(t,\alpha) = a(t) y_l^{-}(t,\alpha) + b_r^{-}(t,\alpha). \end{cases} \tag{2}$$

the linear system (2) can be represented in matrix form by the following equation

$$Y'(t) = A_1(t)Y(t) + B(t). \tag{3}$$

where

$$Y(t) = \begin{pmatrix} y_l^{+}(t,\alpha) \\ y_r^{+}(t,\alpha) \\ y_l^{-}(t,\alpha) \\ y_r^{-}(t,\alpha) \end{pmatrix}, \ A_1(t) = \begin{pmatrix} 0 & a(t) & 0 & 0 \\ a(t) & 0 & 0 & 0 \\ 0 & 0 & 0 & a(t) \\ 0 & 0 & a(t) & 0 \end{pmatrix}$$

and

$$B(t) = \begin{pmatrix} b_l^{+}(t,\alpha) \\ b_r^{+}(t,\alpha) \\ b_l^{-}(t,\alpha) \\ b_r^{-}(t,\alpha) \end{pmatrix}$$

By the variation of constants formula for ordinary differential equations we have

$$Y(t) = \exp\left(\int_0^t A_1(z)dz\right)\left(Y_0 + \exp\left(-\int_0^s A_1(z)dz\right)B(s)ds\right) \tag{4}$$

where,

$$\exp\left(\int_0^t A_1(z)dz\right) = \begin{pmatrix} cosh(\int_0^t a(z)dz) & sinh(\int_0^t a(z)dz) & 0 & 0 \\ sinh(\int_0^t a(z)dz) & cosh(\int_0^t a(z)dz) & 0 & 0 \\ 0 & 0 & cosh(\int_0^t a(z)dz) & sinh(\int_0^t a(z)dz) \\ 0 & 0 & sinh(\int_0^t a(z)dz) & cosh(\int_0^t a(z)dz) \end{pmatrix} \tag{5}$$

and

$$\exp\left(-\int_0^t A_1(z)dz\right) = \begin{pmatrix} cosh(\int_0^s a(z)dz) & -sinh(\int_0^s a(z)dz) & 0 & 0 \\ -sinh(\int_0^s a(z)dz) & cosh(\int_0^s a(z)dz) & 0 & 0 \\ 0 & 0 & cosh(\int_0^s a(z)dz) & -sinh(\int_0^s a(z)dz) \\ 0 & 0 & -sinh(\int_0^s a(z)dz) & cosh(\int_0^s a(z)dz) \end{pmatrix} \tag{6}$$

After replacing the values (5) and (6) in the Eq. (4), then we have

$$\begin{pmatrix} y_l^+(t,\alpha) \\ y_r^+(t,\alpha) \\ y_l^-(t,\alpha) \\ y_r^-(t,\alpha) \end{pmatrix} = \begin{pmatrix} cosh\left(\int_0^t a(z)dz\right) & sinh\left(\int_0^t a(z)dz\right) & 0 & 0 \\ sinh\left(\int_0^t a(z)dz\right) & cosh\left(\int_0^t a(z)dz\right) & 0 & 0 \\ 0 & 0 & cosh\left(\int_0^t a(z)dz\right) & sinh\left(\int_0^t a(z)dz\right) \\ 0 & 0 & sinh\left(\int_0^t a(z)dz\right) & cosh\left(\int_0^t a(z)dz\right) \end{pmatrix}$$

$$\times \begin{pmatrix} y_l^+(0,\alpha) + \int_0^t \left(b_l^+(s,\alpha)cosh\left(\int_0^s a(z)dz\right) - b_r^+(s,\alpha)sinh\left(\int_0^s a(z)dz\right)\right)ds \\ y_r^+(0,\alpha) + \int_0^t \left(-b_l^+(s,\alpha)sinh\left(\int_0^s a(z)dz\right) + b_r^+(s,\alpha)cosh\left(\int_0^s a(z)dz\right)\right)ds \\ y_l^-(0,\alpha) + \int_0^t \left(b_l^-(s,\alpha)cosh\left(\int_0^s a(z)dz\right) - b_r^-(s,\alpha)sinh\left(\int_0^s a(z)dz\right)\right)ds \\ y_r^-(0,\alpha) + \int_0^t \left(-b_l^-(s,\alpha)sinh\left(\int_0^s a(z)dz\right) + b_r^-(s,\alpha)cosh\left(\int_0^s a(z)dz\right)\right)ds \end{pmatrix}$$

Then the solution of the ordinary differential equations system is

$$\begin{cases} y_l^+(t,\alpha) = cosh\left(\int_0^t a(z)dz\right)\left\{y_l^+(0,\alpha) + \int_0^t \left(b_l^+(s,\alpha)cosh\left(\int_0^s a(z)dz\right) - b_r^+(s,\alpha)sinh\left(\int_0^s a(z)dz\right)\right)ds\right\} \\ +sinh\left(\int_0^t a(z)dz\right)\left\{y_r^+(0,\alpha) + \int_0^t \left(-b_l^+(s,\alpha)sinh\left(\int_0^s a(z)dz\right) + b_r^+(s,\alpha)cosh\left(\int_0^s a(z)dz\right)\right)ds\right\}, \\ y_r^+(t,\alpha) = sinh\left(\int_0^t a(z)dz\right)\left\{y_l^+(0,\alpha) + \int_0^t \left(b_l^+(s,\alpha)cosh\left(\int_0^s a(z)dz\right) - b_r^+(s,\alpha)sinh\left(\int_0^s a(z)dz\right)\right)ds\right\} \\ +cosh\left(\int_0^t a(z)dz\right)\left\{y_r^+(0,\alpha) + \int_0^t \left(-b_l^+(s,\alpha)sinh\left(\int_0^s a(z)dz\right) + b_r^+(s,\alpha)cosh\left(\int_0^s a(z)dz\right)\right)ds\right\}, \\ y_l^-(t,\alpha) = cosh\left(\int_0^t a(z)dz\right)\left\{y_l^-(0,\alpha) + \int_0^t \left(b_l^-(s,\alpha)cosh\left(\int_0^s a(z)dz\right) - b_r^-(s,\alpha)sinh\left(\int_0^s a(z)dz\right)\right)ds\right\} \\ +sinh\left(\int_0^t a(z)dz\right)\left\{y_r^-(0,\alpha) + \int_0^t \left(-b_l^-(s,\alpha)sinh\left(\int_0^s a(z)dz\right) + b_r^-(s,\alpha)cosh\left(\int_0^s a(z)dz\right)\right)ds\right\}, \\ y_r^-(t,\alpha) = sinh\left(\int_0^t a(z)dz\right)\left\{y_l^-(0,\alpha) + \int_0^t \left(b_l^-(s,\alpha)cosh\left(\int_0^s a(z)dz\right) - b_r^-(s,\alpha)sinh\left(\int_0^s a(z)dz\right)\right)ds\right\} \\ +cosh\left(\int_0^t a(z)dz\right)\left\{y_r^-(0,\alpha) + \int_0^t \left(-b_l^-(s,\alpha)sinh\left(\int_0^s a(z)dz\right) + b_r^-(s,\alpha)cosh\left(\int_0^s a(z)dz\right)\right)ds\right\}. \end{cases} \tag{7}$$

Then for $a < 0$, the solution of the problem (1) is

$$y(t) = cosh\left(\int_0^t a(z)dz\right)\left\{y(0) \oplus \int_0^t \left(b(s)cosh\left(\int_0^s a(z)dz\right) \ominus b(s)sinh\left(\int_0^s a(z)dz\right)\right)ds\right\}$$
$$\oplus sinh\left(\int_0^t a(z)dz\right)\left\{y(0) \oplus \int_0^t \left(b(s)cosh\left(\int_0^s a(z)dz\right) \ominus b(s)sinh\left(\int_0^s a(z)dz\right)\right)ds\right\}$$

provided the following H-differences

$$b(s)cosh\left(\int_0^s a(z)dz\right) \ominus b(s)sinh\left(\int_0^s a(z)dz\right)$$

exist.

Case 2. $a > 0$: For finding solution we translate problem (1) into the following ODEs system:

$$\begin{cases} y_l'^+(t,\alpha) = a(t)y_l^+(t,\alpha) + b_l^+(t,\alpha), \\ y_r'^+(t,\alpha) = a(t)y_r^+(t,\alpha) + b_r^+(t,\alpha), \\ y_l'^-(t,\alpha) = a(t)y_l^-(t,\alpha) + b_l^-(t,\alpha), \\ y_r'^-(t,\alpha) = a(t)y_r^-(t,\alpha) + b_r^-(t,\alpha). \end{cases} \tag{8}$$

the linear system (8) can be represented in matrix form by the following equation

$$Y'(t) = A_2(t)Y(t) + B(t). \tag{9}$$

where

$$Y(t) = \begin{pmatrix} y_l^+(t,\alpha) \\ y_r^+(t,\alpha) \\ y_l^-(t,\alpha) \\ y_r^-(t,\alpha) \end{pmatrix}, A_2(t) = \begin{pmatrix} a(t) & 0 & 0 & 0 \\ 0 & a(t) & 0 & 0 \\ 0 & 0 & a(t) & 0 \\ 0 & 0 & 0 & a(t) \end{pmatrix}$$

and

$$B(t) = \begin{pmatrix} b_l^+(t,\alpha) \\ b_r^+(t,\alpha) \\ b_l^-(t,\alpha) \\ b_r^-(t,\alpha) \end{pmatrix}$$

By the variation of constants formula for ordinary differential equations we have

$$Y(t) = \exp\left(\int_0^t A_2(z)dz\right)\left(Y_0 + \exp\left(-\int_0^s A_2(z)dz\right)B(s)ds\right) \tag{10}$$

where,

$$\exp\left(\int_0^t A_2(z)dz\right) = \left(cosh\left(\int_0^t a(z)dz\right) + sinh\left(\int_0^t a(z)dz\right)\right)\begin{pmatrix} 1 & 0 & 0 & 0 \\ 0 & 1 & 0 & 0 \\ 0 & 0 & 1 & 0 \\ 0 & 0 & 0 & 1 \end{pmatrix} \tag{11}$$

and

$$\exp\left(-\int_0^t A_2(z)dz\right) = \left(cosh\left(\int_0^t a(z)dz\right) - sinh\left(\int_0^t a(z)dz\right)\right)\begin{pmatrix} 1 & 0 & 0 & 0 \\ 0 & 1 & 0 & 0 \\ 0 & 0 & 1 & 0 \\ 0 & 0 & 0 & 1 \end{pmatrix} \tag{12}$$

After replacing the values (11) and (12) in the Eq. (10), then we have

$$\begin{pmatrix} y_l^+(t,\alpha) \\ y_r^+(t,\alpha) \\ y_l^-(t,\alpha) \\ y_r^-(t,\alpha) \end{pmatrix} = \left(cosh\left(\int_0^t a(z)dz\right) + sinh\left(\int_0^t a(z)dz\right)\right) \begin{pmatrix} 1\,0\,0\,0 \\ 0\,1\,0\,0 \\ 0\,0\,1\,0 \\ 0\,0\,0\,1 \end{pmatrix}$$

$$\times \begin{pmatrix} y_l^+(0,\alpha) + \int_0^t \left(cosh\left(\int_0^s a(z)dz\right) - sinh\left(\int_0^s a(z)dz\right)\right) b_l^+(s,\alpha)ds \\ y_r^+(0,\alpha) + \int_0^t \left(cosh\left(\int_0^s a(z)dz\right) - sinh\left(\int_0^s a(z)dz\right)\right) b_r^+(s,\alpha)ds \\ y_l^-(0,\alpha) + \int_0^t \left(cosh\left(\int_0^s a(z)dz\right) - sinh\left(\int_0^s a(z)dz\right)\right) b_l^-(s,\alpha)ds \\ y_r^-(0,\alpha) + \int_0^t \left(cosh\left(\int_0^s a(z)dz\right) - sinh\left(\int_0^s a(z)dz\right)\right) b_r^-(s,\alpha)ds \end{pmatrix}$$

Then the solution of the ordinary differential equations system is

$$\begin{cases} y_l^+(t,\alpha) = \left(cosh\left(\int_0^t a(z)dz\right) + sinh\left(\int_0^t a(z)dz\right)\right)\Big(y_l^+(0,\alpha) + \\ \qquad \int_0^t \left(cosh\left(\int_0^s a(z)dz\right) - sinh\left(\int_0^s a(z)dz\right)\right) b_l^+(s,\alpha)ds\Big), \\ y_r^+(t,\alpha) = \left(cosh\left(\int_0^t a(z)dz\right) + sinh\left(\int_0^t a(z)dz\right)\right)\Big(y_r^+(0,\alpha) + \\ \qquad \int_0^t \left(cosh\left(\int_0^s a(z)dz\right) - sinh\left(\int_0^s a(z)dz\right)\right) b_r^+(s,\alpha)ds\Big), \\ y_l^-(t,\alpha) = \left(cosh\left(\int_0^t a(z)dz\right) + sinh\left(\int_0^t a(z)dz\right)\right)\Big(y_l^-(0,\alpha) + \\ \qquad \int_0^t \left(cosh\left(\int_0^s a(z)dz\right) - sinh\left(\int_0^s a(z)dz\right)\right) b_l^-(s,\alpha)ds\Big), \\ y_r^-(t,\alpha) = \left(cosh\left(\int_0^t a(z)dz\right) + sinh\left(\int_0^t a(z)dz\right)\right)\Big(y_r^-(0,\alpha) + \\ \qquad \int_0^t \left(cosh\left(\int_0^s a(z)dz\right) - sinh\left(\int_0^s a(z)dz\right)\right) b_r^-(s,\alpha)ds\Big). \end{cases} \tag{13}$$

Then for $a > 0$, the solution of the problem (1) is written as

$$y(t) = \left(cosh\left(\int_0^t a(z)dz\right) + sinh\left(\int_0^t a(z)dz\right)\right)\Big(y(0) \oplus \\ \int_0^t \left(cosh\left(\int_0^s a(z)dz\right) - sinh\left(\int_0^s a(z)dz\right)\right) b(s)ds\Big)$$

Case 3. $a \equiv 0$: For this case, the problem (1) reduces to

$$\begin{cases} y'(t) = b(t) \\ y(0) = \left(y_l^+(0,\alpha), y_r^+(0,\alpha), y_l^-(0,\alpha), y_r^-(0,\alpha)\right) \end{cases} \tag{14}$$

The solution of this problem is

$$y(t) = y(0) \oplus \int_0^t b(s)ds$$

Theorem 2 *1. If $a < 0$, then the solution of the problem* (1) *is written as*

$$y(t) = cosh\left(\int_0^t a(z)dz\right)\left\{y(0) \oplus \int_0^t \left(b(s)cosh\left(\int_0^s a(z)dz\right) \ominus b(s)sinh\left(\int_0^s a(z)dz\right)\right)ds\right\}$$
$$\oplus sinh\left(\int_0^t a(z)dz\right)\left\{y(0) \oplus \int_0^t \left(b(s)cosh\left(\int_0^s a(z)dz\right) \ominus b(s)sinh\left(\int_0^s a(z)dz\right)\right)ds\right\}$$

provided the following H-differences

$$b(s)cosh\left(\int_0^s a(z)dz\right) \ominus b(s)sinh\left(\int_0^s a(z)dz\right)$$

exist.

2. *If $a > 0$, then the solution of the problem* (1) *is written as*

$$y(t) = \left(cosh\left(\int_0^t a(z)dz\right) + sinh\left(\int_0^t a(z)dz\right)\right)\left(y(0) \oplus \right.$$
$$\left.\int_0^t \left(cosh\left(\int_0^s a(z)dz\right) - sinh\left(\int_0^s a(z)dz\right)\right)b(s)ds\right)$$

3. *If $a \equiv 0$, then the solution of the problem* (1) *is written as*

$$y(t) = y(0) \oplus \int_0^t b(s)ds$$

Proof If $a < 0$ and for $\alpha \in [0,1]$ and $s \in I$
we have

$$diam\left(\left[b(s)cosh\left(\int_0^s a(z)dz\right)\right]^\alpha\right) = diam\left(\left[b(s)\right]^\alpha\right)cosh\left(\int_0^s a(z)dz\right)$$

and

$$diam\left(\left[b(s)sinh\left(\int_0^s a(z)dz\right)\right]^\alpha\right) = diam\left(\left[b(s)\right]^\alpha\right)sinh\left(\int_0^s a(z)dz\right)$$

Since $a < 0$ then

$$cosh\left(\int_0^s a(z)dz\right) \geq sinh\left(\int_0^s a(z)dz\right)$$

Therefore,

$$diam\left(\left[b(s)\right]^\alpha\right)cosh\left(\int_0^s a(z)dz\right) \geq diam\left(\left[b(s)\right]^\alpha\right)sinh\left(\int_0^s a(z)dz\right)$$

So

$$diam\left(\left[b(s)cosh\left(\int_0^s a(z)dz\right)\right]^\alpha\right) \geq diam\left(\left[b(s)sinh\left(\int_0^s a(z)dz\right)\right]^\alpha\right) \tag{15}$$

The same idea we prove that

$$diam\left(\left[b(s)cosh\left(\int_0^s a(z)dz\right)\right]_\alpha\right) \geq diam\left(\left[b(s)sinh\left(\int_0^s a(z)dz\right)\right]_\alpha\right) \tag{16}$$

In addition, from (15) and (16) and

$$diam\left(\left[b(s)cosh\left(\int_0^s a(z)dz\right)\right]^\alpha\right)$$

and

$$diam\left(\left[b(s)cosh\left(\int_0^s a(z)dz\right)\right]_\alpha\right)$$

are increasing with respect to α.

Thus the following H-difference

$$b(s)cosh\left(\int_0^s a(z)dz\right) \ominus b(s)sinh\left(\int_0^s a(z)dz\right)$$

exist.

For the cases $a > 0$ and $a \equiv 0$ are discussed in the previous paragraph.

4 Examples

4.1 Example

Case $a > 0$: we consider the linear first order intuitionistic fuzzy differential equation

$$\begin{cases} y'(t) = 2ty(t) + ty(0) \\ y(0) = \Big(1+\alpha, 4-\alpha, 1+2\alpha, 5-2\alpha\Big) \end{cases} \tag{17}$$

The parametric form of the solution is given by the following form:

$$\begin{cases} y_l^+(t,\alpha) = \Big(cosh(t^2) + sinh(t^2)\Big)\Big(1+\alpha + \int_0^t \Big(cosh(s^2) - sinh(s^2)\Big)s(1+\alpha)ds\Big), \\ y_r^+(t,\alpha) = \Big(cosh(t^2) + sinh(t^2)\Big)\Big(4-\alpha + \int_0^t \Big(cosh(s^2) - sinh(s^2)\Big)s(4-\alpha)ds\Big), \\ y_l^-(t,\alpha) = \Big(cosh(t^2) + sinh(t^2)\Big)\Big(1+2\alpha + \int_0^t \Big(cosh(s^2) - sinh(s^2)\Big)s(1+2\alpha)ds\Big), \\ y_r^-(t,\alpha) = \Big(cosh(t^2) + sinh(t^2)\Big)\Big(5-2\alpha + \int_0^t \Big(cosh(s^2) - sinh(s^2)\Big)s(5-2\alpha)ds\Big). \end{cases} \tag{18}$$

Therefore,

$$\begin{cases} y_l^+(t,\alpha) = (1+\alpha)\exp(t^2)\Big(1 + \int_0^t \exp(-s^2)sds\Big), \\ y_r^+(t,\alpha) = (4-\alpha)\exp(t^2)\Big(1 + \int_0^t \exp(-s^2)sds\Big), \\ y_l^-(t,\alpha) = (1+2\alpha)\exp(t^2)\Big(1 + \int_0^t \exp(-s^2)sds\Big), \\ y_r^-(t,\alpha) = (5-2\alpha)\exp(t^2)\Big(1 + \int_0^t \exp(-s^2)sds\Big). \end{cases} \tag{19}$$

Then,

$$\begin{cases} y_l^+(t,\alpha) = \frac{1}{2}(1+\alpha)\Big(3\exp(t^2)-1\Big), \\ y_r^+(t,\alpha) = \frac{1}{2}(4-\alpha)\Big(3\exp(t^2)-1\Big), \\ y_l^-(t,\alpha) = \frac{1}{2}(1+2\alpha)\Big(3\exp(t^2)-1\Big), \\ y_r^-(t,\alpha) = \frac{1}{2}(5-2\alpha)\Big(3\exp(t^2)-1\Big). \end{cases} \tag{20}$$

Thus, the solution of the linear first order intuitionistic fuzzy differential equation (21) is

$$y(t) = \frac{1}{2}y(0)\Big(3\exp(t^2)-1\Big).$$

4.2 Example

Case $a < 0$: we consider the linear first order intuitionistic fuzzy differential equation

$$\begin{cases} y'(t) = -2ty(t) + ty(0) & t > 0 \\ y(0) = \Big(1+\alpha, 4-\alpha, 1+2\alpha, 5-2\alpha\Big) \end{cases} \tag{21}$$

The parametric form of the solution is given by the following form:

$$\begin{cases} y_l^+(t,\alpha) = cosh(-t^2)\Big\{(1+\alpha) + \int_0^t \Big(s(1+\alpha)cosh(-s^2) - s(4-\alpha)sinh(-s^2)\Big)ds\Big\} \\ +sinh(-t^2)\Big\{(4-\alpha) + \int_0^t \Big(-s(1+\alpha)sinh(-s^2) + s(4-\alpha)cosh(-s^2)\Big)ds\Big\}, \\ y_r^+(t,\alpha) = sinh(-t^2)\Big\{(1+\alpha) + \int_0^t \Big(s(1+\alpha)cosh(-s^2) - s(4-\alpha)sinh(-t^2)\Big)ds\Big\} \\ +cosh(-t^2)\Big\{(4-\alpha) + \int_0^t \Big(-s(1+\alpha)sinh(-s^2) + s(4-\alpha)cosh(-s^2)\Big)ds\Big\}, \\ y_l^-(t,\alpha) = cosh(-t^2)\Big\{(1+2\alpha) + \int_0^t \Big(s(1+2\alpha)cosh(-s^2) - s(5-2\alpha)sinh(-s^2)\Big)ds\Big\} \\ +sinh(-t^2)\Big\{(5-2\alpha) + \int_0^t \Big(-s(1+2\alpha)sinh(-s^2) + s(5-2\alpha)cosh(-s^2)\Big)ds\Big\}, \\ y_r^-(t,\alpha) = sinh(-t^2)\Big\{(1+2\alpha) + \int_0^t \Big(s(1+2\alpha)cosh(-s^2) - s(5-2\alpha)sinh(-s^2)\Big)ds\Big\} \\ +cosh(-t^2)\Big\{(5-2\alpha) + \int_0^t \Big(-s(1+2\alpha)sinh(-s^2) + s(5-2\alpha)cosh(-s^2)\Big)ds\Big\}. \end{cases} \tag{22}$$

Therefore

$$\begin{cases} y_l^+(t,\alpha) = cosh\Big(-t^2\Big)\Big\{\frac{1}{4}\Big((3-2\alpha)\exp(-t^2)+5\exp(t^2)+6\alpha-4\Big)\Big\} \\ +sinh\Big(-t^2\Big)\Big\{\frac{1}{4}\Big((2\alpha-3)\exp(-t^2)+5\exp(t^2)-6\alpha+14\Big)\Big\}, \\ \\ y_r^+(t,\alpha) = sinh\Big(-t^2\Big)\Big\{\frac{1}{4}\Big((3-2\alpha)\exp(-t^2)+5\exp(t^2)+6\alpha-4\Big)\Big\} \\ +cosh\Big(-t^2\Big)\Big\{\frac{1}{4}\Big((2\alpha-3)\exp(-t^2)+5\exp(t^2)-6\alpha+14\Big)\Big\}, \\ \\ y_l^-(t,\alpha) = cosh\Big(-t^2\Big)\Big\{\frac{1}{2}\Big(2(1-\alpha)\exp(-t^2)+3\exp(t^2)+6\alpha-3\Big)\Big\} \\ +sinh\Big(-t^2\Big)\Big\{\frac{1}{2}\Big(2(\alpha-1)\exp(-t^2)+3\exp(t^2)-6\alpha+9\Big)\Big\}, \\ \\ y_r^-(t,\alpha) = sinh\Big(-t^2\Big)\Big\{\frac{1}{2}\Big(2(1-\alpha)\exp(-t^2)+3\exp(t^2)+6\alpha-3\Big)\Big\} \\ +cosh\Big(-t^2\Big)\Big\{\frac{1}{2}\Big(2(\alpha-1)\exp(-t^2)+3\exp(t^2)-6\alpha+9\Big)\Big\}. \end{cases} \tag{23}$$

Case 3. $a \equiv 0$: For this case, we consider the linear first order intuitionistic fuzzy differential equation

$$\begin{cases} y'(t) = ty(0) \\ y(0) = \Big(1+\alpha, 4-\alpha, 1+2\alpha, 5-2\alpha\Big) \end{cases} \tag{24}$$

The solution of this problem is

$$y(t) = y(0) \oplus \int_0^t sy(0)ds$$

Then,

$$\begin{cases} y_l^+(t,\alpha) = (1+\alpha) + \int_0^t s(1+\alpha)ds, \\ \\ y_r^+(t,\alpha) = (4-\alpha) + \int_0^t s(4-\alpha)ds, \\ \\ y_l^-(t,\alpha) = (1+2\alpha) + \int_0^t s(1+2\alpha)ds, \\ \\ y_r^-(t,\alpha) = (5-2\alpha) + \int_0^t s(5-2\alpha)ds. \end{cases}$$

Thus, the solution of the linear first order intuitionistic fuzzy differential equation (24) is

$$y(t) = \left(\frac{2 + t^2}{2}\right) y(0)$$

References

1. Atanassov, K.: Intuitionistic fuzzy sets. Fuzzy Sets Syst. **20**, 87–96 (1986)
2. Atanassov, K.: Intuitionistic Fuzzy Sets: Theory and Applications. Physica-Verlag (1999)
3. Ettoussi, R., Melliani, S., Chadli, L.S.: Differential equation with intuitionistic fuzzy parameters. Notes Intuit. Fuzzy Sets **23**(4), 46–61 (2017)
4. Ettoussi, R., Melliani, S., Elomari, M., Chadli, L.S.: Solution of intuitionistic fuzzy differential equations by successive approximations method. Notes Intuit. Fuzzy Sets **21**(2), 51–62 (2015)
5. Keyanpour, M., Akbarian, T.: Solving intuitionistic fuzzy nonlinear equations. J. Fuzzy Set Valued Anal. **2014**, 1–6 (2014)
6. Melliani, S., Elomari, M., Chadli, L.S., Ettoussi, R.: Intuitionistic fuzzy metric spaces. Notes Intuit. Fuzzy Sets **21**(1), 43–53 (2015)
7. Melliani, S., Chadli, L.S.: Introduction to intuitionistic fuzzy partial differential equations. Notes Intuit. Fuzzy Sets **7**(3), 39–42 (2001)
8. Melliani, S., Chadli, L.S.: Introduction to intuitionistic fuzzy differential equations. Notes Intuit. Fuzzy Sets **6**(2), 31–41 (2000)
9. Royden, H.L.: Real Analysis, 2nd edn. Macmillan, New York (1968)
10. Zadeh, L.A.: Fuzzy set. Inform. Control **8**(3), 338–353 (1965)

A Secure Variant of the Fiat and Shamir Authentication Protocol Using Gaussian Integers

Leila Zahhafi and Omar Khadir

Abstract In this paper we propose a cryptographic topic. It's a new identification protocol. We use the Gaussian integers to ameliorate the Fiat and Shamir scheme. The method is based on the hardness to factor large composite numbers. We analyze security and complexity of the proposed protocol.

1 Introduction

Cryptography is a strong tool to secure computer systems. Indeed, this science verifies several properties as confidentiality, integrity and authenticity. When a user tries to connect to a system, he will have to prove his identity. In this case, authenticity allows to ensure that there is no third entity trying to identify himself as the user. This secures access to computer systems.

There are several methods for identification [1, p. 56]. Among these methods we quote the zero knowledge proof initiated by Goldwasser, Micali and Rackoff [5] in 1985. It means that the prover can prove to the verifier that he knows a secret x without sharing any information about x. Then the verifier can check the received identification using public keys of the prover.

In the public key cryptography, there are several identification protocols. We distinguish methods based on the discrete logarithm problem [2, p. 53] and those based on factorization of large composite numbers [9, 11].

In 1991, Schnorr [12] created a new identification system. Okamoto [8] proposed on 1993 an other identification scheme. The security of their methods is related to the hardness of solving a discrete logarithm problem. In 1987, Fiat and Shamir [3]

L. Zahhafi (✉) · O. Khadir
Laboratory of Mathematics, Cryptography, Mechanics and Numerical Analysis, University Hassan II of Casablanca, fstm, BP 146, Mohammedia, Morocco
e-mail: leila.zahhafi@gmail.com

O. Khadir
e-mail: khadir@hotmail.com

S. Melliani and O. Castillo (eds.), *Recent Advances in Intuitionistic Fuzzy Logic Systems and Mathematics*, Studies in Fuzziness and Soft Computing 395,
https://doi.org/10.1007/978-3-030-53929-0_14

published an identification protocol inspired by the cryptosystem of Rabin [10]. Their method relies on the factorization of large composite integers.

In this work, we present a new identification protocol inspired by the work of Fiat and Shamir. Indeed, We have improved this sheme to make it stronger by using Gaussian integers ([4, p. 27] and [7, p. 427])

The paper is organized as follows: We present in Sect. 2 the Fiat and Shamir identification method. In Sect. 3, we recall Gaussian integers. We describe steps of our identification protocol in Sect. 4. And we end by a conclusion in Sect. 5.

We denote by $\mathbf{Z}/n\mathbf{Z}$ the finite ring of modular integers for every natural number n. We write $a \equiv b\,[n]$ if n divides the difference $a - b$ where: a, b and n are three integers.

2 Fiat and Shamir Identification Scheme [3]

A trusted center chooses an integer n, product of two large and distinct primes (p, q). The parameter n is known, but p and q are destroyed for security reasons.

In the Fiat and Shamir scheme, Alice proves to Bob that she knows the square root $1/M$ modulo n of a given number $C \in Z/nZ$, in other words: $(1/M)^2 \equiv C[n]$. Where $M \in Z/nZ$.

The protocol works as follow:

1. Alice chooses a random number $r \in \{1, 2, .., n\}$ and computes $X \equiv r^2[n]$. She sends the result X to the verifier, Bob.
2. Bob chooses a random number $\varepsilon \in \{0, 1\}$ and sends it to Alice.
3. Alice replies by sending $Y \equiv r.M^{\varepsilon}[n]$.

Proposition 1 *We have* $Y^2C^{\varepsilon} \equiv X[n]$.

As application of proposition 1: Bob accepts Alices identification if and only if $Y^2C^{\varepsilon} \equiv X[n]$.

3 The Gaussian Integers

Definition 1 The Gaussian integers ([4, p. 27] and [7, p. 427]) are the set $\mathbf{Z}[i] = \{a + bi \mid a, b \in \mathbf{Z}\}$, where $i^2 = -1$.

In other words, a Gaussian integer is a complex number such that its real and imaginary parts are both integers. Since the Gaussian integers are closed under addition and multiplication, they form a commutative ring, which is a sub-ring of the field of complex numbers.

The norm of a Gaussian integer is its product with its conjugate.
$N(a + bi) = (a + bi)(a - bi) = a^2 + b^2$.

Definition 2 We define a modular reduction of a Gaussian integer $(a + bi)$ as the fallowing expression:
$(a, b)\ mod\ n = (a\ mod\ n, b\ mod\ n)$

Definition 3 Let (a, b) and (c, d) be two Gaussian integers. Their product is as follow:
$(a, b)(c, d)\ mod\ n = ((ac - bd)\ mod\ n, (bc + ad)\ mod\ n)$.

Definition 4 Let (a, b), (c, d) be two Gaussian integers and n an integer. (a, b) is the square root of (c, d) if:
$(c, d) = (a, b)^2\ mod\ n$.

Definition 5 Let (a, b), (c, d) be two Gaussian integers and n an integer. If the square root of (c, d) exists, then (c, d) is called a Gaussian quadratic residue modulo n (GQR). Otherwise (c, d) is called a Gaussian quadratic nonresidue modulo n (GQNR).

4 Our Contribution

In this section, we describe the different steps of our proposed identification method.

4.1 *Description of the Protocol*

We first have an authority trusted by everyone. All what she says is true. She distributes to all interested parties a secret based on their identity.

The authority chooses an integer n, product of two large and distinct primes p and q. She publishes n and destroys p and q for security reasons.

Alice selects two Gaussian integers (a_1, b_1) and (a_2, b_2) then she computes $(c_1, d_1) = (a_1, b_1)^2 (mod\ n)$ and $(c_2, d_2) = (a_2, b_2)^2 (mod\ n)$. Alice publishes (c_1, d_1) and (c_2, d_2) and keeps (a_1, b_1) and (a_2, b_2) secret.

The protocol works as follows:

1. Alice selects two Gaussian integers (s_1, v_1) and (s_2, v_2) at random, then she computes $(x_1, y_1) = (s_1, v_1)^2 (mod\ n)$ and $(x_2, y_2) = (s_2, v_2)^2 (mod\ n)$. She sends (x_1, y_1) and (x_2, y_2) to Bob.
2. Bob chooses two numbers $(\varepsilon_1, \varepsilon_2)$ at random and sends them to Alice.
3. Alice computes $(z_1, z_2) \equiv (s_1, v_1)(s_2, v_2)(a_1, b_1)^{\varepsilon_1} (a_2, b_2)^{\varepsilon_2} mod\ n$. She sends it to Bob.

Verification of the Alice's identification is done as follows:

1. Bob receives (z_1, z_2) and checks that $(z_1, z_2)^2 = (x_1, y_1)(x_2, y_2)(c_1, d_1)^{\varepsilon 1} (c_2, d_2)^{\varepsilon 2}\ mod\ n$, then he asks Alice to send him (s_1, v_1) or (s_2, v_2).

2. Bob checks that $(x_1, y_1) = (s_1, v_1)^2 (mod\ n)$ or $(x_2, y_2) = (s_2, v_2)^2 (mod\ n)$.

Remark 1 No one can propose a (z_1, z_2) that checks the previous equation except Alice, because she uses her secret parameters.

Example 1 Let $n = 223.127 = 28321$, Alice's secret keys are $(a_1, b_1) = (24, 42)$ and $(a_2, b_2) = (76, 12)$. She calculates its public keys: $(c_1, d_1) = (a_1, b_1)^2 (mod\ n =$ $(27133, 2016)$ and $(c_2, d_2) = (a_2, b_2)^2 (mod\ n) = (5632, 1824)$ Alice wants to identify herself to Bob. She follows these steps:

- Alice chooses two numbers $(s_1, v_1) = (24, 32)$ and $(s_2, v_2) = (174, 72)$, then she computes $(x_1, y_1) = (s_1, v_1)^2 (mod\ n) = (27873, 1536)$ and $(x_2, y_2) = (s_2, v_2)^2$ $(mod\ n) = (25092, 25056)$. She sends (x_1, y_1) and (x_2, y_2) to Bob.
- Bob submits to Alice a challenge by generating a couple $(\varepsilon_1 = 1, \varepsilon_2 = 1$ which he sends to Alice.
- Alice computes $(z_1, z_2) \equiv (s_1, v_1)(s_2, v_2)(a_1, b_1)^{\varepsilon 1}(a_2, b_2)^{\varepsilon 2} mod\ n = (20970,$ $2310)$.

Verification of Alice's identification is done as follow:

1. Bob receives $(z_1, z_2) = (20970, 2310)$ and checks that: $(z_1, z_2)^2 = (x_1, y_1)(x_2, y_2)(c_1, d_1)^{\varepsilon 1}(c_2, d_2)^{\varepsilon 2} mod\ n = (17302, 23580)$, then he asks Alice to send to him (s_1, v_1).
2. Alice sends $(s_1, v_1) = (24, 32)$ to Bob.
3. Bob checks that $(x_1, y_1) = (s_1, v_1)^2 (mod\ n) = (27873, 1536)$.

So, He accepts Alice's identification.

4.2 Security Analysis

Assume that Oscar is an attacker.
Attack 1: If Oscar intercepts the value of (z_1, z_2), then he will not be able to find Alice's secrets keys. Indeed, he must solve the equation:
$(z_1, z_2) \equiv (s_1, v_1)(s_2, v_2)(a_1, b_1)^{\varepsilon 1}(a_2, b_2)^{\varepsilon 2} mod\ n$ with four unknown variables: (s_1, v_1), (s_2, v_2), (a_1, b_1), and (a_2, b_2).
Attack 2: If Oscar intercepts values of (x_1, y_1), (x_2, y_2) and (z_1, z_2) and tries to imitate the identification of Alice then he will be blocked by the challenge numbers $\varepsilon 1$ and $\varepsilon 2$ which are changeable with every trial of identification.
Attack 3 (man in the middle): If Oscar tries to identify himself as Alice, then he takes any two numbers: (x_1, y_1), (z_1, z_2) and computes (x_2, y_2) using the equation: $(z_1, z_2)^2 = (x_1, y_1)(x_2, y_2)(c_1, d_1)^{\varepsilon_1}(c_2, d_2)^{\varepsilon_2}\ mod\ n$. The bank, or Bob, poses a challenge to Oscar by proposing him to send one of the two numbers (s_1, v_1) or (s_2, v_2) such that: $(x_1, y_1) = (s_1, v_1)^2 (mod\ n)$ and $(x_2, y_2) = (s_2, v_2)^2 (mod\ n)$. So we put $\delta \in \{1, 2\}$ with:

- If $\delta = 1$, Bob asks Oscar the number (s_1, v_1).
- If $\delta = 2$, Bob asks Oscar the number (s_2, v_2).

4.3 Complexity

In this paragraph, we discuss the complexity of our method.

Let T_{exp} and T_{mult} be appropriately the time to calculate an exponentiation and a modular multiplication. We neglect the time needed for modular substraction, additions and comparisons.

- To generate her keys (c_1, d_1) and (c_2, d_2), Alice needs to execute 2 modular exponentiation.
- In the identification step, she performs 4 modular exponentiation and 3 multiplication.
- To verify Alice's identification, Bob calculates 4 modular exponentiation and 3 multiplication.

So, there are 10 modular exponentiations and 6 multiplications.

In other words, the total time T_{tot} required to execute all the identification's operations is:

$$T_{tot} = 1T_{exp} + 6T_{mult}$$

We have: $T_{exp} = O((\log n)^3)$ and $T_{mult} = O((\log n)^2)$ [6]. So, the final complexity of our identification scheme is as follow:

$$T_{tot} = O((\log n)^2 + (\log n)^3)$$

Finally, we assume that the protocol works on a polylogarithmic time.

5 Conclusion

In this paper, we presented a new variant of the Fiat and Shamir identification protocol. The work is based on the zero knowledge proof concept using Gaussian integers. The purpose of the method is to allow an entity to identify with a system without revealing its secret parameters. We analyzed the security of our scheme.

References

1. Brassard, J.: Cryptologie Contemporaine. Masson, Paris (1993)
2. Dubertret, G.: Initiation à la Cryptographie, 2nd edn. VUIBERT (2000)
3. Fiatm, A., Shamir, A.: How to prove yourself: practical solutions to identification and signature problems. Advances in Cryptology, Proceedings of Crypto '86. Lecture Notes in Computer Science, No 263, pp. 186–194. Springer (1987)

4. Goldstein, C., Schappacher, N., Schwermer, J.: The Shaping of Arithmetic after C.F. Gauss's Disquisitiones Arithmeticae. Springer, Berlin (2007)
5. Goldwasser, S., Micali, S., Rackoff, C.: The knowledge of interactive proof systems. In: 17 ACM Symposium on Theory of Computing, pp. 291–304 (1985)
6. Menezes, A.J., van Oorschot, P.C., Vanstone, S.A.: Handbook of Applied Cryptography, pp. 72 (1996)
7. Niven, I., Zuckerman, H.S., Montgomery H.L.: An Introduction to the Theory of Numbers, 5th edn. Wiley, Hoboken (1991)
8. Okamoto, T.: Provably secure and practical identification schemes and corresponding signature schemes. In: Brickell, E.F. (eds.) Advances in Cryptology Crypto '92. Lecture Notes in Computer Science, vol. 740. Springer, Berlin (1992)
9. Pollard, J.M.: A Monte Carlo method for factorization. BIT Numer. Math., 331–334 (1975)
10. Rabin, M.O.: Digital signatures and public-key functions as intractable as factorization. Technical report MIT/LCS/TR-212 (1978)
11. Shor., P.: Polynomial-time algorithms for prime factorization and discrete logarithms on a quantum computer. SIAM J. Comput., 1484–1509 (1997)
12. Schnorr, C.P.: Efficient signature generation by smart cards. J. Crypt., 161–174 (1991)

Resolution of a System of the Max-Product Fuzzy Relation Equations Using $B \circ B^t$-Factorization

Hamid Sadiki, Lalla Saadia Chadli, Said Melliani, and Idris Bakhadach

Abstract In this paper, the Cholesky-factorization is extended to the square fuzzy symmetric matrix with respect to the max-product composition operator called $B \circ B^t$-factorization. Equivalently, we will find two fuzzy triangular matrices B and B^t for a fuzzy square symmetric matrix A such that $A = B \circ B^t$, where "$\circ$" is the max-product composition. An algorithm is presented to find the matrix B. Furthermore, some necessary and sufficient conditions are proposed for the existence and uniqueness of the $B \circ B^t$-factorization for a given fuzzy square symmetric matrix A. An algorithm is also proposed to find the solution set of a square system of Fuzzy Relation Equations (FRE) using the $B \circ B^t$-factorization.

1 Introduction

The notion of FRE is associated with the composition of fuzzy binary relations. The FRE have been intensively investigated both from a theoretical standpoint and in view of applications since they were first introduced by Sanchez [1, 9]. The FRE play important roles in many applications, such as intelligence technology, image reconstruction etc. Therefore, how to compute the solutions of FRE is a fundamental problem.

H. Sadiki · L. S. Chadli · S. Melliani (✉) · I. Bakhadach
LMACS, Laboratory of Applied Mathematics and Computing Sciences, Sultan Moulay Slimane University, PO Box 523, 23000 Beni Mellal, Morocco
e-mail: saidmelliani@gmail.com

H. Sadiki
e-mail: sadiki.info@gmail.com

L. S. Chadli
e-mail: sa.chadli@yahoo.fr

I. Bakhadach
e-mail: i.bakhadach@usms.ma

S. Melliani and O. Castillo (eds.), *Recent Advances in Intuitionistic Fuzzy Logic Systems and Mathematics*, Studies in Fuzziness and Soft Computing 395,
https://doi.org/10.1007/978-3-030-53929-0_15

The Cholesky-factorization method in the linear algebra with the summation and product operations is very interesting and useful for the reduction of a non-sparse square matrix to the product of two sparse square matrices called lower and upper triangular matrices. The method is applied to solve a square linear system have a symmetric matrix . Using the $B \circ B^t$-factorization method, the non-sparse square linear system is easily reduced to two sparse square linear systems as lower and upper triangular systems. Hence, we can easily solve the systems by the forward and backward substitutions. This procedure is called the resolution of the system by the $B \circ B^t$-factorization.

2 $B \circ B^t$-Factorization of a Fuzzy Matrix with Respect to the Max-Product Composition Operator

Let $A = [a_{ij}]$, $0 \leq a_{ij} \leq 1$,be an $n \times n$ dimensional fuzzy symmetric matrix .We will show how to decompose a fuzzy square matrix A into two fuzzy (lower and upper) triangular matrices B and B^t such that $A = B \circ B^t$, where the symbol "o" denotes the max-product composition operator. In other words, we try to find one fuzzy (lower) triangular matrix $B = [b_{ij}]_{n \times n}$ such that $0 \leq b_{ij} \leq 1$

$$b_{ij} = \begin{cases} b_{ij} & i \geq j \\ 0 & i < j \end{cases} \tag{1}$$

and

$$\max_{t \in \underline{n}} \{b_{it}.b_{jt}\} = a_{ij} \qquad \forall\ 1 \leq i, j \leq n \tag{2}$$

where $\underline{n} := 1, 2, \ldots, n$. With regard to the special structure of the fuzzy lower matrix B and the max-product composition operator, we can write:

(S_1) $a_{11} = b_{11}^2, a_{i1} = b_{i1} b_{11} \forall\ i \in \{2, \ldots, n\}$

(S_2) $a_{22} = \max\{b_{21}^2, b_{22}^2\}$; $a_{32} = \max\{b_{31} b_{21}, b_{32} b_{22}\}$;
$a_{i2} = \max\{b_{i1} b_{21}, b_{i2} b_{22}\} \forall\ i \in \{3, \ldots, n\}$

(S_3) $a_{33} = \max\{b_{31}^2, b_{32}^2, b_{33}^2\}$; $a_{43} = \max\{b_{41} b_{31}, b_{42} b_{32}, b_{43} b_{33}\}$;
$a_{i3} = \max\{b_{i1} b_{31}, b_{i2} b_{32}, b_{i3} b_{33}\} \forall\ i \in \{4, \ldots, n\}$

(S_4) $a_{44} = \max\{b_{41}^2, b_{42}^2, b_{43}^2, b_{44}^2\}$;
$a_{54} = \max\{b_{51} b_{41}, b_{52} b_{42}, b_{53} b_{43}, b_{54} b_{44}\}$;
$a_{i4} = \max\{b_{i1} b_{41}, b_{i2} b_{42}, b_{i3} b_{43}, b_{i4} b_{44}\} \forall\ i \in \{5, \ldots, n\}$

$\vdots$

(S_j) $a_{jj} = \max\{b_{j1}^2, b_{j2}^2, \ldots, b_{jj}^2\}$;
$a_{ij} = \max\{b_{i1} b_{j1}, b_{i2} b_{j2}, b_{i3} b_{j3} \ldots, b_{ij} b_{jj}\} \forall\ i \in \{j+1, \ldots, n\}$;

(S_n) $a_{nn} = \max\{b_{n1}^2, b_{n2}^2, \ldots, b_{nn}^2\}$;

If we can solve the system $(S_1) - (S_n)$ with respect to the components b_{ij} for $i < j$ where $i, j \in \underline{n}$ then at least an $B \circ B^t$–factorization is found for the square fuzzy symmetric matrix $A = [a_{ij}]$. Now, an algorithm is designed to find the components b_{ij} for $i \geq j$, where $i, j \in \underline{n}$.

An algorithm for computing the matrix B

Algorithm 2.1 Let $A = [a_{ij}], 0 \leq a_{ij} \leq 1$, be an $n \times n$-dimensional fuzzy symmetric matrix.

Step 1. Let $b_{11} = \sqrt{a_{11}}$;
For $i = 2$ to n
{Run **Proc2**($a_{i1}, 0, b_{11}$); Let b_{i1}=**Proc2**($a_{i1}, 0, b_{11}$);}

Step 2. Run **Proc1**(a_{22}, b_{21}^2); Let b_{22}=**Proc1**(a_{22}, b_{21}^2);
For $i = 3$ to n
{Run **Proc2**($a_{i2}, b_{i1}b_{21}, b_{22}$); Let b_{i2}=**Proc2**($a_{i2}, b_{i1}b_{21}, b_{22}$);}

Step 3. Let max=$\max\{b_{31}^2, b_{32}^2\}$;
Run **Proc1**(a_{33},max); Let b_{33}=**Proc1**(a_{33}, max);
For $i = 4$ to n
{ Let max= $\max\{b_{i1}b_{31}, b_{i2}b_{32}\}$;
Run **Proc2**(a_{i3},max, b_{33}); Let b_{i3}=**Proc2**(a_{i3},max, b_{33}); }

Step 4. Let max=$\max\{b_{41}^2, b_{42}^2, b_{43}^2\}$;
Run **Proc1**(a_{44},max); Let b_{44}=**Proc1**(a_{44},max);
For $i = 5$ to n
{ Let max= $\max\{b_{i1}b_{41}, b_{i2}b_{42}, b_{i3}b_{43}\}$;
Run **Proc2**(a_{i4},max, b_{44}); Let b_{i4}=**Proc2**(a_{i4},max, b_{44}); }

⋮

Step j. Let max=$\max\{b_{j1}^2, b_{j2}^2, b_{j3}^2, ..., b_{j,j-1}^2\}$;
Run **Proc1**(a_{jj},max); Let b_{jj}=**Proc1**(a_{jj},max);
For $i = j + 1$ to n
{ Let max= $\max\{b_{i1}b_{j1}, b_{i2}b_{j2}, b_{i3}b_{j3}, ..., b_{i,j-1}b_{j,j-1}\}$;
Run **Proc2**(a_{ij},max, b_{jj}); Let b_{ij}=**Proc2**(a_{ij},max, b_{jj}); }

⋮

Step n. Let max=$\max\{b_{n,1}^2, b_{n,2}^2, b_{n,3}^2, ..., b_{n,n-1}^2\}$;
Run **Proc1**($a_{n,n}$,max); Let $b_{n,n}$=**Proc1**($a_{n,n}$,max);

End.

Furthermore, the used procedures in the algorithm are presented below:

Procedure Proc1(a, b^2)
Begin
if $a > b^2$ then let $d = \sqrt{a}$;
if $a = b^2$ then choose the value d from $[0, \sqrt{a}]$, arbitrarily;
if $a < b^2$ then the matrix A has no $B \circ B^t$–factorization and $d = \emptyset$;
Return(d);
End.

Procedure Proc2(a, bb^t, d)
Begin
if $bb^t < a \leq d$ then let $b = \frac{a}{d}$ and Return(b);
if $bb^t = a \leq d$ then let $b \in [0, \frac{a}{d}]$ and Return(b);
Else the matrix A has no $B \circ B^t$–factorization and $b = \emptyset$; and Return(b);
End.

Example Consider the fuzzy matrix A as follows:

$$A = \begin{pmatrix} 0.3 & 0.2 & 0.14 & 0.2 & 0.2 \\ 0.2 & 0.4 & 0.26 & 0.3 & 0.3 \\ 0.14 & 0.26 & 1 & 0.75 & 0.85 \\ 0.2 & 0.3 & 0.75 & 0.85 & 0.7 \\ 0.2 & 0.3 & 0.85 & 0.7 & 0.9 \end{pmatrix}$$

The resulted matrice B is as follows:

$$B = \begin{pmatrix} 0.55 & 0 & 0 & 0 & 0 \\ 0.37 & 0.63 & 0 & 0 & 0 \\ 0.26 & 0.41 & 1 & 0 & 0 \\ 0.37 & 0.47 & 0.75 & 0.92 & 0 \\ 0.37 & 0.47 & 0.85 & 0.76 & 0.95 \end{pmatrix}$$

2.1 *Time Computational Complexity of Algorithm 2.1*

We now study the computational complexity of Algorithm 2.1. We might view the computational time for Algorithm 2.1 as allocated to the two basic operations:

1. **Running Proc 1 and Proc 2**: the algorithm performs the two procedures in step k for $k \geq 2$ with different times. If we assume the required time to run Proc1 and Proc2 as $O(P1)$ and $O(P2)$, respectively, then the sum of allocated time to run each step including Proc 1 and Proc 2 of Algorithm 2.1 for a fuzzy matrix $A_{n\times n}$ is as follows:

   - **Time in Step 1.** There are no Proc 1 and $(n-1) \times O(P2)$.
   - **Time in Step k.** $1 \times O(P1) + (n-(k+1)+1) \times O(P2)$, for each $k = 2, \ldots, n-1$.

   The sum of above time (T_1) is as follows:
   $T_1 = \left(\sum_{k=2}^{n} 1\right) \times O(P1) + \left(\sum_{k=1}^{n-1} (n-(k+1)+1)\right) \times O(P2) = (n-1)$
   $\left(O(P1) + \frac{n}{2} \times O(P2)\right)$.
   If we consider the required time for the comparison of two numbers as $O(1)$, then with attention to the procedures of Proc 1 and Proc 2, at most running

time of Proc 1 and Proc 2 are as $O(P1) = 3 \times O(1)$ and $O(P2) = 6 \times O(1)$, respectively. Hence, we can write: $T_1 = 3(n^2 - 1) \times O(1)$

2. **Maximization Operations:** the algorithm performs the maximization operations in step k for $k \geq 3$. The required time to find the maximum element in a set depends on the number of elements of the set. If we apply Algorithm 2.1 for a fuzzy matrix $A_{n \times n}$, then the maximization operations are divided into two groups.

   - **Group 1:** the maximization operations which are done inside of the "for loops".
   - **Group 2:** the maximization operations which are done outside of the "for loops". In steps 1 and 2, there are no maximization operations. Therefore, the sum of the required time for Group 1 in step k, for each $k \geq 3$, is computed as follows:
     Time in Step k. $(n - (k + 1) + 1) \times (k - 2) \times O(1)$,
     where $(n - (k + 1) + 1)$ is the number of the iterations in a "for loop" , and $(k - 2)$ is the number of required comparisons to find the maximum element in the k-element sets and $O(1)$ denotes the required time to do a comparison operation.
     Therefore, the sum of required time for the maximization operations in Group 1 is as follows:

     T(Group 1)=$\sum_{k=3}^{n-1} (n - (k + 1) + 1) \times (k - 2) \times O(P1)$.
     The sum of the required time for Group 2 is computed as follows. First of all, we remind that the sum of the required time to find the maximum element of an k-element set in Group 2 from step k to step n is equal to $(n - k + 1) \times (k - 2) \times O(1)$, where $(n - k + 1)$ is the number of k-element sets outside of the "for loops" in Algorithm 2.1 that we require to find their maximum elements and $(k - 2)$ is the number of required comparisons to find the maximum element in an k-element set. Therefore, the sum of required time for the maximization operations in Group 2 is as follows:

     T(Group 2)=$\sum_{k=3}^{n} (n - k + 1) \times (k - 2) \times O(P1)$.

   Therefore, we can compute the sum of the required time for the maximization operations (T_2) as follows:

   $$T_2 = T(Group1) + T(Group2) = \sum_{k=3}^{n-1} (n - (k + 1) + 1) \times (k - 2) \times O(P1)$$
   $$+ \sum_{k=3}^{n} (n - k + 1) \times (k - 2) \times O(P1)$$
   $$= \left(\tfrac{1}{3}n^3 + \tfrac{1}{2}n^2 - \tfrac{23}{6}n - 1\right) \times O(1).$$

   With attention to the algorithm structure, the time computational complexity of Algorithm 2.1 (T) is as follows:

$$T = T_1 + T_2 = (3(n^2-1)) \times O(1) + \left(\frac{1}{3}n^3 + \frac{1}{2}n^2 - \frac{23}{6}n - 1\right) \times O(1) = \left(\frac{1}{3}n^3 + \frac{7}{2}n^2 - \frac{23}{6}n - 4\right) \times O(1)$$

For large values n, we have $T \cong \frac{1}{3}n^3 \times O(1) = O(n^3)$

Therefore, we have established the following result.

Theorem 1 *Algorithm 2.1 finds an Cholesky-factorization for a $n \times n$ fuzzy symmetric matrix A in $O(n^3)$ time.*

3 Existence and Uniqueness of $B \circ B^t$-Factorization

It is very important to answer the following question about the $B \circ B^t$–factorization of a fuzzy square symmetric matrix."When does there exist an$B \circ B^t$-factorization for a fuzzy square symmetric matrix A"?. In this section, we try to answer the above question. If the system $(S_1) - (S_n)$ has at least an solution, then there is at least an $B \circ B^t$-factorization for the fuzzy square symmetric matrix A. Also, if the system has no solutions, then there is no $B \circ B^t$-factorization for the fuzzy square symmetric matrix A. The uniqueness of the solution set of the system $(S_1) - (S_n)$ causes that the $B \circ B^t$-factorization of a fuzzy square symmetric be unique. The following theorems express the necessary and sufficient conditions for the existence and uniqueness of the $B \circ B^t$-factorization for a given fuzzy square symmetric matrix.

Theorem 2 *Let $A = [a_{i,j}], 0 \le a_{i,j} \le 1$,be an $n \times n$ dimensional symmetric fuzzy matrix. Then there exist(s) at least an $B \circ B^t$-factorization for the fuzzy matrix A with respect to the max-product composition if and only if there exist at least a $n \times n$-dimensional fuzzy (lower) triangular matrix as $B = [b_{i,j}]$ such that their components satisfy the following conditions.*
(C_1) $b_{1,1} \ge a_{i,1} = b_{i,1}b_{1,1} \forall\ i \in \{2, \ldots, n\}$
(C_2) $a_{2,2} \ge b_{2,1}^2$; *and* $b_{2,2} \ge a_{i,2} \ge b_{i,1}b_{2,1} \forall\ i \in \{3, \ldots, n\}$
(C_3) $a_{3,3} \ge \max(b_{3,1}^2, b_{3,2}^2)$; *and* $b_{3,3} \ge a_{i,3} \ge \max(b_{i,1}b_{3,1}, b_{i,2}b_{3,2}) \forall\ i \in \{4, \ldots, n\}$

.

.

(C_j) $a_{j,j} \ge \max(b_{j,1}^2, b_{j,2}^2, \ldots b_{j,j-1}^2)$; *and* $b_{jj} \ge a_{i,j} \ge \max(b_{i,1}b_{j,1}, b_{i,2}b_{j,2}, \ldots. b_{i,j-1}b_{j,j-1}) \forall\ i \in \{j+1, \ldots, n\}$

.

.

.

(C_n) $a_{nn} \ge \max(b_{n,1}^2, b_{n,2}^2, \ldots b_{n,n-1}^2)$;

Proof If the component B satisfy the relations $(C_1) - (C_n)$, then it is easily shown that the solution set of the system $(S_1) - (S_n)$ is non-empty and the fuzzy matrix $A = [a_{i,j}]_{n\times n}$ can be decomposed into two fuzzy lower and upper triangular matrices(B and B^t) $B = [b_{i,j}]_{n\times n}$ and $B^t = [b_{j,i}]_{n\times n}$ such that $A = B \circ B^t$. Conversely, if the fuzzy matrix $A = [a_{i,j}]_{n\times n}$ is decomposed into two fuzzy lower and upper triangular matrices B and B^t such that $A = B \circ B^t$, then by extending the relation $A = B \circ B^t$ with the max-product composition,we can obtain the relations $(C_1) - (C_n)$. Since there exist(s) at least an $B \circ B^t$-factorization for the fuzzy matrix A, the solution set of the system $(S_1) - (S_n)$ is non-empty. On other hand, since the matrices B and B^t are fuzzy matrices, the existence conditions of solution for the system $(S_1) - (S_n)$ are as $(C_1) - (C_n)$.

Theorem 3 *Let $A = [a_{i,j}], 0 \leq a_{i,j} \leq 1$, be an $n \times n$ dimensional fuzzy symmetric matrix and has at least an $B \circ B^t$-factorization with respect to the max-product composition. If the following conditions are satisfied, then the $B \circ B^t$-factorization is unique.*

(Q_1) $\quad a_{22} > b_{21}^2 and b_{22} \geq a_{i2} > b_{i1}b_{21} \forall \ i \in \{3, \ldots, n\}$ *and*

(Q_2) $\quad a_{33} > \max\{b_{31}^2, b_{32}^2\}$; *and*
$b_{33} \geq a_{i3} > \max\{b_{i1}b_{31}, b_{i2}b_{32}\} \forall \ i \in \{4, \ldots, n\}$ *and*

.
.
.

(Q_{j-1}) $\quad a_{jj} > \max\{b_{j1}^2, b_{j2}^2, \ldots, b_{j,j-1}^2\}$; *and*
$b_{jj} \geq a_{ij} > \max\{b_{i1}b_{j1}, b_{i2}b_{j2}, b_{i3}b_{j3}, b_{i,j-1}b_{j,j-1}\} \forall \ i \in \{j+1, \ldots, n\}$; *and*

.
.
.

(Q_n) $\quad a_{nn} > \max\{b_{n1}^2, b_{n2}^2, \ldots, b_{n,n-1}^2\}$;

Proof Since the matrix A has at least an $B \circ B^t$-factorization and the conditions $(Q_1) - (Q_{n-1})$ are satisfied, the system $(S_1) - (S_n)$ has an $B \circ B^t$-factorization. To show it, suppose that the condition (Q_1) is satisfied, i.e.,$a_{22} > b_{21}^2$ and $b_{22} \geq a_{i2} > b_{i1}b_{21} \forall \ i \in \{3, \ldots, n\}$ and then the statement $a_{22} = \max\{b_{21}^2, b_{22}^2\}$, is reduced to $a_{22} = b_{22}^2$. The equality results that $b_{22} = \sqrt{a_{22}}$ is only value for b_{22}. It is necessary to note that the component a_{22} is known,and the statement $b_{22} \geq a_{i2} > b_{i1}b_{21} \forall \ i \in \{3, \ldots, n\}$ is reduced to $b_{22} \geq a_{i2} = b_{i2}b_{22}$ i.e.($b_{i2} = \frac{a_{i2}}{b_{22}}$ The equality results that b_{i2} is only value). Similarly, if each of the conditions $(Q_1) - (Q_{n-1})$ is satisfied, then we can show that the values $b_{ij}, fori \geq j$, are determined uniquely.

Therefore, under the conditions $(Q_1) - (Q_{n-1})$, the matrix B is determined uniquely.

4 Resolution of the System of the Max-Product Fuzzy Relation Equations Using the $B \circ B^t$-Factorization

Let $A = [a_{ij}], 0 \leq a_{ij} \leq 1$,be an $n \times n$ dimensional symmetric fuzzy matrix and $b = [b_1, b_2, \ldots, b_n]^t$,$0 \leq b_i \leq 1$, be an n-dimensional fuzzy vector. Then the following system of fuzzy relation equations is defined by A and b as follows:

$$A \circ x = b, \tag{3}$$

where "o" denotes the max-product composition operator. In other words, we try to find a solution vector $x = [x_1, x_2, \ldots, x_n]^t$, with $0 \leq x_i \leq 1$ such that

$$\max_{j \in \underline{n}} \{a_{ij}.x_j\} = b_i, \forall i \in \underline{n} \tag{4}$$

According to [5, 8], when $X(A, b) = \{x \in [0, 1]^n \mid A \circ x = b\} \neq \emptyset$ it can be completely determined by the maximum solution and a finite number of minimal solutions. Chen and Wang [6, 10], Markovskii [4, 7], and Lin [2], showed that finding the minimal solutions of a system of fuzzy relation equations is an NP-hard problem. In this section, we propose an algorithm to find the solution set of the system (3) without finding the minimal solutions of the system. First of all, we present a property of the max-product composition operator in the following remark.

Remark 1 Let $A = [a_{ij}]_{m \times n}$, $B = [b_{ij}]_{n \times s}$ and $C = [c_{ij}]_{s \times k}$ be three fuzzy matrices and the operator "o" is the max-product composition. Then we have

$$(A \circ B) \circ C = A \circ (B \circ C) \tag{5}$$

i.e., the operator "o" has the associative property.

We now focus on the resolution of the system of fuzzy relation equations with the fuzzy square matrix A. Assume that the fuzzy matrix A has at least an $B \circ B^t$-factorization. Then we can rewrite the system (3) as follows:

$$(B \circ B^t) \circ x = b \tag{6}$$

Using **Remark** 4.1, we have

$$B \circ (B^t \circ x) = b \tag{7}$$

Now, let $B^t \circ x = y$ where $y = [y_1, y_2, \ldots, y_n]^t$ Then the system (3) is reduced to the following system:

$$B \circ y = b \tag{8}$$

with regard to the special structure of the matrix B, we can easily solve the system (8). Here the component y_1 is obtained from the first equation and then inserting its value in the second equation, y_2 is obtained, and so on. This process is called the forward substitution. Then the system $B^t \circ x = y$ is similarly solved. At first, the component x_n is computed from the last equation. Then by inserting its value in the $(n-1)$th equation, the component x_{n-1} is obtained and so on. This process is called the backward substitution. Using the two substitutions, the solution set of the system (3) is found. Now, we express this method by the following algorithm.

4.1 An Algorithm for the Resolution of the System (3)

Algorithm 1

Step 1. Compute the maximum solution of the system (3) by the following formula:

$$\hat{x} = A \Diamond b = [\wedge_{i=1}^{n} (a_{ij} \Diamond b_i)]_{j \in N} \text{ where } \wedge_{i=1}^{n} (a_{ij} \Diamond b_i) = \begin{cases} 1 & if\ a_{ij} \leq b_i, \\ & a \wedge b = \min(a, b) \text{ and } \wedge \emptyset = 0, \\ \frac{b_i}{a_{ij}} & \text{otherwise,} \end{cases}$$

Step 2. Check its feasibility by verifying whether $A \circ \hat{x} = b$. If it is infeasible, then stop and $X(A, b) = \emptyset$.

Step 3. Run Algorithm 2.1 to find an$B \circ B^t$-factorization for the fuzzy matrix A.
If there is no $B \circ B^t$-factorization for the fuzzy matrix A, then stop!

Step 4. Find the solution set of the system $B \circ y = b$ by the forward substitution.

Step 5. Find the solution set of the system $B^t \circ x = y$ by the backward substitution.

Step 6. End.

4.2 Time Computational Complexity of Algorithm 1

We now analyze the time computational complexity of Algorithm 1.

We now study the computational complexity of Algorithm 1. To do this, we compute the required time to run each step. According to [3], the time computational complexity for establishing the consistency of the system and computing $\hat{x}$ for a fuzzy symmetric matrix $A_{n \times n}$ (steps 1 and 2) is from order $n^2 (= O(n^2))$. As we saw in Sect. 2.1, the time computational complexity of step 3 is as $O(n^3)$. Furthermore, the required time for resolution of a triangular system as $B \circ y = b$ s as follows: the kth equation in the system has a maximization operation among k elements. Hence, it requires $(k - 1)$ comparison operations with $(k - 1) \times O(1)$ time. Therefore, the sum of required time for this step is computed as: $\sum_{k=1}^{n} (k - 1) \times O(1) = \frac{n(n-1)}{2} \times O(1)$. For large values n, we can write $\frac{n(n-1)}{2} \times O(1) \cong O(n^2)$.

Similar to step 4, the required time for the resolution of a triangular system as $B^t \circ x = y$ is as $\frac{n(n-1)}{2} \times O(1) \cong O(n^2)$.

With attention to the above points, the required time T for the resolution of a square fuzzy system of dimension $n \times n$ is as follows: $T = O(n^2) + O(n^2) + O(n^3) + O(n^2) + O(n^2)$. For large values n, we can write $T \cong O(n^3)$. Therefore, we have the following result.

Theorem 4 *Algorithm 1 solves the system (3) in $O(n^3)$ time.*

4.3 Numerical Example

Now, the algorithm is illustrated by an example. As it is seen, the solution set of the system (3) is found without using the maximum solution and the minimal solutions of the original system.We only apply the maximum solution to check the feasibility of the system. Since the maximum solution has a special formula, its computation is easily done.

Example Consider the system of fuzzy relation equations as follows.
$A \circ x = b$,

$$\text{Where } A = \begin{pmatrix} 0.3 & 0.2 & 0.14 & 0.2 & 0.2 \\ 0.2 & 0.4 & 0.26 & 0.3 & 0.3 \\ 0.14 & 0.26 & 1 & 0.75 & 0.85 \\ 0.2 & 0.3 & 0.75 & 0.85 & 0.7 \\ 0.2 & 0.3 & 0.85 & 0.7 & 0.9 \end{pmatrix} \text{ and } b = \begin{bmatrix} 0.3 \\ 0.4 \\ 0.6 \\ 0.6 \\ 0.51 \end{bmatrix}$$

Algorithm 1 is applied to solve the example.

Step 1. The maximum solution of the system is as $\hat{x} = [1, 1, 0.6, 0.71, 0.57]'$

Step 2. Since $A \circ \hat{x} = b$, the system is feasible.Hence, go to **Step 3**.

Step 3. Applying Algorithm 2.1, the $B \circ B'$-factorization of the matrix A is as follows:
$A = B \circ B'$,
where

$$B = \begin{pmatrix} \frac{\sqrt{30}}{10} & 0 & 0 & 0 & 0 \\ \frac{\sqrt{30}}{15} & \frac{\sqrt{10}}{5} & 0 & 0 & 0 \\ \frac{7\sqrt{13}}{65} & \frac{13\sqrt{10}}{100} & 1 & 0 & 0 \\ \frac{\sqrt{30}}{15} & \frac{3\sqrt{10}}{20} & \frac{\sqrt{3}}{4} & \frac{\sqrt{17}}{20} & 0 \\ \frac{\sqrt{30}}{15} & \frac{3\sqrt{10}}{20} & \frac{\sqrt{17}}{20} & \frac{7\sqrt{85}}{85} & \frac{3\sqrt{10}}{10} \end{pmatrix}$$

Step 4. In this step, we find the solution set of the system $B \circ y = b$ by the forward substitution.

1. $y_1 = \frac{0.3}{0.55} = 0.55$.

2. $\max\{0.37y_1, 0.63y_2\} = 0.4 \Rightarrow \max\{0.2, 0.63y_2\} = 0.4 \Rightarrow y_2 = \frac{0.4}{0.63} = 0.63$.

3. $\max\{0.26y_1, 0.41y_2, y_3\} = 0.6 \Rightarrow \max\{0.14, 0.26, y_3\} = 0.6 \Rightarrow y_3 = 0.6$.

4. $\max\{0.37y_1, 0.47y_2, 0.75y_3, 0.92y_4\} = 0.6 \Rightarrow \max\{0.2, 0.3, 0.45, 0.92y_4\} = 0.6$
$\Rightarrow y_4 = \frac{0.6}{0.92} = 0.65$.

5. $\max\{0.37y_1, 0.47y_2, 0.85y_3, 0.76y_4, 0.95y_5\} = 0.51 \Rightarrow \max\{0.2, 0.3, 0.51, 0.95y_5\} = 0.51$
$\Rightarrow y_5 \in [0, \frac{0.51}{0.95}] = [0, 0.54]$.

Therefore, the solution set of the system $B \circ y = b$ is as follows:
$X(B, b) = \{y = (y_1, y_2, y_3, y_4, y_5) \in [0, 1]^5 | y \in \{0.55\} \times \{0.63\} \times \{0.6\} \times \{0.65\} \times [0, 0.54]\}$.

Step 5. In this step, we find the solution set $B' \circ x = y$ by the backward substitution.

1. $0.95x_5 = [0, 0.54] \Rightarrow 0 \leq x_5 \leq 0.57$.

2. $\max\{0.92x_4, 0.76x_5\} = 0.65 \Rightarrow \max\{0.92x_4, [0, 0.43]\} = 0.65 \Rightarrow x_4 = \frac{0.65}{0.92} = 0.71$.

3. $\max\{x_3, 0.75x_4, 0.85x_5\} = 0.6 \Rightarrow \max\{x_3, 0.53, [0, 0.48]\} = 0.6 \Rightarrow x_3 = 0.6$.

4. $\max\{0.63x_2, 0.41x_3, 0.47x_4, 0.47x_5\} = 0.63 \Rightarrow \max\{0.63x_2, 0.25, 0.34, [0, 0.27]\} = 0.63$
$\Rightarrow x_2 = \frac{0.63}{0.63} = 1$.

5. $\max\{0.55x_1, 0.37x_2, 0.26x_3, 0.37x_4, 0.37x_5\} = 0.55$
$\Rightarrow \max\{0.55x_1, 0.37, 0.16, 0.26, [0, 0.21]\} = 0.55 \Rightarrow x_1 = 1$.

With regard to the above items, we conclude that the solution set $A \circ x = b$ is as follows:
$X(A, b) = \{x = (x_1, x_2, x_3, x_4, x_5) \in [0, 1]^5 | x \in \{1\} \times \{1\} \times \{0.6\} \times \{0.71\} \times [0, \frac{13}{19}]\}$.

Therefore, the solution set of the system was calculated without computing minimal solutions of the fuzzy relation equations and without using the maximum solution in the determination of the solution set.

5 Conclusions

In this paper, the Cholesky-factorization was extended to a fuzzy square symmetric matrix with respect to the max-product composition operator called $B \circ B^t$-factorization. The sufficient and necessary conditions of its existence and uniqueness for a fuzzy matrix were presented. Moreover, an algorithm was proposed to find the fuzzy lower and upper triangular matrices B and B^t , respectively.

Finally, applying the $B \circ B^t$-factorization of a fuzzy matrix, an algorithm was suggested to solve a system of fuzzy relation equations. This algorithm uses the forward and backward substitution to solve the system without finding the minimal solutions of the system.

References

1. Molai, A.A.: Resolution of a system of the max-product fuzzy relation equations using $L \circ U$-factorization
2. Cechlárová, K.: Unique solvability of maxmin fuzzy equations and strong regularity of matrices over fuzzy algebra. Fuzzy Sets Syst. **75**, 165–177 (1995)
3. Peeva, K., Kyosev, Y.: Algorithm for solving max-product fuzzy relational equations. Soft Comput. **11**, 593–605 (2007)
4. Bour, L., Lamotte, M.: Solutions minimales d'quations de relations floues avec la composition max t-norme. BUSEFAL **31**, 24–31 (1987)
5. Luoh, L., Wang, W.J., Liaw, Y.K.: Matrix-pattern-based computer algorithm for solving fuzzy relation equations. IEEE Trans. Fuzzy Syst. **11**, 100–108 (2003)
6. Allame, M., Vatankhahan, B.: Iteration algorithm for solving $Ax = b$ in maxmin algebra. Appl. Math. Comput. **175**, 269–276 (2006)
7. Bourke, M.M., Fisher, D.G.: Solution algorithms for fuzzy relational equations with max-product composition. Fuzzy Sets Syst. **94**, 61–69 (1998)
8. Miyakoshi, M., Shimbo, M.: Solutions of composite fuzzy relational equations with triangular norms. Fuzzy Sets Syst. **16**, 53–63 (1985)
9. Arnould, T., Tano, S.: A rule-based method to calculate the widest solution sets of a maxmin fuzzy relational equation. Int. J. Uncertain., Fuzziness Knowl.-Based Syst. **2**, 247–256 (1994)
10. Arnould, T., Tano, S.: A rule-based method to calculate exactly the widest solution sets of a maxmin fuzzy relational inequality. Fuzzy Sets Syst. **64**, 39–58 (1994)

On an Infinite Family of Imaginary Triquadratic Number Fields

M. M. Chems-Eddin, A. Azizi, and A. Zekhnini

Abstract Let p be a prime integer and $L = \mathbb{Q}(\zeta_8, \sqrt{p})$. In this paper, we determine the fields L for which the 2-class group is of type $(2, 8)$ and those for which the 16-rank of the 2-class group equals 1.

1 Introduction

Let k be a number field and $\mathrm{Cl}(k)$ its class group in the wide sense. Denote by $\mathrm{Cl}_2(k)$ the 2-class group of k, i.e. the 2-Sylow subgroup of $\mathrm{Cl}(k)$. The determination of the structure of $\mathrm{Cl}_2(k)$ is a classical and difficult problem of algebraic number theory that have many uses. In fact, it serves to give answers to many other problems related to the number field k, as the determination of the Hilbert 2-class field, the structure of the Galois group of the second Hilbert 2-class field and the 2-class group of the unramified extensions of k within its Hilbert 2-class field. Note that many authors interested in this problem for quadratic fields and biquadratic number fields (cf. [1–3, 6, 10]). In the present paper we shall determine the structure of the 2-class group of an infinite family of triquadratic number fields of the form $L := \mathbb{Q}(\zeta_8, \sqrt{p}) = \mathbb{Q}(\sqrt{-1}, \sqrt{2}, \sqrt{p})$, where p is a prime number. More precisely, we determine all the prime numbers for which the 2-class group of L is of type $(2, 8)$ and those for which the 16-rank $\mathrm{Cl}_2(L)$ equals 1. The following notations will be used for the rest of this article.

M. M. Chems-Eddin (✉) · A. Azizi · A. Zekhnini
Mathematics Department, Sciences Faculty, Mohammed First University, Oujda, Morocco
e-mail: 2m.chemseddin@gmail.com

A. Azizi
e-mail: abdelmalekazizi@yahoo.fr

A. Zekhnini
e-mail: zekha1@yahoo.fr

S. Melliani and O. Castillo (eds.), *Recent Advances in Intuitionistic Fuzzy Logic Systems and Mathematics*, Studies in Fuzziness and Soft Computing 395,
https://doi.org/10.1007/978-3-030-53929-0_16

2 Notations

- d: A positive odd square free integer,
- K: $\mathbb{Q}(\zeta_8)$,
- L: $K(\sqrt{d})$,
- $\mathrm{Cl}_2(L)$: The 2-class group of L,
- $h(k)$: The class number of k,
- $h_2(k)$: The 2-class number of k,
- $h_2(m)$: The 2-class number of the quadratic field $\mathbb{Q}(\sqrt{m})$,
- E_k: The unit group of k,
- W_k: The set of roots of unity contained in k,
- ω_k: The cardinality of W_k,
- k^+: The maximal real subfield of an imaginary number field k,
- Q_k: The Hasse's index, that is $[E_k : W_k k^+]$, if k/k^+ is CM,
- $q(k) := [E_L : \prod_i E_{k_i}]$, with k_i are the quadratic subfields of k,
- $\left(\frac{\cdot}{\cdot}\right)$: The Legendre symbol,
- $\left(\frac{\cdot}{\cdot}\right)_4$: The rational biquadratic symbol.

3 Some Preliminary Results

In this section we shall recall some results that we will need later. Let $K := \mathbb{Q}(\zeta_8)$ and $L := K(\sqrt{d})$.

Lemma 1 ([5]) *Let k/k' be a quadratic extension. If the class number of k' is odd, then the rank of the 2-class group of k is given by*

$$2\text{-}rank(\mathrm{Cl}(k)) = t - 1 - e,$$

where t is the number of ramified primes (finite or infinite) in the extension k/k' and e is defined by: $2^e = [E_{k'} : E_{k'} \cap N_{k/k'}(k^*)]$.

Lemma 2 *Let $d = p$ be a prime integer. Then, there are exactly four prime ideals that ramify in L/K if $p \equiv 1 \pmod 8$ and there are two prime ideals that ramify in L/K, if $p \not\equiv 1 \pmod 8$.*

Proof Assume that $p \equiv 1 \pmod 8$. Since 2 is unramified in $\mathbb{Q}(\sqrt{p})$, then ramified primes of L/K are the prime ideals of K dividing p. If $p \equiv 1 \pmod 8$, then by the cyclotomic reciprocity law (see [14, Theorem 2.13]) one can verify that there are exactly 4 prime ideals of K above p. We similarly complete the proof.

Lemma 3 *The unit group of K is given by $E_K = \langle \zeta_8, \varepsilon_2 \rangle$.*

Proof The field $K = \mathbb{Q}(\zeta_8) = \mathbb{Q}(\sqrt{-1}, \sqrt{2})$ is a Galois extension of $\mathbb{Q}$ whose Galois group is an elementary 2-group of order 4. We have, $\mathbb{Q}(\sqrt{2})$ is the real quadratic subfield of K with fundamental unit $\varepsilon_2 = 1+\sqrt{2}$, which is also a fundamental unit of K. Thus by [11, Theorem 42], the Hasse's unit index is equal to 1, i.e., $[E_K : E_{\mathbb{Q}(\sqrt{2})}W_K] = 1$, where W_K is the set of roots of unity contained in K.

4 The Main Result

Let p be a prime integer. We say that p takes the form (1) if

$$p \equiv 9 \pmod{16},\ \left(\frac{2}{p}\right)_4 = -1 \text{ and } \left(\frac{g}{p}\right)_4 \neq \left(\frac{2h}{g}\right), \tag{1}$$

We say that p takes the form (2) if

$$p \equiv 9 \pmod{16},\ \left(\frac{2}{p}\right)_4 = -1 \text{ and } \left(\frac{g}{p}\right)_4 = \left(\frac{2h}{g}\right), \tag{2}$$

where h and g are odd positive integers such that $p = 2g^2 - h^2$. Now we state the main result of this paper.

Theorem 1 *Let p be a prime integer and $L = \mathbb{Q}(\zeta_8, \sqrt{p})$. Then we have*

$$\mathrm{Cl}_2(L) \simeq (2, 8) \textit{ if and only if } p \textit{ takes the form } (1).$$

Furthermore, the 16*-rank of* $\mathrm{Cl}_2(L)$ *equals* 1 *if and only if* p *takes the form* (2).

Proof • We eliminate all the fields L such that the 2-$rank(\mathrm{Cl}(L)) \neq 2$, as follows. Note that the class group of $K = \mathbb{Q}(\zeta_8)$ is trivial. So by Lemma 1, the rank of the 2-class group of L equals $t-1-e$, where e is defined by: $2^e = [E_L : E_K \cap N_{L/K}(L^*)]$. If $p \not\equiv 1 \pmod 8$, then by Lemma 2, we have 2-$rank(\mathrm{Cl}(L)) = 1 - e < 2$. So this case is eliminated.
Now suppose that $p \equiv 1 \pmod 8$. Denote by $\mathfrak{p}$ any prime ideal of K above p and by $\left(\frac{\zeta_8, p}{\mathfrak{p}}\right)$ the quadratic norm residue symbol. We have

$$\left(\frac{\zeta_8, p}{\mathfrak{p}}\right) = (-1)^{\frac{p-1}{8}} \text{ and } \left(\frac{\varepsilon_2, p}{\mathfrak{p}}\right) = \left(\frac{2}{p}\right)_4 \left(\frac{p}{2}\right)_4.$$

Thus ε_2 (resp. ζ_8) is a norm in L/K if and only if $\left(\frac{2}{p}\right)_4 = \left(\frac{p}{2}\right)_4$ (resp. $p \equiv 1 \pmod{16}$). It follows by Lemma 2, that the rank of the 2-class group of L equals 2 if and only if $p \equiv 9 \pmod{16}$ or $\left(\frac{2}{p}\right)_4 \neq \left(\frac{p}{2}\right)_4$.

• Let us eliminate the case $p \equiv 1 \pmod{16}$ and $\left(\frac{2}{p}\right)_4 \neq \left(\frac{p}{2}\right)_4$. Consider the following diagram:

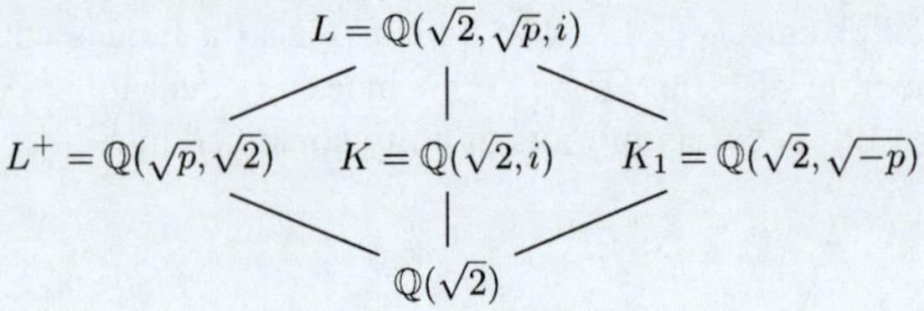

By the class number formula (see [8]), we have:

$$h(L) = \frac{Q_L}{Q_{K_1} Q_K} \frac{\omega_L}{\omega_{K_1} \omega_K} \frac{h(L^+) h(K_1) h(K)}{h(\mathbb{Q}(\sqrt{2}))}.$$

Then $h(L) = \frac{Q_L}{Q_{K_1} Q_K} \frac{1}{2} h(L^+) h(K_1) h(K)$. As $Q_L = 1$ (cf. [3]) and $Q_K = 1$ (see Lemma 3), so:

$$\begin{aligned} h(L) &= \tfrac{1}{2Q_{K_1}} h(L^+) h(K_1), \\ &= \tfrac{1}{2Q_{K_1}} h(L^+) \tfrac{1}{2} Q_{K_1} h_2(-p) h_2(-2p), \\ &= \tfrac{1}{4} h(L^+) h_2(-p) h_2(-2p). \end{aligned}$$

Recall that $h_2(-p) = 4$ if and only if $\left(\frac{2}{p}\right)_4 \neq \left(\frac{p}{2}\right)_4$ (cf. [4]).

So assuming that $p \equiv 1 \pmod{16}$ and $\left(\frac{2}{p}\right)_4 \neq \left(\frac{p}{2}\right)_4$ (i.e., $\left(\frac{2}{p}\right)_4 = -1$ and $\left(\frac{p}{2}\right)_4 = 1$), we get $h(L^+)$ is odd (see [7, Theorem 2]). Thus:

$$h_2(L) = \frac{1}{4} h_2(-p) h_2(-2p).$$

As $h_2(-p) = 4$, then $h_2(L) = h_2(-2p)$. Since $h_2(-2p) = 4$ if and only if $\left(\frac{p}{2}\right)_4 = 1$ (cf. [12]). So this case is eliminated.

- Assume that $\left(\frac{2}{p}\right)_4 = \left(\frac{p}{2}\right)_4 = -1$ (i.e. $p \equiv 9 \pmod{16}$ and $\left(\frac{2}{p}\right)_4 = -1$). By the class number formula for a multiquadratic number field (see [13]), we have:

$$\begin{aligned} h_2(L) &= \frac{1}{2^5} q(L) h_2(-p) h_2(p) h_2(-2p) h_2(2p) \\ &= \frac{1}{2^5} .4.h_2(-p).1.4.4 = 2.h_2(-p). \end{aligned}$$

Since $h_2(-p)$ is divisible by 8, then by [9, Theorem 1], $h_2(-p) = 8$ if and only if $\left(\frac{g}{p}\right)_4 \neq \left(\frac{2h}{g}\right)$. As the rank of the 2-class group of $\mathrm{Cl}_2(L)$ equals 2 and 4-rank of $\mathrm{Cl}_2(L)$ equals 1 (cf. [3, Théorème 10]), then we get the result.

Remark 1 By the previous proof, if p takes the form (2) then $\mathrm{Cl}_2(L) \simeq (2, h_2(-p))$ and $h(L)$ is divisible by 32.

Remark 2 Let d be a square-free integer. Note that we determined the triquadratic fields $\mathbb{Q}(\sqrt{-1}, \sqrt{2}, \sqrt{d})$ for which the 2-class group is of type $(2, 2)$, $(2, 4)$ or $(2, 2, 2)$ in submitted papers.

Using Pari/gp, we close this section by some numerical examples illustrating our theorem.

p	$p \pmod{16}$	$\left(\frac{2}{p}\right)_4$	g	h	$\left(\frac{g}{p}\right)_4$	$\left(\frac{2g}{h}\right)$	$\mathrm{Cl}_2(L)$
41	9	-1	5	3	-1	1	$(2, 8)$
137	9	-1	9	5	-1	1	$(2, 8)$
313	9	-1	13	5	-1	1	$(2, 8)$
409	9	-1	17	13	1	1	$(2, 16)$
521	9	-1	21	19	1	1	$(2, 32)$

References

1. Azizi, A. et Benhamza, I.: *Sur la capitulation des 2-classes d'idéaux de* $\mathbb{Q}(\sqrt{d}, \sqrt{-2})$, Ann. Sci. Math. Québec., 29 (2005), 1-20
2. Azizi, A., Mouhib, A.: Sur le rang du 2-groupe de classes de $\mathbb{Q}(\sqrt{m}, \sqrt{d})$ *où* $m = 2$ ou un premier $p \equiv 1 \pmod 4$. Trans. Amer. Math. Soc. **353**, 2741–2752 (2001)
3. Azizi, A., Taous, M.: Capitulation des 2-classes d'idéaux de $k = \mathbb{Q}(\sqrt{2p}, i)$. Acta Arith. **131**, 103–123 (2008)
4. Barrucand, P., Cohn, H.: Note on primes of type $x^2 + 32y^2$, class number, and residuacity. J. Reine. Angew. Math. **238**, 67–70 (1969)
5. Gras, G.: Sur les l-classes d'idéaux dans les extensions cycliques relatives de degré premier l. Ann. Inst. Fourier (Grenoble) **23**, 1–48 (1973)
6. Kaplan, P.: Sur le 2 -groupe de classes d'idéaux des corps quadratiques. J. Reine angew. Math. **283**(284), 313–363 (1976)
7. Kučera, R.: On the parity of the class number of a biquadratic field. J. Number Theory **52**, 43–52 (1995)
8. Lemmermeyer, F.: Ideal class groups of cyclotomic number fields I. Acta Arith. **72**, 347–359 (1995)
9. Leonard, P.A., Williams, K.S.: On the divisibility of the class numbers of $\mathbb{Q}(\sqrt{-d})$ em and $\mathbb{Q}(\sqrt{-2d})$ by 16. Can. Math. Bull. **25**, 200–206 (1982)
10. McCall, T.M., Parry, C.J., Ranalli, R.R.: Imaginary bicyclic biquadratic fields with cyclic 2 -class group. J. Number Theory **53**, 88–99 (1995)
11. Fröhlich, A., Taylor, M.J.: Algebraic Number Theory. Cambridge Studies in Advanced Mathematics, vol. 27. Cambridge University Press, Cambridge (1993)
12. Scholz, A.: Über die lösbarkeit der gleichung $t^2 - Du^2 = -4$. Math. Z. **39**, 95–111 (1935)
13. Wada, H.: On the class number and the unit group of certain algebraic number fields. J. Fac. Sci. Univ. Tokyo **13**, 201–209 (1966)
14. Washington, L.C.: Introduction to Cyclotomic Fields. Graduate Texts in Mathematics, vol. 83. Springer, New York (1982)

Computational Methods for Solving Intuitionistic Fuzzy Linear Systems

Hafida Atti, Bouchra Ben Amma, Said Melliani, Mohamed Oukessou, and Lalla Saadia Chadli

Abstract In this paper, we propose a method for solving intuitionistic fuzzy systems of linear equations. The method is discussed in detail and considered in two cases, namely with the right hand side as an intuitionistic fuzzy vector and as an intuitionistic fuzzy symmetric vector. Finally, we solve numerical examples in order to demonstrate the applicability of the proposed concept.

Keywords Intuitionistic fuzzy set · Intuitionistic fuzzy number vector

1 Introduction

Fuzzy set theory is a useful tool to describe the situation in which data are imprecise or vague or uncertain. A membership function of a classical fuzzy set assigns to each element of the universe of discourse a number from the unit interval to indicate the degree of belongingness to the set under consideration. The degree of non belongingness is just automatically the complement to 1 of the membership degree. However, a human being who expresses the degree of membership of given element

H. Atti · B. Ben Amma · S. Melliani (✉) · M. Oukessou · L. S. Chadli
Laboratory of Applied Mathematics and Scientific Computing, Faculty of Sciences and Technologies, Sultan Moulay Slimane University, 523, 23000 Beni Mellal, Morocco
e-mail: saidmelliani@gmail.com

H. Atti
e-mail: hafida.atti@gmail.com

B. Ben Amma
e-mail: bouchrabenamma@gmail.com

M. Oukessou
e-mail: ouk.mohamed@yahoo.fr

L. S. Chadli
e-mail: sa.chadli@yahoo.fr

S. Melliani and O. Castillo (eds.), *Recent Advances in Intuitionistic Fuzzy Logic Systems and Mathematics*, Studies in Fuzziness and Soft Computing 395,
https://doi.org/10.1007/978-3-030-53929-0_17

in a fuzzy set very often does not express corresponding degree of non membership as the complement to 1. This reflects a well known psychological fact that the linguistic negation not always identifies with logical negation. There may be some hesitation about the belongingness and nonbelongingness. To handle such situations, Atanassov [1] extended the concept of fuzzy set theory introduced by Zadeh [8], by intuitionistic fuzzy set (IFS) theory. Atanassov [2] explored the concept of fuzzy set theory by intuitionistic fuzzy set (IFS) theory. Now-a-days, IFSs are being studied extensively and being used in different disciplines of Science and Technology we can include [5–7].

Systems of simultaneous linear equations play a major role in various areas as such as mathematics, physics, statistics, neural network and etc and in most of the problems, the system's parameters and measurements are vague or imprecise. In that situation we can represent the systems with given data as intuitionistic fuzzy numbers rather than crisp numbers. Intuitionistic fuzzy linear systems are the linear systems whose parameters are all or partially represented by intuitionistic fuzzy numbers. According to our understanding, research on the property of solutions of intuitionistic fuzzy linear systems are very limited. However, a very few results [3, 4] of intuitionistic fuzzy linear systems, the authors have studied a some models for solving an intuitionistic fuzzy linear system using different concepts, in this paper we present a method for solving $n \times n$ intuitionistic fuzzy linear system where the right hand side is an intuitionistic fuzzy number vector (general case) or a symmetric intuitionistic fuzzy number vector (special case).

The structure of the paper is organized as follows. In Sect. 2, we give some basic definitions and notations. In Sect. 3 we propose a procedure to solve intuitionistic fuzzy linear equations. A numerical examples are provided to illustrate the efficiency of the method in Sect. 4 and finally some conclusions in Sect. 5.

2 Preliminaries

Definition 1 An intuitionistic fuzzy set (IFS) A in X is defined as an object of the following form

$$A = \{(x, \mu_A(x), \nu_A(x)) \mid x \in X\}$$

where

- $\mu_A : X \to [0, 1]$ degree of membership
- $\nu_A : X \to [0, 1]$ degree of non-membership

and $0 \leq \mu_A(x) + \nu_A(x) \leq 1$ for every $x \in X$.

$\pi_A(x) = 1 - \mu_A(x) - \nu_A(x)$ is called the degree of non-determinacy (or uncertainty) of the element $x \in X$ to the intuitionistic fuzzy set A and $\pi_A(x) \in [0, 1]$.

We denote by

$$\mathrm{IF}_1 = \mathrm{IF}(\mathbb{R}) = \left\{\langle u, v\rangle \,:\, \mathbb{R} \to [0, 1]^2, \,|\forall\, x \in \mathbb{R} 0 \leq u(x) + v(x) \leq 1\right\}$$

An element $\langle u, v\rangle$ of IF_1 is said an intuitionistic fuzzy number if it satisfies the following conditions

(i) $\langle u, v\rangle$ is normal i.e there exists $x_0, x_1 \in \mathbb{R}$ such that $u(x_0) = 1$ and $v(x_1) = 1$.
(ii) u is fuzzy convex and v is fuzzy concave.
(iii) u is upper semi-continuous and v is lower semi-continuous
(iv) $supp\ \langle u, v\rangle = \mathrm{cl}\{x \in \mathbb{R}\ :|\ v(x) < 1\}$ is bounded.

so we denote the collection of all intuitionistic fuzzy number by IF_1

A Triangular Intuitionistic Fuzzy Number (TIFN) $\langle u, v\rangle$ is an intuitionistic fuzzy set in $\mathbb{R}$ with the following membership function u and non-membership function v:

$$u(y) = \begin{cases} \dfrac{y-a_1}{a_2-a_1} & \text{if } a_1 \le y \le a_2 \\ \dfrac{a_3-y}{a_3-a_2} & \text{if } a_2 \le y \le a_3, \\ 0 & \text{otherwise} \end{cases}$$

$$v(y) = \begin{cases} \dfrac{a_2-y}{a_2-a_1^{'}} & \text{if } a_1^{'} \le y \le a_2 \\ \dfrac{y-a_2}{a_3^{'}-a_2} & \text{if } a_2 \le y \le a_3^{'}, \\ 1 & otherwise. \end{cases}$$

where $a_1^{'} \le a_1 \le a_2 \le a_3 \le a_3^{'}$.

We define $0_{(1,0)} \in \mathrm{IF}_1$ as

$$0_{(1,0)}(t) = \begin{cases} (1,0) & t = 0 \\ (0,1) & t \neq 0 \end{cases}$$

Definition 2 ([3]) An intuitionistic fuzzy number x in parametric form is a pair $x = \Big((\underline{x}^+, \overline{x}^+), (\underline{x}^-, \overline{x}^-)\Big)$ of functions $\underline{x}^-(\alpha)$, $\overline{x}^-(\alpha)$, $\underline{x}^+(\alpha)$ and $\overline{x}^+(\alpha)$ which satisfies the following requirements:

1. $\underline{x}^+(\alpha)$ is a bounded monotonic increasing left continuous function,
2. $\overline{x}^+(\alpha)$ is a bounded monotonic decreasing left continuous function,
3. $\underline{x}^-(\alpha)$ is a bounded monotonic increasing left continuous function,
4. $\overline{x}^-(\alpha)$ is a bounded monotonic decreasing left continuous function,
5. $\underline{x}^-(\alpha) \le \overline{x}^-(\alpha)$ and $\underline{x}^+(\alpha) \le \overline{x}^+(\alpha)$, for all $0 \le r \le 1$.

This TIFN is denoted by $x = \langle a_1, a_2, a_3; a_1^{'}, a_2, a_3^{'}\rangle$.

Its parametric form is

$$\underline{x}^{+}(\alpha) = a_1 + \alpha(a_2 - a_1),\ \overline{x}^{+}(\alpha) = a_3 - \alpha(a_3 - a_2)$$
$$\underline{x}^{-}(\alpha) = a_1' + \alpha(a_2 - a_1'),\ \overline{x}^{-}(\alpha) = a_3' - \alpha(a_3' - a_2)$$

For two intuitionistic fuzzy numbers $x = \left(\underline{x}^{+}(r), \overline{x}^{+}(r), \underline{x}^{-}(r), \overline{x}^{-}(r)\right)$ and $y = \left(\underline{y}^{+}(r), \overline{y}^{+}(r), \underline{y}^{-}(r), \overline{y}^{-}(r)\right)$ and $k \in \mathbb{R}$, the equality, the addition and scaler-multiplication are defined as follows:

- $x = y$ if and only if $\underline{x}^{+}(r) = \underline{y}^{+}(r), \overline{x}^{+}(r) = \overline{y}^{+}(r), \underline{x}^{-}(r) = \underline{y}^{-}(r)$ and $\overline{x}^{-}(r) = \overline{y}^{-}(r)$
- $x + y = \left(\underline{x}^{+}(r) + \underline{y}^{+}(r), \overline{x}^{+}(r) + \overline{y}^{+}(r), \underline{x}^{-}(r) + \underline{y}^{-}(r), \overline{x}^{-}(r) + \overline{y}^{-}(r)\right)$
- $$kx = \begin{cases} (k\underline{x}^{+}(r), k\overline{x}^{+}(r), k\underline{x}^{-}(r), k\overline{x}^{-}(r)) \text{ si } 0 \leq k \\ (k\overline{x}^{+}(r), k\underline{x}^{+}(r), k\overline{x}^{-}(r), k\underline{x}^{-}(r)) \text{ si } k < 0 \end{cases}$$

3 Intuitionistic Fuzzy Linear System

3.1 Intuitionistic Fuzzy Linear System in General Case

Definition 3 ([3]) The $n \times n$ linear system of equations

$$\begin{cases} a_{11}x_1 + a_{12}x_2 + \ldots + a_{1n}x_n = y_1, \\ a_{21}x_1 + a_{22}x_2 + \ldots + a_{2n}x_n = y_2, \\ \vdots \\ a_{n1}x_1 + a_{n2}x_2 + \ldots + a_{nn}x_n = y_n, \end{cases} \tag{1}$$

where the coefficients matrix $A = (a_{ij})$,$1 \leq i \leq n, 1 \leq j \leq n$ is a crisp $n \times n$ matrix and $y_i \in \mathrm{IF}_1$, $1 \leq i \leq n$ and the unknown $x_j \in \mathrm{IF}_1$, $1 \leq j \leq n$, is called an intuitionistic fuzzy linear system (IFLS).

Definition 4 ([3]) An intuitionistic fuzzy vector $\left(\tilde{x}_1, \tilde{x}_2, ..., \tilde{x}_n\right)^t$ given by $\tilde{x}_j = \left(\underline{x_j^{+}}(r), \overline{x_j^{+}}(r), \underline{x_j^{-}}(r), \overline{x_j^{-}}(r)\right)$, $1 \leq j \leq n$, $0 \leq r \leq 1$, is called solution of (1) if:

$$\underline{\sum_{j=1}^{n} a_{ij}x_j^+} = \sum_{j=1}^{n} \underline{a_{ij}x_j^+} = \underline{y_i^+}, i = 1, 2, ..., n$$

$$\overline{\sum_{j=1}^{n} a_{ij}x_j^+} = \sum_{j=1}^{n} \overline{a_{ij}x_j^+} = \overline{y_i^+}, i = 1, 2, ..., n$$

$$\underline{\sum_{j=1}^{n} a_{ij}x_j^-} = \sum_{j=1}^{n} \underline{a_{ij}x_j^-} = \underline{y_i^-}, i = 1, 2, ..., n$$

$$\overline{\sum_{j=1}^{n} a_{ij}x_j^-} = \sum_{j=1}^{n} \overline{a_{ij}x_j^-} = \overline{y_i^-}, i = 1, 2, ..., n$$

if, for a particular i, $a_{ij} > 0, 1 \leq j \leq n$, we get

$$\underline{\sum_{j=1}^{n} a_{ij}x_j^+} = \sum_{j=1}^{n} \underline{a_{ij}x_j^+} = \sum_{j=1}^{n} a_{ij}\underline{x_j^+} = \underline{y_i^+},$$

$$\overline{\sum_{j=1}^{n} a_{ij}x_j^+} = \sum_{j=1}^{n} \overline{a_{ij}x_j^+} = \sum_{j=1}^{n} a_{ij}\overline{x_j^+} = \overline{y_i^+},$$

$$\underline{\sum_{j=1}^{n} a_{ij}x_j^-} = \sum_{j=1}^{n} \underline{a_{ij}x_j^-} = \sum_{j=1}^{n} a_{ij}\underline{x_j^-} = \underline{y_i^-},$$

$$\overline{\sum_{j=1}^{n} a_{ij}x_j^-} = \sum_{j=1}^{n} \overline{a_{ij}x_j^-} = \sum_{j=1}^{n} a_{ij}\overline{x_j^-} = \overline{y_i^-}.$$

Let matrix B contains the positive entries of A and matrix C contains the absolute value of the negative entries of A, hence $A = B - C$.

Now using matrix notation for (1) we get $AX = Y$ or $(B - C)X = Y$.

Theorem 1 ([3]) *if $(B + C)$ and $(B - C)$ are both nonsingular and y is an arbitrary intuitionistic fuzzy number vector. Then the system (1) has an unique intuitionistic fuzzy solution if $(B + C)^{-1}$ is a nonnegative matrix.*

Theorem 2 *Suppose the inverse of matrix A in Eq. 1 exists and $(\tilde{x}_1, \tilde{x}_2, ..., \tilde{x}_n)^t$ is an intuitionistic fuzzy solution of this equation. Then $\underline{x^+} + \overline{x^+} = (\underline{x_1^+} + \overline{x_1^+}, \underline{x_2^+} + \overline{x_2^+}, \ldots, \underline{x_n^+} + \overline{x_n^+})^t$ and $\underline{x^-} + \overline{x^-} = (\underline{x_1^-} + \overline{x_1^-}, \underline{x_2^-} + \overline{x_2^-}, \ldots, \underline{x_n^-} + \overline{x_n^-})^t$ are the solutions of the following systems respectively*

$$A(\underline{x^+} + \overline{x^+}) = \underline{y^+} + \overline{y^+}$$

and

$$A(\underline{x^-} + \overline{x^-}) = \underline{y^-} + \overline{y^-}$$

where $\underline{y}^{+}+\overline{y^{+}}=(\underline{y_1^{+}}+\overline{y_1^{+}}, \underline{y_2^{+}}+\overline{y_2^{+}}, \ldots, \underline{y_n^{+}}+\overline{y_n^{+}})^{t}$ *and* $\underline{y}^{-}+\overline{y^{-}}=(\underline{y_1^{-}}+\overline{y_1^{-}}, \underline{y_2^{-}}+\overline{y_2^{-}}, \ldots, \underline{y_n^{-}}+\overline{y_n^{-}})^{t}$.

Proof Suppose the parametric form of x_j be $x_j = (\underline{x_j^{+}}, \overline{x_j^{+}}, \underline{x_j^{-}}, \overline{x_j^{-}})$, $1 \le j \le n$ and let $a_{ij} = b_{ij} - c_{ij}$ such that b_{ij} et c_{ij} are positive and $b_{ij}.c_{ij} = 0$. If we consider Eq. 1 in parametric form then for $i = 1, 2, ..., m$ we have:

$$(b_{i1} - c_{i1})(\underline{x_1^{+}}, \overline{x_1^{+}}, \underline{x_1^{-}}, \overline{x_1^{-}}) + \cdots + (b_{in} - c_{in})(\underline{x_n^{+}}, \overline{x_n^{+}}, \underline{x_n^{-}}, \overline{x_n^{-}}) = (\underline{y_i^{+}}, \overline{y_i^{+}}, \underline{y_i^{-}}, \overline{y_i^{-}}) \quad (2)$$

hence

$$b_{i1}\underline{x_1^{+}} - c_{i1}\overline{x_1^{+}} + b_{i2}\underline{x_1^{+}} - c_{i2}\overline{x_2^{+}} \cdots + b_{in}\underline{x_n^{+}} - c_{in}\overline{x_n^{+}} = \underline{y_i^{+}} \quad (3)$$

$$b_{i1}\overline{x_1^{+}} - c_{i1}\underline{x_1^{+}} + b_{i2}\overline{x_2^{+}} - c_{i2}\underline{x_2^{+}} \cdots + b_{in}\overline{x_n^{+}} - c_{in}\underline{x_n^{+}} = \overline{y_i^{+}} \quad (4)$$

and

$$b_{i1}\underline{x_1^{-}} - c_{i1}\overline{x_1^{-}} + b_{i2}\underline{x_1^{-}} - c_{i2}\overline{x_2^{-}} \cdots + b_{in}\underline{x_n^{-}} - c_{in}\overline{x_n^{-}} = \underline{y_i^{-}} \quad (5)$$

$$b_{i1}\overline{x_1^{-}} - c_{i1}\underline{x_1^{-}} + b_{i2}\overline{x_2^{-}} - c_{i2}\underline{x_2^{-}} \cdots + b_{in}\overline{x_n^{-}} - c_{in}\underline{x_n^{-}} = \overline{y_i^{-}} \quad (6)$$

By addition of (3) and (4) we have:

$$(b_{i1} - c_{i1})(\underline{x_1^{+}} + \overline{x_1^{+}}) + \cdots + (b_{in} - c_{in})(\underline{x_n^{+}} + \overline{x_n^{+}}) = (\underline{y_i^{+}} + \overline{y_i^{+}}) \quad (7)$$

and same for (5) and (6) we have:

$$(b_{i1} - c_{i1})(\underline{x_1^{-}} + \overline{x_1^{-}}) + \cdots + (b_{in} - c_{in})(\underline{x_n^{-}} + \overline{x_n^{-}}) = (\underline{y_i^{-}} + \overline{y_i^{-}}) \quad (8)$$

and hence

$$a_{i1}(\underline{x_1^{+}} + \overline{x_1^{+}}) + \cdots + a_{in}(\underline{x_n^{+}} + \overline{x_n^{+}}) = (\underline{y_i^{+}} + \overline{y_i^{+}}) \quad (9)$$

and

$$a_{i1}(\underline{x_1^{-}} + \overline{x_1^{-}}) + \cdots + a_{in}(\underline{x_n^{-}} + \overline{x_n^{-}}) = (\underline{y_i^{-}} + \overline{y_i^{-}}) \quad (10)$$

Then, $X^{+}=(\underline{x_1^{+}} + \overline{x_1^{+}}, \underline{x_2^{+}} + \overline{x_2^{+}}, \ldots, \underline{x_n^{+}} + \overline{x_n^{+}})^{t}$ and $X^{-}=(\underline{x_1^{-}} + \overline{x_1^{-}}, \underline{x_2^{-}} + \overline{x_2^{-}}, \ldots, \underline{\underline{x_n^{-}}} + \overline{x_n^{-}})^{t}$ are the solutions of $A(\underline{x^{+}} + \overline{x^{+}}) = \underline{y^{+}} + \overline{y^{+}}$ and $A(\underline{x^{-}} + \overline{x^{-}}) = \underline{y^{-}} + \overline{y^{-}}$.

For solving (1), we first solve the following systems:

$$
\begin{cases}
a_{11}(\underline{x_1^+}(r)+\overline{x_1^+}(r))+\cdots+a_{1n}(\underline{x_n^+}(r)+\overline{x_n^+}(r))=(\underline{y_1^+}(r)+\overline{y_1^+}(r))\\
a_{21}(\underline{x_1^+}(r)+\overline{x_1^+}(r))+\cdots+a_{2n}(\underline{x_n^+}(r)+\overline{x_n^+}(r))=(\underline{y_2^+}(r)+\overline{y_2^+}(r))\\
\vdots\\
a_{n1}(\underline{x_1^+}(r)+\overline{x_1^+}(r))+\cdots+a_{nn}(\underline{x_n^+}(r)+\overline{x_n^+}(r))=(\underline{y_n^+}(r)+\overline{y_n^+}(r))
\end{cases}
\tag{11}
$$

and

$$
\begin{cases}
a_{11}(\underline{x_1^-}(r)+\overline{x_1^-}(r))+\cdots+a_{1n}(\underline{x_n^-}(r)+\overline{x_n^-}(r))=(\underline{y_1^-}(r)+\overline{y_1^-}(r))\\
a_{21}(\underline{x_1^+}(r)+\overline{x_1^+}(r))+\cdots+a_{2n}(\underline{x_n^+}(r)+\overline{x_n^+}(r))=(\underline{y_2^+}(r)+\overline{y_2^+}(r))\\
\vdots\\
a_{n1}(\underline{x_1^-}(r)+\overline{x_1^-}(r))+\cdots+a_{nn}(\underline{x_n^-}(r)+\overline{x_n^-}(r))=(\underline{y_n^-}(r)+\overline{y_n^-}(r))
\end{cases}
\tag{12}
$$

and suppose the solution of (11) be as

$$
d^+ = \begin{pmatrix} d_1^+ \\ d_2^+ \\ \vdots \\ d_n^+ \end{pmatrix} = \begin{pmatrix} \underline{x_1^+}(r)+\overline{x_1^+}(r) \\ \underline{x_2^+}(r)+\overline{x_2^+}(r) \\ \vdots \\ \underline{x_n^+}(r)+\overline{x_n^+}(r) \end{pmatrix}
$$

and the solution of (12) be as

$$
d^- = \begin{pmatrix} d_1^- \\ d_2^- \\ \vdots \\ d_n^- \end{pmatrix} = \begin{pmatrix} \underline{x_1^-}(r)+\overline{x_1^-}(r) \\ \underline{x_2^-}(r)+\overline{x_2^-}(r) \\ \vdots \\ \underline{x_n^-}(r)+\overline{x_n^-}(r) \end{pmatrix}
$$

Let matrix B contains the positive entries of A and matrix C contains the absolute of the negative entries of A, hence $A = B - C$.

Now using matrix notation for Eq. 1 we get $Ax = y$ or $(B - C)x = y$ and in parametric form:

$$
(B-C)(\underline{x^+}(r), \overline{x^+}(r)) = (\underline{y^+}(r), \overline{y^+}(r))
$$

$$
(B-C)(\underline{x^-}(r), \overline{x^-}(r)) = (\underline{y^-}(r), \overline{y^-}(r))
$$

We can write this system as follows:

$$\begin{cases} B\underline{x}^{+}(r) - C\overline{x^{+}}(r) = \underline{y}^{+}(r), \\ B\overline{x^{+}}(r) - C\underline{x}^{+}(r) = \overline{y^{+}}(r), \\ B\underline{x}^{-}(r) - C\overline{x^{-}}(r) = \underline{y}^{-}(r), \\ B\overline{x^{-}}(r) - C\underline{x}^{-}(r) = \overline{y^{-}}(r). \end{cases}$$

By substituting of $\overline{x^{+}(r)} = d^{+} - \underline{x}^{+}(r)$, $\underline{x}^{+}(r) = d^{+} - \overline{x^{+}}(r)$, $\overline{x^{-}}(r) = d^{-} - \underline{x}^{-}(r)$, and $\underline{x}^{-}(r) = d^{-} - \overline{x^{-}}(r)$ in equations of above system, respectively, we have:

$$\begin{cases} (B + C)\underline{x}^{+}(r) = \underline{y}^{+}(r) + Cd^{+} \\ (B + C)\overline{x^{+}}(r) = \overline{y^{+}}(r) + Cd^{+} \\ (B + C)\underline{x}^{-}(r) = \underline{y}^{-}(r) + Cd^{-} \\ (B + C)\overline{x^{-}}(r) = \overline{y^{-}}(r) + Cd^{-} \end{cases} \tag{13}$$

If the inverse of matrix $(B + C)$ exist then:

$$\underline{x}^{+}(r) = (B + C)^{-1}(\underline{y}^{+}(r) + Cd^{+}),$$
$$\overline{x^{+}}(r) = (B + C)^{-1}(\overline{y^{+}}(r) + Cd^{+}).$$
$$\underline{x}^{-}(r) = (B + C)^{-1}(\underline{y}^{-}(r) + Cd^{-}),$$
$$\overline{x^{-}}(r) = (B + C)^{-1}(\overline{y^{-}}(r) + Cd^{-}).$$

Therefore, we can solve intuitionistic fuzzy linear system Eq. 1 by solving Eqs. (11)–(13). The solution vector is unique but may still not be an appropriate intuitionistic fuzzy vector.

3.2 Intuitionistic Fuzzy Linear System in Special Case

Definition 5 A triangular intuitionistic fuzzy number $x = (\underline{x}^{+}(r), \overline{x}^{+}(r), \underline{x}^{-}(r), \underline{x}^{-}(r))$ is symmetric if $\frac{\underline{x}^{+}(r)+\overline{x}^{+}(r)}{2}$ and $\frac{\underline{x}^{-}(r)+\overline{x}^{-}(r)}{2}$ are constants.

Consider the problem Eq. 1 but we consider y_i as triangular symmetric intuitionistic fuzzy number.

Remark 1 ([3]) Let X be an intuitionistic fuzzy solution of system (1) where A is nonsingular matrix and Y is an intuitionistic fuzzy number. If Y is symmetric intuitionistic fuzzy number vector then X is symmetric intuitionistic fuzzy number vector.

For solving (1), we first solve the following system:

$$\begin{cases} a_{11}(\underline{x_1^+}(0) + \overline{x_1^+}(0)) + \cdots + a_{1n}(\underline{x_n^+}(0) + \overline{x_n^+}(0)) = (\underline{y_1^+}(0) + \overline{y_1^+}(0)) \\ a_{21}(\underline{x_1^+}(0) + \overline{x_1^+}(0)) + \cdots + a_{2n}(\underline{x_n^+}(0) + \overline{x_n^+}(0)) = (\underline{y_2^+}(0) + \overline{y_2^+}(0))) \\ \vdots \\ a_{n1}(\underline{x_1^+}(0) + \overline{x_1^+}(0)) + \cdots + a_{nn}(\underline{x_n^+}(0) + \overline{x_n^+}(0)) = (\underline{y_n^+}(0) + \overline{y_n^+}(0)) \end{cases} \tag{14}$$

and

$$\begin{cases} a_{11}(\underline{x_1^-}(1) + \overline{x_1^-}(1)) + \cdots + a_{1n}(\underline{x_n^-}(1) + \overline{x_n^-}(1)) = (\underline{y_1^-}(1) + \overline{y_1^-}(1)) \\ a_{21}(\underline{x_1^-}(1) + \overline{x_1^-}(1)) + \cdots + a_{2n}(\underline{x_n^-}(1) + \overline{x_n^-}(1)) = (\underline{y_2^-}(1) + \overline{y_2^-}(1)) \\ \vdots \\ a_{n1}(\underline{x_1^-}(1) + \overline{x_1^-}(1)) + \cdots + a_{nn}(\underline{x_n^-}(1) + \overline{x_n^-}(1)) = (\underline{y_n^-}(1) + \overline{y_n^-}(1)) \end{cases} \tag{15}$$

and suppose the solution of (14) be as

$$d^+ = \begin{pmatrix} d_1^+ \\ d_2^+ \\ \vdots \\ d_n^+ \end{pmatrix} = \begin{pmatrix} \underline{x_1^+}(0) + \overline{x_1^+}(0) \\ \underline{x_2^+}(0) + \overline{x_2^+}(0) \\ \vdots \\ \underline{x_n^+}(0) + \overline{x_n^+}(0) \end{pmatrix}$$

and the solution of (15) be as

$$d^- = \begin{pmatrix} d_1^- \\ d_2^- \\ \vdots \\ d_n^- \end{pmatrix} = \begin{pmatrix} \underline{x_1^-}(1) + \overline{x_1^-}(1) \\ \underline{x_2^-}(1) + \overline{x_2^-}(1) \\ \vdots \\ \underline{x_n^-}(1) + \overline{x_n^-}(1) \end{pmatrix}$$

Now using matrix notation for (1) we get $Ax = y$ or $(B - C)x = y$, then we have:

$$(B - C)(\underline{x^+}(0), \overline{x^+}(0)) = (\underline{y^+}(0), \overline{y^+}(0))$$

$$(B - C)(\underline{x^-}(1), \overline{x^-}(1)) = (\underline{y^-}(1), \overline{y^-}(1))$$

We can write this system as follows:

$$\begin{cases} B\underline{x}^{+}(0) - C\overline{x^{+}}(0) = \underline{y}^{+}(0), \\ B\overline{x^{+}}(0) - C\underline{x}^{+}(0) = \overline{y^{+}}(0), \\ B\underline{x}^{-}(1) - C\overline{x^{-}}(1) = \underline{y}^{-}(1), \\ B\overline{x^{-}}(1) - C\underline{x}^{-}(1) = \overline{y^{-}}(1). \end{cases}$$

By substituting of $\overline{x^{+}}(0) = d^{+} - \underline{x}^{+}(0)$, $\underline{x}^{+}(0) = d^{+} - \overline{x^{+}}(0)$, $\overline{x^{-}}(1) = d^{-} - \underline{x}^{-}(1)$ and $\underline{x}^{-}(1) = d^{-} - \overline{x^{-}}(1)$ in equations of above system, respectively, we have:

$$\begin{cases} (B + C)\underline{x}^{+}(0) = \underline{y}^{+}(0) + Cd^{+} \\ (B + C)\overline{x^{+}}(0) = \overline{y^{+}}(0) + Cd^{+} \\ (B + C)\underline{x}^{-}(1) = \underline{y}^{-}(1) + Cd^{-} \\ (B + C)\overline{x^{-}}(1) = \overline{y^{-}}(1) + Cd^{-} \end{cases} \tag{16}$$

If the matrix $B + C$ is nonsingular then:

$$\underline{x}^{+}(0) = (B + C)^{-1}(\underline{y}^{+}(0) + Cd^{+}),$$
$$\overline{x^{+}}(0) = (B + C)^{-1}(\overline{y^{+}}(0) + Cd^{+}).$$
$$\underline{x}^{-}(1) = (B + C)^{-1}(\underline{y}^{-}(1) + Cd^{-}),$$
$$\overline{x^{-}}(1) = (B + C)^{-1}(\overline{y^{-}}(1) + Cd^{-}).$$

4 Examples

Example 1 Consider the following intuitionistic fuzzy linear system

$$\begin{cases} 2x_1 + 3x_2 = & (2 + 2r; 8 - 4r; 4 - 3r; 4 + 5r) \\ 5x_1 - x_2 = & (4r; 6 - 2r; 4 - 5r; 4 + 3r) \end{cases} \tag{17}$$

By using the Eqs. (11)–(12) we have (Fig. 1):

$$d^{+} = \begin{pmatrix} \frac{28}{17} + \frac{4r}{17} \\ \frac{38}{17} - \frac{14r}{17} \end{pmatrix}$$

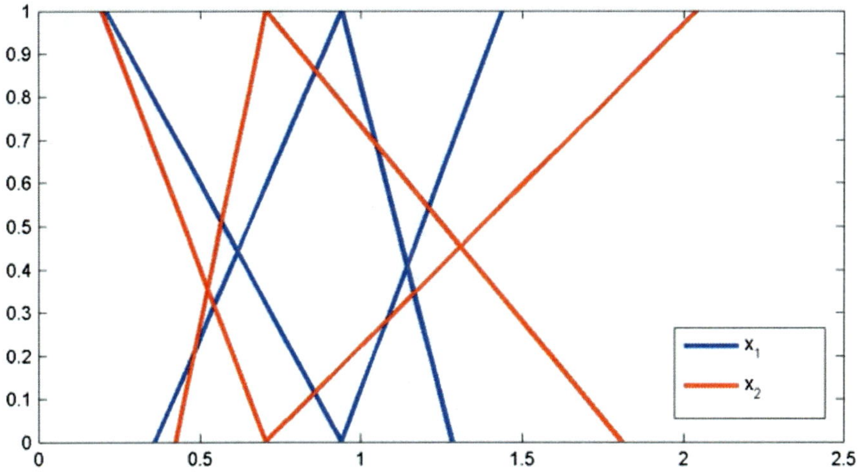

Fig. 1 The intuitionistic fuzzy numbers x_1 and x_2

and

$$d^- = \begin{pmatrix} \frac{32}{17} - \frac{4r}{17} \\ \frac{24}{17} + \frac{14r}{17} \end{pmatrix}$$

From Eq. (13) we have:

$$\underline{X^+} = \begin{pmatrix} \frac{80}{221} + \frac{128r}{221} \\ \frac{94}{221} + \frac{62r}{221} \end{pmatrix}, \quad \overline{X^+} = \begin{pmatrix} \frac{284}{221} - \frac{76r}{221} \\ \frac{400}{221} - \frac{244r}{221} \end{pmatrix}$$

and

$$\underline{X^-} = \begin{pmatrix} \frac{16}{17} - \frac{162r}{221} \\ \frac{12}{17} - \frac{113r}{221} \end{pmatrix}, \quad \overline{X^-} = \begin{pmatrix} \frac{16}{17} + \frac{110r}{221} \\ \frac{12}{17} + \frac{295r}{221} \end{pmatrix}$$

Hence

$$\begin{cases} x_1 = & \left(\frac{80}{221} + \frac{128r}{221}; \frac{284}{221} - \frac{76r}{221}; \frac{16}{17} - \frac{162r}{221}; \frac{16}{17} + \frac{110r}{221}\right) \\ x_2 = & \left(\frac{94}{221} + \frac{62r}{221}; \frac{400}{221} - \frac{244r}{221}; \frac{12}{17} - \frac{113r}{221}; \frac{12}{17} + \frac{295r}{221}\right) \end{cases} \tag{18}$$

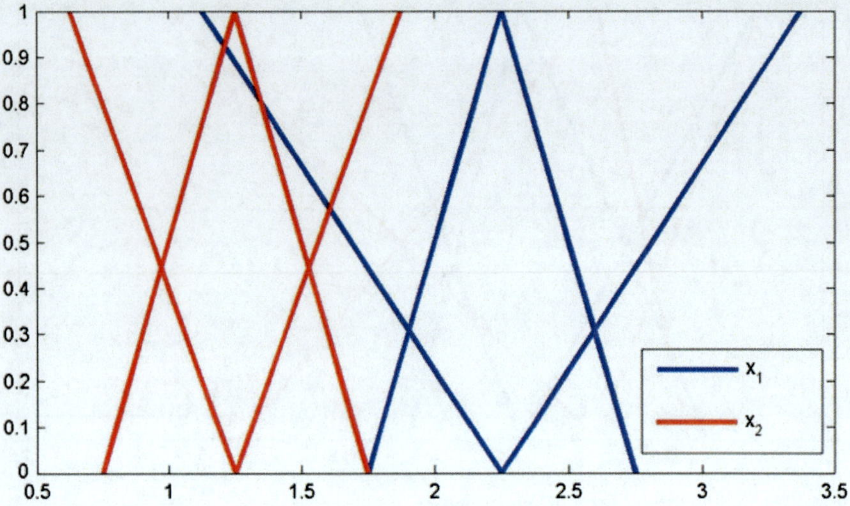

Fig. 2 The intuitionistic fuzzy numbers x_1 and x_2

Example 2 Consider the following intuitionistic fuzzy linear system:

$$\begin{cases} x_1 - x_2 = & (r; 2 - r; 1 - 1.75r; 1 + 1.75r) \\ \\ x_1 + 3x_2 = & (4 + 2r; 8 - 2r; 6 - 3r; 6 + 3r) \end{cases} \tag{19}$$

By using the Eqs. (14)–(15) we have:

$$d^{+}(0) = \begin{pmatrix} 4.5 \\ \\ 2.5 \end{pmatrix}$$

and

$$d^{-}(1) = \begin{pmatrix} 4.5 \\ \\ 2.5 \\ \\ \end{pmatrix}$$

From Eq. (16) we have (Fig. 2):

$$\underline{X^{+}}(0) = \begin{pmatrix} 1.75 \\ \\ 0.75 \end{pmatrix}, \quad \overline{X^{+}}(0) = \begin{pmatrix} 2.75 \\ \\ 1.75 \end{pmatrix}$$

and

$$\underline{X^-}(1) = \begin{pmatrix} 1.125 \\ 0.625 \end{pmatrix}, \quad \overline{X^-}(1) = \begin{pmatrix} 3.375 \\ 1.875 \end{pmatrix}$$

Hence

$$\begin{cases} x_1 = & (0.5r + 1.75; -0.5r + 2.75; -1.125r + 2.25; 1.125r + 2.25) \\ x_2 = & (0.5r + 0.75; -0.5r + 1.75; -0.625r + 1.25; 0.625r + 1.25) \end{cases} \tag{20}$$

5 Conclusion

In this paper, we have presented a procedure for solving intuitionistic fuzzy linear systems where the right hand side is an intuitionistic fuzzy number vector (general case) or a symmetric intuitionistic fuzzy number vector (special case). The original system is replaced by four $n \times n$ crisp systems. The solution vector is a symmetric solution if the right hand side vector is symmetric. These results are illustrated by some computational examples.

References

1. Atanassov, K.T.: Intuitionistic fuzzy sets. VII ITKR's session, Sofia (deposited in Central Science and Technical Library of the Bulgarian Academy of Sciences 1697/84) (1983)
2. Atanassov, K.: Intuitionistic fuzzy sets. Fuzzy Sets Syst. **20**, 87–96 (1986)
3. Atti, H., Amma, B.B., Melliani, S., Chadli, L.S.: Intuitionistic fuzzy linear systems. In: Intuitionistic and Type-2 Fuzzy Logic Enhancements in Neural and Optimization Algorithms: Theory and Applications, vol. 862. Springer International Publishing, Berlin (2019)
4. Pradhan, R., Pal, M.: Solvability of system of intuitionistic fuzzy linear equations. Int. J. Fuzzy Logic Syst. (IJFLS) **4**(3), 13–324 (2014)
5. Sotirov, S., Sotirova, E., Melin, P., Castillo, O., Atanassov.: Modular neural network preprocessing procedure with intuitionistic fuzzy intercriteria analysis method. Flexib. Query Answ. Syst. **2015**, 175–186 (2015)
6. Sotirov, S., Sotirova, E., Atanassova, V., Atanassov, K., Castillo, O., Melin, P., Petkov, T., Surchev, S.: A hybrid approach for modular neural network design using intercriteria analysis and intuitionistic fuzzy logic. Complexity **2018**, (2018)
7. Wang, Z., Li, K.W., Wang, W.: An approach to multiattribute decision making with interval-valued intuitionistic fuzzy assessments and incomplete weights. Inf. Sci. **179**(7), 3026–3040 (2009)
8. Zadeh, L.A.: Fuzzy sets. Inf. Control **8**(3), 338–353 (1965)

On the Principal Minors Assignment Problem for Skew-Symmetric Matrices

R. Matoui and K. Driss

Abstract Let A be an $n \times n$ real or complex skew-symmetric matrix. Define an *elementary transformation* $\mathcal{T}$: A $\rightarrow T \circ A$/where $(\circ)$ is the Hadamard product, t an $n \times n$ matrix with elements in $\{-1, 1\}$ such A and $T \circ A$ are skew-symmetric with equal corresponding principal minors of all orders. The aim of this work is to present an algorithm that returns all elementary transformations of an arbitrary skew-symmetric matrix of order n, preserving the form of the matrix and it's principal minors, further more, the conjecture of Boussairi and Chergui regarding an equivalence relation between skew-symmetric matrices with no zeros off diagonal and having equal corresponding principal minors is discussed.

1 Introduction

In Algebra, the study of the principal minors has been raised in many areas of interest, more precisely the relationship between a square matrix of order n and a vector of 2^n values representing a set of principal minors, or the dependency relationship between the elements of this vector. For dense symmetric matrices, it has been shown in particular that the main minors up to order 3 defines the rest of the principal minors and analogically for the skew-symmetric matrices, that the main minors up to order 4 defines the rest of the principal minors. Part of this work is to answer the question of generating the transformations of an skew-symmetric matrices leading to matrices having equal principal minors of all order through an algorithm.

R. Matoui (✉) · K. Driss
Laboratoire de Mathématique, Département de Mathématique,
Faculté des Sciences et Techniques de Mohammedia, 20800 Mohammedia, Morocco
e-mail: rachid.matoui@gmail.com

K. Driss
e-mail: driss.kariim@gmail.com

S. Melliani and O. Castillo (eds.), *Recent Advances in Intuitionistic Fuzzy Logic Systems and Mathematics*, Studies in Fuzziness and Soft Computing 395,
https://doi.org/10.1007/978-3-030-53929-0_19

2 Definitions and Examples

Definition 1 Let $\mathscr{M}_n$ be the set of complex matrices of order n and $A \in \mathscr{M}_n$ a known or arbitrary matrix.

We define a set of transformations $\mathscr{T}$ by the set $\left\{t, t \circ A \overset{pm}{=} A\right\}$, where ($\circ$) is the Hadamard product [3] and t an $n \times n$ real or complex matrix such that for all matrices $t \in \mathscr{T}$, the matrix $t \circ A$ have the same principal minors of all orders. We use the notation $A \overset{pm}{=} t \circ A$ to specify principal minors of all orders equality.

A transformation is said to be *elementary*, if A and $t \circ A$ are both symmetric or both skew-symmetric accordingly and t having elements in $\{-1, 0, 1\}$.

Remark 1 Some remarks about the previous definition

1. Let $A = [t_{ij}]_{1 \le i,j \le n}$ be a matrix with some null elements, clearly, the corresponding elements in all elements of $\mathscr{T}$ can be any numbers $\in \mathbb{C}$. By convention, we choose those elements to be null as well.
2. For an elementary transformation $\mathscr{T}$, all the elements $t \in \mathscr{T}$ are symmetric matrices.
3. A skew-symmetric transformation element $t \in \mathscr{T}$ switches the form of a symmetric matrix to a skew-symmetric matrix and vice-versa.
4. Clearly, the set of transformations of an arbitrary matrix is included but not always equal to the set of transformations of a specific matrix of the same size, see Example 3.

3 Examples

Below are some examples of transformations preserving principal minors.

Example 1 An example of a transformation not preserving the form of the initial matrix. The resulting matrix is not symmetric nor skew-symmetric.

$$\begin{pmatrix} 0 & -i & i \\ i & 0 & 1 \\ -i & 1 & 0 \end{pmatrix} \circ \begin{pmatrix} 0 & -x_{12} \cdot i & x_{13} \cdot i \\ -x_{12} \cdot i & 0 & x_{23} \\ x_{13} \cdot i & -x_{23} & 0 \end{pmatrix} \overset{pm}{=} \begin{pmatrix} 0 & x_{12} & x_{13} \\ -x_{12} & 0 & x_{23} \\ -x_{13} & -x_{23} & 0 \end{pmatrix}$$

Example 2 An example of an *elementary transformation*.

$$\begin{pmatrix} 0 & 1 & -1 & 1 \\ 1 & 0 & -1 & 1 \\ -1 & -1 & 0 & -1 \\ 1 & 1 & -1 & 0 \end{pmatrix} \circ \begin{pmatrix} 0 & x_{12} & x_{13} & x_{14} \\ -x_{12} & 0 & x_{23} & x_{24} \\ -x_{13} & -x_{23} & 0 & x_{34} \\ -x_{14} & -x_{24} & -x_{34} & 0 \end{pmatrix} \overset{pm}{=} \begin{pmatrix} 0 & x_{12} & -x_{13} & x_{14} \\ -x_{12} & 0 & -x_{23} & x_{24} \\ x_{13} & x_{23} & 0 & -x_{34} \\ -x_{14} & -x_{24} & x_{34} & 0 \end{pmatrix}$$

Example 3 An example of an elementary transformation switching a single element's sign for a specific matrix, such transformation is not a valid for arbitrary matrices.

$$\begin{pmatrix} 0 & -1 & 1 & 1 & 1 & 1 \\ -1 & 0 & 1 & 1 & 1 & 1 \\ 1 & 1 & 0 & 1 & 1 & 1 \\ 1 & 1 & 1 & 0 & 1 & 1 \\ 1 & 1 & 1 & 1 & 0 & 1 \\ 1 & 1 & 1 & 1 & 1 & 0 \end{pmatrix} \circ \begin{pmatrix} 0 & -1 & 1 & 1 & 1 & -1 \\ 1 & 0 & 1 & 1 & 1 & -1 \\ -1 & -1 & 0 & 1 & 1 & -1 \\ -1 & -1 & -1 & 0 & 1 & -1 \\ -1 & -1 & -1 & -1 & 0 & -1 \\ 1 & 1 & 1 & 1 & 1 & 0 \end{pmatrix} \overset{pm}{=} \begin{pmatrix} 0 & 1 & 1 & 1 & 1 & -1 \\ -1 & 0 & 1 & 1 & 1 & -1 \\ -1 & -1 & 0 & 1 & 1 & -1 \\ -1 & -1 & -1 & 0 & 1 & -1 \\ -1 & -1 & -1 & -1 & 0 & -1 \\ 1 & 1 & 1 & 1 & 1 & 0 \end{pmatrix}$$

4 The Conjecture

In order to state the *BC conjecture* named after Boussairi and Chergui [1] specifying an equivalence relation between skew-symmetric matrices having equal corresponding principal minors, we need the following definitions and notations. Let $A = [a_{ij}]$ be an $n \times n$ matrix and let X; Y be two nonempty subsets of [n] (where [n] := 1, ..., n). We denote by A[X; Y] the sub-matrix of A having row indices in X and column indices in Y. If X = Y, then A[X; X] is a principal sub-matrix of A and we abbreviate this to A[X].

Following [1], a subset X of [n] is an HL-clan of A if both of matrices $A[\bar{X}; X]$ and $A[X; \bar{X}]$ have rank of at most 1 (where $\bar{X} := [n]\backslash X$). By definition, $\emptyset$, $[n]$ and singletons are HL-clans. Considering the particular case when A is skew-symmetric, let X be a subset of [n]. We denote by $Inv(X; A) := [t_{ij}]$ the matrix obtained from A as follows. For any $i; j \in [n]$, $t_{ij} = -a_{ij}$ if $i; j \in X$ and $t_{ij} = a_{ij}$, otherwise. More generally, let A and B be two skew-symmetric matrices, assume that there exists a sequence $A_0 = A, \ldots, A_m = B$ of $n \times n$ skew-symmetric matrices such that for $k = 0, \ldots, m$; $A_{k+1} = Inv(X_k; A_k)$, where X_k is an HL-clan of A_k. Two matrices A, B obtained in this way are called HL-clan-reversal-equivalent.

The conjecture of Boussairi and Chergui is stated as follow.

Two $n \times n$ skew-symmetric real matrices have equal corresponding principal minors of all orders if and only if they are HL-clan-reversal-equivalent.

5 Determining the Principal Minors of Skew-Symmetric Matrices

For dense symmetric matrices, Oeding pointed in [2] that the principal minors of order at most 3 defines the rest of the principal minors, but since the odd principal minors of skew-symmetric matrices are null, we need to use the forth principal minors. In the following, we define an equivalence relation between *pairwise different* matrices having equal corresponding principal minors and the transformation results of the program in Sect. 6.1. The pairwise different condition is specified to preserve the general form of matrices.

Theorem 1 *For two dense skew-symmetric $n \times n$ matrices pairwise different, the following statements are equivalent:*

1. *A and B have equal corresponding principal minors of all orders;*
2. *A and B have equal corresponding principal minors of order at most 4;*
3. *An elementary transformation t is returned by the SMPM program Sect. 6.1 such that $A \overset{pm}{=} t \circ B$.*

Definition 2 A skew-symmetric matrix A such

$$A = \begin{pmatrix} 0 & a & b & c \\ -a & 0 & d & e \\ -b & -d & 0 & f \\ -c & -e & -f & 0 \end{pmatrix}$$

is said pairwise different if the scalars af, $-be$ and dc are *pairwise different*.

Proposition 1 *Let $A = [a_{ij}]$ and $B = [b_{ij}]$ two 4×4 dense skew-symmetric matrices such $a_{ij} = \pm b_{ij}, i, j \in \{1, \ldots, 4\}$ have the same determinant if and only if one of the above statements is true:*

1. *The couple's elements $\{a_{12}a_{34}, b_{12}b_{34}\}$, $\{a_{13}a_{24}, b_{13}, b_{24}\}$ and $\{a_{23}a_{14}nb_{23}, b_{14}\}$ have respectively and independently a positive or negative but the same signs;*
2. *The couple's elements$\{a_{12}a_{34}, b_{12}b_{34}\}$, $\{a_{13}a_{24}, b_{13}, b_{24}\}$ and $\{a_{23}a_{14}nb_{23}, b_{14}\}$ have all respectively the opposite sign;*
3. *The determinant is null;*
4. *At least one couple from the set $\{a_{12}a_{34}, a_{13}a_{24}, -a_{23}a_{14}\}$ are equals. In that case the matrices A and B are said to be pairwise equal otherwise they are pairwise different.*

Proof Let A and B two dense skew-symmetric matrices such

$$A = \begin{pmatrix} 0 & a_{12} & a_{13} & a_{14} \\ -a_{12} & 0 & a_{23} & a_{24} \\ -a_{13} & -a_{23} & 0 & a_{34} \\ -a_{14} & -a_{24} & -a_{34} & 0 \end{pmatrix}, B = \begin{pmatrix} 0 & b_{12} & b_{13} & b_{14} \\ -b_{12} & 0 & b_{23} & b_{24} \\ -b_{13} & -b_{23} & 0 & b_{34} \\ -b_{14} & -b_{24} & -b_{34} & 0 \end{pmatrix}$$

since $det(A) = (a_{12}a_{34} - a_{13}a_{24} + a_{23}a_{14})^2$, $det(B) = (b_{12}b_{34} - b_{13}b_{24} + b_{23}b_{14})^2$, and $a_{ij} = \pm b_{ij}, i, j \in \{1, \ldots, 4\}$, and since there are no zeros off the diagonal and $a_{ij} = \pm b_{ij}$, there exists three unknowns $\{x, y, z\}$ having values in $\{-1, 1\}$ such $det(B) = (xa_{14}a_{23} - ya_{13}a_{24} + za_{12}a_{34})^2$. We can easily check that solving $det(A) = det(B)$ leads to the previous statements.

Remark 2 As consequence of the Proposition 1, the elementary transformations for 4×4 pairwise different matrices are:

$$1.\begin{pmatrix} 0 & 1 & 1 & 1 \\ -1 & 0 & 1 & 1 \\ -1 & -1 & 0 & 1 \\ -1 & -1 & -1 & 0 \end{pmatrix} \quad 2.\begin{pmatrix} 0 & -1 & 1 & 1 \\ 1 & 0 & 1 & 1 \\ -1 & -1 & 0 & -1 \\ -1 & -1 & 1 & 0 \end{pmatrix} \quad 3.\begin{pmatrix} 0 & 1 & -1 & 1 \\ -1 & 0 & 1 & -1 \\ 1 & -1 & 0 & 1 \\ -1 & 1 & -1 & 0 \end{pmatrix}$$

$$4.\begin{pmatrix} 0 & 1 & 1 & -1 \\ -1 & 0 & -1 & 1 \\ -1 & 1 & 0 & 1 \\ 1 & -1 & -1 & 0 \end{pmatrix} \quad 5.\begin{pmatrix} 0 & -1 & -1 & 1 \\ 1 & 0 & 1 & -1 \\ 1 & -1 & 0 & -1 \\ -1 & 1 & 1 & 0 \end{pmatrix} \quad 6.\begin{pmatrix} 0 & -1 & 1 & -1 \\ 1 & 0 & -1 & 1 \\ -1 & 1 & 0 & -1 \\ 1 & -1 & 1 & 0 \end{pmatrix}$$

$$7.\begin{pmatrix} 0 & 1 & -1 & -1 \\ -1 & 0 & -1 & -1 \\ 1 & 1 & 0 & 1 \\ 1 & 1 & -1 & 0 \end{pmatrix} \quad 8.\begin{pmatrix} 0 & -1 & -1 & -1 \\ 1 & 0 & -1 & -1 \\ 1 & 1 & 0 & -1 \\ 1 & 1 & 1 & 0 \end{pmatrix} \quad 9.\begin{pmatrix} 0 & -1 & -1 & 1 \\ 1 & 0 & -1 & 1 \\ 1 & 1 & 0 & 1 \\ -1 & -1 & -1 & 0 \end{pmatrix}$$

$$10.\begin{pmatrix} 0 & 1 & 1 & -1 \\ -1 & 0 & 1 & -1 \\ -1 & -1 & 0 & -1 \\ 1 & 1 & 1 & 0 \end{pmatrix} \quad 11.\begin{pmatrix} 0 & -1 & -1 & -1 \\ 1 & 0 & 1 & 1 \\ 1 & -1 & 0 & 1 \\ 1 & -1 & -1 & 0 \end{pmatrix} \quad 12.\begin{pmatrix} 0 & 1 & 1 & 1 \\ -1 & 0 & -1 & -1 \\ -1 & 1 & 0 & -1 \\ -1 & 1 & 1 & 0 \end{pmatrix}$$

$$13.\begin{pmatrix} 0 & -1 & 1 & 1 \\ 1 & 0 & -1 & -1 \\ -1 & 1 & 0 & 1 \\ -1 & 1 & -1 & 0 \end{pmatrix} \quad 14.\begin{pmatrix} 0 & 1 & -1 & -1 \\ -1 & 0 & 1 & 1 \\ 1 & -1 & 0 & -1 \\ 1 & -1 & 1 & 0 \end{pmatrix} \quad 15.\begin{pmatrix} 0 & 1 & -1 & 1 \\ -1 & 0 & -1 & 1 \\ 1 & 1 & 0 & -1 \\ -1 & -1 & 1 & 0 \end{pmatrix}$$

$$16.\begin{pmatrix} 0 & -1 & 1 & -1 \\ 1 & 0 & 1 & -1 \\ -1 & -1 & 0 & 1 \\ 1 & 1 & -1 & 0 \end{pmatrix}$$

Lemma 1 *Let A and B two 6×6 dense skew-symmetric matrices equals up to a sign pairwise different. A and B have equal corresponding principal minors of order at most 4 if and only if they have equal corresponding principal minors of all orders.*

Proof This result is proved under weaker conditions by Boussairi and Chergui in [1]

Definition 3 Let A be a skew-symmetric matrix, and v a 2^{n-1} vector representing the set of it's principal minors, for conventional notation, v is said to be in *combinatorial order* if it is ordered by blocks of principal minors order, then by the combinatorial order of chosen rows and columns for each order. In other words, for a specific principal minor order k, there are $c = \binom{n}{k}$ vectors of size k representing the selected rows/columns forming the principal sub-matrix, such the rows/columns number of the vectors are in ascending order in the vectors individually and for the corresponding order in the set.

Remark 3 As a consequence of the definition of combinatorial order, the entries of v for a 3×3 matrix are as follows:

$$v = \left(\underbrace{\underbrace{A[1]}_{PM1_1}, \underbrace{A[2]}_{PM1_2}, \underbrace{A[3]}_{PM1_3}}_{\text{PM1}}, \underbrace{\underbrace{A[1,2]}_{PM2_1}, \underbrace{A[1,3]}_{PM2_2}, \underbrace{A[2,3]}_{PM2_3}}_{\text{PM2}}, \underbrace{\underbrace{A[1,2,3]}_{PM3_1}}_{\text{PM3}} \right)$$

Proof (*the main Theorem*) The equivalence between the second and last statements holds by design, the proof of the equivalence between the first and second statements based on the Lemma 1.

6 The Algorithm and Computational Results

Throughout this section we explain the algorithm of the program SMPM generating the elementary transformations for an $n \times n$ arbitrary dense matrix. We present then computational results and analysis.

6.1 The Program SMPM

Using the results of the Proposition 1, we initiate the algorithm with the matrices obtained in Remark 2. The matrices can be hard coded in the implementation because this same set is used abstraction of the size of the transformations. The algorithm uses the principle of the Theorem 1 by building recursively matrices of equal corresponding principal minors of 4th order. The algorithm proceeds as the following:

```
%Generate Elementary transformations of an $n\times n$
skew-symmetric matrix
\begin{algorithm2e}[H]
  \SetAlgoLined
  \KwResult{Generate Elementary transformations of an
  $n\times n $ skew-symmetric matrix}
initialization of $\mathcal{E}_{1\leq j \leq16}$
\tcp{the set of 16 elementary transformation matrices
of order 4}
initialization of $\mathcal{R}_{j=1}$ \tcp{Initiate
the result set with one element formed by unknowns
elements}
initialization of $PM_{1 \leq i \leq 2^n-1}$
\tcp{Initiate the set of the principal minors}
  \For{All principal minors combinations $PM_i$}{
  \For{All $\mathcal{R}_j$ elements of $\mathcal{R}$ }{
```

```
    \For{All $\mathcal{E}_k$ elements of  $\mathcal{E}$}{
      \uIf{the elements of $\mathcal{R}_i$ are still unknown}{
       Fill the unknown elements and add the new matrix
       to $\mathcal{R}$ \;     }
      \uElseIf{no element of $E$ is equal to $PM_i$}{
        Remove the matrix from $\mathcal{R}$\;}
  }
    }
  }
   Return $\mathcal{R}$ \;
    \caption{Generate Elementary transformations of an
    $n\times n $ skew-symmetric matrix}
  \end{algorithm2e}
```

The Matlab®implementation is given in the appendix.

6.2 *Elementary Transformations for* 6 × 6 *Matrices*

In this section we present the elementary transformations returned by the program for a 6 × 6 matrix. The result consists of 64 skew-symmetric dense matrices. The Hadamard products of any skew-symmetric matrix by the resulting matrices share the same principal minors of all orders.

$$\begin{pmatrix} 0 & 1 & 1 & 1 & 1 & 1 \\ -1 & 0 & 1 & 1 & 1 & 1 \\ -1 & -1 & 0 & 1 & 1 & 1 \\ -1 & -1 & -1 & 0 & 1 & 1 \\ -1 & -1 & -1 & -1 & 0 & 1 \\ -1 & -1 & -1 & -1 & -1 & 0 \end{pmatrix}$$
$$\begin{pmatrix} 0 & 1 & 1 & 1 & 1 & -1 \\ -1 & 0 & 1 & 1 & 1 & -1 \\ -1 & -1 & 0 & 1 & 1 & -1 \\ -1 & -1 & -1 & 0 & 1 & -1 \\ -1 & -1 & -1 & -1 & 0 & -1 \\ 1 & 1 & 1 & 1 & 1 & 0 \end{pmatrix}$$
$$\begin{pmatrix} 0 & 1 & 1 & 1 & -1 & 1 \\ -1 & 0 & 1 & 1 & -1 & 1 \\ -1 & -1 & 0 & 1 & -1 & 1 \\ -1 & -1 & -1 & 0 & -1 & 1 \\ 1 & 1 & 1 & 1 & 0 & -1 \\ -1 & -1 & -1 & -1 & 1 & 0 \end{pmatrix}$$
$$\begin{pmatrix} 0 & 1 & 1 & 1 & -1 & -1 \\ -1 & 0 & 1 & 1 & -1 & -1 \\ -1 & -1 & 0 & 1 & -1 & -1 \\ -1 & -1 & -1 & 0 & -1 & -1 \\ 1 & 1 & 1 & 1 & 0 & 1 \\ 1 & 1 & 1 & 1 & -1 & 0 \end{pmatrix}$$

$$\begin{pmatrix} 0 & -1 & 1 & 1 & 1 & 1 \\ 1 & 0 & 1 & 1 & 1 & 1 \\ -1 & -1 & 0 & -1 & -1 & -1 \\ -1 & -1 & 1 & 0 & -1 & -1 \\ -1 & -1 & 1 & 1 & 0 & -1 \\ -1 & -1 & 1 & 1 & 1 & 0 \end{pmatrix}$$

$$\begin{pmatrix} 0 & -1 & 1 & 1 & 1 & -1 \\ 1 & 0 & 1 & 1 & 1 & -1 \\ -1 & -1 & 0 & -1 & -1 & 1 \\ -1 & -1 & 1 & 0 & -1 & 1 \\ -1 & -1 & 1 & 1 & 0 & 1 \\ 1 & 1 & -1 & -1 & -1 & 0 \end{pmatrix}$$

$$\begin{pmatrix} 0 & -1 & 1 & 1 & -1 & 1 \\ 1 & 0 & 1 & 1 & -1 & 1 \\ -1 & -1 & 0 & -1 & 1 & -1 \\ -1 & -1 & 1 & 0 & 1 & -1 \\ 1 & 1 & -1 & -1 & 0 & 1 \\ -1 & -1 & 1 & 1 & -1 & 0 \end{pmatrix}$$

$$\begin{pmatrix} 0 & -1 & 1 & 1 & -1 & -1 \\ 1 & 0 & 1 & 1 & -1 & -1 \\ -1 & -1 & 0 & -1 & 1 & 1 \\ -1 & -1 & 1 & 0 & 1 & 1 \\ 1 & 1 & -1 & -1 & 0 & -1 \\ 1 & 1 & -1 & -1 & 1 & 0 \end{pmatrix}$$

$$\begin{pmatrix} 0 & 1 & -1 & 1 & 1 & 1 \\ -1 & 0 & 1 & -1 & -1 & -1 \\ 1 & -1 & 0 & 1 & 1 & 1 \\ -1 & 1 & -1 & 0 & -1 & -1 \\ -1 & 1 & -1 & 1 & 0 & -1 \\ -1 & 1 & -1 & 1 & 1 & 0 \end{pmatrix}$$

$$\begin{pmatrix} 0 & 1 & -1 & 1 & 1 & -1 \\ -1 & 0 & 1 & -1 & -1 & 1 \\ 1 & -1 & 0 & 1 & 1 & -1 \\ -1 & 1 & -1 & 0 & -1 & 1 \\ -1 & 1 & -1 & 1 & 0 & 1 \\ 1 & -1 & 1 & -1 & -1 & 0 \end{pmatrix}$$

$$\begin{pmatrix} 0 & 1 & -1 & 1 & -1 & 1 \\ -1 & 0 & 1 & -1 & 1 & -1 \\ 1 & -1 & 0 & 1 & -1 & 1 \\ -1 & 1 & -1 & 0 & 1 & -1 \\ 1 & -1 & 1 & -1 & 0 & 1 \\ -1 & 1 & -1 & 1 & -1 & 0 \end{pmatrix}$$

$$\begin{pmatrix} 0 & 1 & -1 & 1 & -1 & -1 \\ -1 & 0 & 1 & -1 & 1 & 1 \\ 1 & -1 & 0 & 1 & -1 & -1 \\ -1 & 1 & -1 & 0 & 1 & 1 \\ 1 & -1 & 1 & -1 & 0 & -1 \\ 1 & -1 & 1 & -1 & 1 & 0 \end{pmatrix}$$

$$\begin{pmatrix} 0 & 1 & 1 & -1 & -1 & -1 \\ -1 & 0 & -1 & 1 & 1 & 1 \\ -1 & 1 & 0 & 1 & 1 & 1 \\ 1 & -1 & -1 & 0 & -1 & -1 \\ 1 & -1 & -1 & 1 & 0 & -1 \\ 1 & -1 & -1 & 1 & 1 & 0 \end{pmatrix}$$

$$\begin{pmatrix} 0 & 1 & 1 & -1 & -1 & 1 \\ -1 & 0 & -1 & 1 & 1 & -1 \\ -1 & 1 & 0 & 1 & 1 & -1 \\ 1 & -1 & -1 & 0 & -1 & 1 \\ 1 & -1 & -1 & 1 & 0 & 1 \\ -1 & 1 & 1 & -1 & -1 & 0 \end{pmatrix}$$

$$\begin{pmatrix} 0 & 1 & 1 & -1 & 1 & -1 \\ -1 & 0 & -1 & 1 & -1 & 1 \\ -1 & 1 & 0 & 1 & -1 & 1 \\ 1 & -1 & -1 & 0 & 1 & -1 \\ -1 & 1 & 1 & -1 & 0 & 1 \\ 1 & -1 & -1 & 1 & -1 & 0 \end{pmatrix}$$

$$\begin{pmatrix} 0 & 1 & 1 & -1 & 1 & 1 \\ -1 & 0 & -1 & 1 & -1 & -1 \\ -1 & 1 & 0 & 1 & -1 & -1 \\ 1 & -1 & -1 & 0 & 1 & 1 \\ -1 & 1 & 1 & -1 & 0 & -1 \\ -1 & 1 & 1 & -1 & 1 & 0 \end{pmatrix}$$

$$\begin{pmatrix} 0 & -1 & -1 & 1 & 1 & 1 \\ 1 & 0 & 1 & -1 & -1 & -1 \\ 1 & -1 & 0 & -1 & -1 & -1 \\ -1 & 1 & 1 & 0 & 1 & 1 \\ -1 & 1 & 1 & -1 & 0 & 1 \\ -1 & 1 & 1 & -1 & -1 & 0 \end{pmatrix}$$

$$\begin{pmatrix} 0 & -1 & -1 & 1 & 1 & -1 \\ 1 & 0 & 1 & -1 & -1 & 1 \\ 1 & -1 & 0 & -1 & -1 & 1 \\ -1 & 1 & 1 & 0 & 1 & -1 \\ -1 & 1 & 1 & -1 & 0 & -1 \\ 1 & -1 & -1 & 1 & 1 & 0 \end{pmatrix}$$

$$\begin{pmatrix} 0 & -1 & -1 & 1 & -1 & 1 \\ 1 & 0 & 1 & -1 & 1 & -1 \\ 1 & -1 & 0 & -1 & 1 & -1 \\ -1 & 1 & 1 & 0 & -1 & 1 \\ 1 & -1 & -1 & 1 & 0 & -1 \\ -1 & 1 & 1 & -1 & 1 & 0 \end{pmatrix}$$

$$\begin{pmatrix} 0 & -1 & -1 & 1 & -1 & -1 \\ 1 & 0 & 1 & -1 & 1 & 1 \\ 1 & -1 & 0 & -1 & 1 & 1 \\ -1 & 1 & 1 & 0 & -1 & -1 \\ 1 & -1 & -1 & 1 & 0 & 1 \\ 1 & -1 & -1 & 1 & -1 & 0 \end{pmatrix}$$

$$\begin{pmatrix} 0 & -1 & 1 & -1 & -1 & -1 \\ 1 & 0 & -1 & 1 & 1 & 1 \\ -1 & 1 & 0 & -1 & -1 & -1 \\ 1 & -1 & 1 & 0 & 1 & 1 \\ 1 & -1 & 1 & -1 & 0 & 1 \\ 1 & -1 & 1 & -1 & -1 & 0 \end{pmatrix}$$

$$\begin{pmatrix} 0 & -1 & 1 & -1 & -1 & 1 \\ 1 & 0 & -1 & 1 & 1 & -1 \\ -1 & 1 & 0 & -1 & -1 & 1 \\ 1 & -1 & 1 & 0 & 1 & -1 \\ 1 & -1 & 1 & -1 & 0 & -1 \\ -1 & 1 & -1 & 1 & 1 & 0 \end{pmatrix}$$

$$\begin{pmatrix} 0 & -1 & 1 & -1 & 1 & -1 \\ 1 & 0 & -1 & 1 & -1 & 1 \\ -1 & 1 & 0 & -1 & 1 & -1 \\ 1 & -1 & 1 & 0 & -1 & 1 \\ -1 & 1 & -1 & 1 & 0 & -1 \\ 1 & -1 & 1 & -1 & 1 & 0 \end{pmatrix}$$

$$\begin{pmatrix} 0 & -1 & 1 & -1 & 1 & 1 \\ 1 & 0 & -1 & 1 & -1 & -1 \\ -1 & 1 & 0 & -1 & 1 & 1 \\ 1 & -1 & 1 & 0 & -1 & -1 \\ -1 & 1 & -1 & 1 & 0 & 1 \\ -1 & 1 & -1 & 1 & -1 & 0 \end{pmatrix}$$

$$\begin{pmatrix} 0 & 1 & -1 & -1 & -1 & -1 \\ -1 & 0 & -1 & -1 & -1 & -1 \\ 1 & 1 & 0 & 1 & 1 & 1 \\ 1 & 1 & -1 & 0 & 1 & 1 \\ 1 & 1 & -1 & -1 & 0 & 1 \\ 1 & 1 & -1 & -1 & -1 & 0 \end{pmatrix}$$

$$\begin{pmatrix} 0 & 1 & -1 & -1 & -1 & 1 \\ -1 & 0 & -1 & -1 & -1 & 1 \\ 1 & 1 & 0 & 1 & 1 & -1 \\ 1 & 1 & -1 & 0 & 1 & -1 \\ 1 & 1 & -1 & -1 & 0 & -1 \\ -1 & -1 & 1 & 1 & 1 & 0 \end{pmatrix}$$

$$\begin{pmatrix} 0 & 1 & -1 & -1 & 1 & -1 \\ -1 & 0 & -1 & -1 & 1 & -1 \\ 1 & 1 & 0 & 1 & -1 & 1 \\ 1 & 1 & -1 & 0 & -1 & 1 \\ -1 & -1 & 1 & 1 & 0 & -1 \\ 1 & 1 & -1 & -1 & 1 & 0 \end{pmatrix}$$

$$\begin{pmatrix} 0 & 1 & -1 & -1 & 1 & 1 \\ -1 & 0 & -1 & -1 & 1 & 1 \\ 1 & 1 & 0 & 1 & -1 & -1 \\ 1 & 1 & -1 & 0 & -1 & -1 \\ -1 & -1 & 1 & 1 & 0 & 1 \\ -1 & -1 & 1 & 1 & -1 & 0 \end{pmatrix}$$

$$\begin{pmatrix} 0 & -1 & -1 & -1 & -1 & -1 \\ 1 & 0 & -1 & -1 & -1 & -1 \\ 1 & 1 & 0 & -1 & -1 & -1 \\ 1 & 1 & 1 & 0 & -1 & -1 \\ 1 & 1 & 1 & 1 & 0 & -1 \\ 1 & 1 & 1 & 1 & 1 & 0 \end{pmatrix}$$

$$\begin{pmatrix} 0 & -1 & -1 & -1 & -1 & 1 \\ 1 & 0 & -1 & -1 & -1 & 1 \\ 1 & 1 & 0 & -1 & -1 & 1 \\ 1 & 1 & 1 & 0 & -1 & 1 \\ 1 & 1 & 1 & 1 & 0 & 1 \\ -1 & -1 & -1 & -1 & -1 & 0 \end{pmatrix}$$

$$\begin{pmatrix} 0 & -1 & -1 & -1 & 1 & -1 \\ 1 & 0 & -1 & -1 & 1 & -1 \\ 1 & 1 & 0 & -1 & 1 & -1 \\ 1 & 1 & 1 & 0 & 1 & -1 \\ -1 & -1 & -1 & -1 & 0 & 1 \\ 1 & 1 & 1 & 1 & -1 & 0 \end{pmatrix}$$

$$\begin{pmatrix} 0 & -1 & -1 & -1 & 1 & 1 \\ 1 & 0 & -1 & -1 & 1 & 1 \\ 1 & 1 & 0 & -1 & 1 & 1 \\ 1 & 1 & 1 & 0 & 1 & 1 \\ -1 & -1 & -1 & -1 & 0 & -1 \\ -1 & -1 & -1 & -1 & 1 & 0 \end{pmatrix}$$

$$\begin{pmatrix} 0 & -1 & -1 & 1 & -1 & -1 \\ 1 & 0 & -1 & 1 & -1 & -1 \\ 1 & 1 & 0 & 1 & -1 & -1 \\ -1 & -1 & -1 & 0 & 1 & 1 \\ 1 & 1 & 1 & -1 & 0 & -1 \\ 1 & 1 & 1 & -1 & 1 & 0 \end{pmatrix}$$

$$\begin{pmatrix} 0 & -1 & -1 & 1 & -1 & 1 \\ 1 & 0 & -1 & 1 & -1 & 1 \\ 1 & 1 & 0 & 1 & -1 & 1 \\ -1 & -1 & -1 & 0 & 1 & -1 \\ 1 & 1 & 1 & -1 & 0 & 1 \\ -1 & -1 & -1 & 1 & -1 & 0 \end{pmatrix}$$

$$\begin{pmatrix} 0 & -1 & -1 & 1 & 1 & -1 \\ 1 & 0 & -1 & 1 & 1 & -1 \\ 1 & 1 & 0 & 1 & 1 & -1 \\ -1 & -1 & -1 & 0 & -1 & 1 \\ -1 & -1 & -1 & 1 & 0 & 1 \\ 1 & 1 & 1 & -1 & -1 & 0 \end{pmatrix}$$

$$\begin{pmatrix} 0 & -1 & -1 & 1 & 1 & 1 \\ 1 & 0 & -1 & 1 & 1 & 1 \\ 1 & 1 & 0 & 1 & 1 & 1 \\ -1 & -1 & -1 & 0 & -1 & -1 \\ -1 & -1 & -1 & 1 & 0 & -1 \\ -1 & -1 & -1 & 1 & 1 & 0 \end{pmatrix}$$

$$\begin{pmatrix} 0 & 1 & 1 & -1 & 1 & 1 \\ -1 & 0 & 1 & -1 & 1 & 1 \\ -1 & -1 & 0 & -1 & 1 & 1 \\ 1 & 1 & 1 & 0 & -1 & -1 \\ -1 & -1 & -1 & 1 & 0 & 1 \\ -1 & -1 & -1 & 1 & -1 & 0 \end{pmatrix}$$

$$\begin{pmatrix} 0 & 1 & 1 & -1 & 1 & -1 \\ -1 & 0 & 1 & -1 & 1 & -1 \\ -1 & -1 & 0 & -1 & 1 & -1 \\ 1 & 1 & 1 & 0 & -1 & 1 \\ -1 & -1 & -1 & 1 & 0 & -1 \\ 1 & 1 & 1 & -1 & 1 & 0 \end{pmatrix}$$

$$\begin{pmatrix} 0 & 1 & 1 & -1 & -1 & 1 \\ -1 & 0 & 1 & -1 & -1 & 1 \\ -1 & -1 & 0 & -1 & -1 & 1 \\ 1 & 1 & 1 & 0 & 1 & -1 \\ 1 & 1 & 1 & -1 & 0 & -1 \\ -1 & -1 & -1 & 1 & 1 & 0 \end{pmatrix}$$

$$\begin{pmatrix} 0 & 1 & 1 & -1 & -1 & -1 \\ -1 & 0 & 1 & -1 & -1 & -1 \\ -1 & -1 & 0 & -1 & -1 & -1 \\ 1 & 1 & 1 & 0 & 1 & 1 \\ 1 & 1 & 1 & -1 & 0 & 1 \\ 1 & 1 & 1 & -1 & -1 & 0 \end{pmatrix}$$

$$\begin{pmatrix} 0 & -1 & -1 & -1 & 1 & 1 \\ 1 & 0 & 1 & 1 & -1 & -1 \\ 1 & -1 & 0 & 1 & -1 & -1 \\ 1 & -1 & -1 & 0 & -1 & -1 \\ -1 & 1 & 1 & 1 & 0 & 1 \\ -1 & 1 & 1 & 1 & -1 & 0 \end{pmatrix}$$

$$\begin{pmatrix} 0 & -1 & -1 & -1 & 1 & -1 \\ 1 & 0 & 1 & 1 & -1 & 1 \\ 1 & -1 & 0 & 1 & -1 & 1 \\ 1 & -1 & -1 & 0 & -1 & 1 \\ -1 & 1 & 1 & 1 & 0 & -1 \\ 1 & -1 & -1 & -1 & 1 & 0 \end{pmatrix}$$

$$\begin{pmatrix} 0 & -1 & -1 & -1 & -1 & 1 \\ 1 & 0 & 1 & 1 & 1 & -1 \\ 1 & -1 & 0 & 1 & 1 & -1 \\ 1 & -1 & -1 & 0 & 1 & -1 \\ 1 & -1 & -1 & -1 & 0 & -1 \\ -1 & 1 & 1 & 1 & 1 & 0 \end{pmatrix}$$

$$\begin{pmatrix} 0 & -1 & -1 & -1 & -1 & -1 \\ 1 & 0 & 1 & 1 & 1 & 1 \\ 1 & -1 & 0 & 1 & 1 & 1 \\ 1 & -1 & -1 & 0 & 1 & 1 \\ 1 & -1 & -1 & -1 & 0 & 1 \\ 1 & -1 & -1 & -1 & -1 & 0 \end{pmatrix}$$

$$\begin{pmatrix} 0 & 1 & 1 & 1 & -1 & -1 \\ -1 & 0 & -1 & -1 & 1 & 1 \\ -1 & 1 & 0 & -1 & 1 & 1 \\ -1 & 1 & 1 & 0 & 1 & 1 \\ 1 & -1 & -1 & -1 & 0 & -1 \\ 1 & -1 & -1 & -1 & 1 & 0 \end{pmatrix}$$

$$\begin{pmatrix} 0 & 1 & 1 & 1 & -1 & 1 \\ -1 & 0 & -1 & -1 & 1 & -1 \\ -1 & 1 & 0 & -1 & 1 & -1 \\ -1 & 1 & 1 & 0 & 1 & -1 \\ 1 & -1 & -1 & -1 & 0 & 1 \\ -1 & 1 & 1 & 1 & -1 & 0 \end{pmatrix}$$

$$\begin{pmatrix} 0 & 1 & 1 & 1 & 1 & -1 \\ -1 & 0 & -1 & -1 & -1 & 1 \\ -1 & 1 & 0 & -1 & -1 & 1 \\ -1 & 1 & 1 & 0 & -1 & 1 \\ -1 & 1 & 1 & 1 & 0 & 1 \\ 1 & -1 & -1 & -1 & -1 & 0 \end{pmatrix}$$

$$\begin{pmatrix} 0 & 1 & 1 & 1 & 1 & 1 \\ -1 & 0 & -1 & -1 & -1 & -1 \\ -1 & 1 & 0 & -1 & -1 & -1 \\ -1 & 1 & 1 & 0 & -1 & -1 \\ -1 & 1 & 1 & 1 & 0 & -1 \\ -1 & 1 & 1 & 1 & 1 & 0 \end{pmatrix}$$

$$\begin{pmatrix} 0 & -1 & 1 & 1 & -1 & -1 \\ 1 & 0 & -1 & -1 & 1 & 1 \\ -1 & 1 & 0 & 1 & -1 & -1 \\ -1 & 1 & -1 & 0 & -1 & -1 \\ 1 & -1 & 1 & 1 & 0 & 1 \\ 1 & -1 & 1 & 1 & -1 & 0 \end{pmatrix}$$

$$\begin{pmatrix} 0 & -1 & 1 & 1 & -1 & 1 \\ 1 & 0 & -1 & -1 & 1 & -1 \\ -1 & 1 & 0 & 1 & -1 & 1 \\ -1 & 1 & -1 & 0 & -1 & 1 \\ 1 & -1 & 1 & 1 & 0 & -1 \\ -1 & 1 & -1 & -1 & 1 & 0 \end{pmatrix}$$

$$\begin{pmatrix} 0 & -1 & 1 & 1 & 1 & -1 \\ 1 & 0 & -1 & -1 & -1 & 1 \\ -1 & 1 & 0 & 1 & 1 & -1 \\ -1 & 1 & -1 & 0 & 1 & -1 \\ -1 & 1 & -1 & -1 & 0 & -1 \\ 1 & -1 & 1 & 1 & 1 & 0 \end{pmatrix}$$

$$\begin{pmatrix} 0 & -1 & 1 & 1 & 1 & 1 \\ 1 & 0 & -1 & -1 & -1 & -1 \\ -1 & 1 & 0 & 1 & 1 & 1 \\ -1 & 1 & -1 & 0 & 1 & 1 \\ -1 & 1 & -1 & -1 & 0 & 1 \\ -1 & 1 & -1 & -1 & -1 & 0 \end{pmatrix}$$

$$\begin{pmatrix} 0 & 1 & -1 & -1 & 1 & 1 \\ -1 & 0 & 1 & 1 & -1 & -1 \\ 1 & -1 & 0 & -1 & 1 & 1 \\ 1 & -1 & 1 & 0 & 1 & 1 \\ -1 & 1 & -1 & -1 & 0 & -1 \\ -1 & 1 & -1 & -1 & 1 & 0 \end{pmatrix}$$

$$\begin{pmatrix} 0 & 1 & -1 & -1 & 1 & -1 \\ -1 & 0 & 1 & 1 & -1 & 1 \\ 1 & -1 & 0 & -1 & 1 & -1 \\ 1 & -1 & 1 & 0 & 1 & -1 \\ -1 & 1 & -1 & -1 & 0 & 1 \\ 1 & -1 & 1 & 1 & -1 & 0 \end{pmatrix}$$

$$\begin{pmatrix} 0 & 1 & -1 & -1 & -1 & 1 \\ -1 & 0 & 1 & 1 & 1 & -1 \\ 1 & -1 & 0 & -1 & -1 & 1 \\ 1 & -1 & 1 & 0 & -1 & 1 \\ 1 & -1 & 1 & 1 & 0 & 1 \\ -1 & 1 & -1 & -1 & -1 & 0 \end{pmatrix}$$

$$\begin{pmatrix} 0 & 1 & -1 & -1 & -1 & -1 \\ -1 & 0 & 1 & 1 & 1 & 1 \\ 1 & -1 & 0 & -1 & -1 & -1 \\ 1 & -1 & 1 & 0 & -1 & -1 \\ 1 & -1 & 1 & 1 & 0 & -1 \\ 1 & -1 & 1 & 1 & 1 & 0 \end{pmatrix}$$

$$\begin{pmatrix} 0 & 1 & -1 & 1 & -1 & -1 \\ -1 & 0 & -1 & 1 & -1 & -1 \\ 1 & 1 & 0 & -1 & 1 & 1 \\ -1 & -1 & 1 & 0 & -1 & -1 \\ 1 & 1 & -1 & 1 & 0 & 1 \\ 1 & 1 & -1 & 1 & -1 & 0 \end{pmatrix}$$

$$\begin{pmatrix} 0 & 1 & -1 & 1 & -1 & 1 \\ -1 & 0 & -1 & 1 & -1 & 1 \\ 1 & 1 & 0 & -1 & 1 & -1 \\ -1 & -1 & 1 & 0 & -1 & 1 \\ 1 & 1 & -1 & 1 & 0 & -1 \\ -1 & -1 & 1 & -1 & 1 & 0 \end{pmatrix}$$

$$\begin{pmatrix} 0 & 1 & -1 & 1 & 1 & -1 \\ -1 & 0 & -1 & 1 & 1 & -1 \\ 1 & 1 & 0 & -1 & -1 & 1 \\ -1 & -1 & 1 & 0 & 1 & -1 \\ -1 & -1 & 1 & -1 & 0 & -1 \\ 1 & 1 & -1 & 1 & 1 & 0 \end{pmatrix}$$

$$\begin{pmatrix} 0 & 1 & -1 & 1 & 1 & 1 \\ -1 & 0 & -1 & 1 & 1 & 1 \\ 1 & 1 & 0 & -1 & -1 & -1 \\ -1 & -1 & 1 & 0 & 1 & 1 \\ -1 & -1 & 1 & -1 & 0 & 1 \\ -1 & -1 & 1 & -1 & -1 & 0 \end{pmatrix}$$

$$\begin{pmatrix} 0 & -1 & 1 & -1 & 1 & 1 \\ 1 & 0 & 1 & -1 & 1 & 1 \\ -1 & -1 & 0 & 1 & -1 & -1 \\ 1 & 1 & -1 & 0 & 1 & 1 \\ -1 & -1 & 1 & -1 & 0 & -1 \\ -1 & -1 & 1 & -1 & 1 & 0 \end{pmatrix}$$

$$\begin{pmatrix} 0 & -1 & 1 & -1 & 1 & -1 \\ 1 & 0 & 1 & -1 & 1 & -1 \\ -1 & -1 & 0 & 1 & -1 & 1 \\ 1 & 1 & -1 & 0 & 1 & -1 \\ -1 & -1 & 1 & -1 & 0 & 1 \\ 1 & 1 & -1 & 1 & -1 & 0 \end{pmatrix}$$

$$\begin{pmatrix} 0 & -1 & 1 & -1 & -1 & 1 \\ 1 & 0 & 1 & -1 & -1 & 1 \\ -1 & -1 & 0 & 1 & 1 & -1 \\ 1 & 1 & -1 & 0 & -1 & 1 \\ 1 & 1 & -1 & 1 & 0 & 1 \\ -1 & -1 & 1 & -1 & -1 & 0 \end{pmatrix}$$

$$\begin{pmatrix} 0 & -1 & 1 & -1 & -1 & -1 \\ 1 & 0 & 1 & -1 & -1 & -1 \\ -1 & -1 & 0 & 1 & 1 & 1 \\ 1 & 1 & -1 & 0 & -1 & -1 \\ 1 & 1 & -1 & 1 & 0 & -1 \\ 1 & 1 & -1 & 1 & 1 & 0 \end{pmatrix}$$

We illustrate all HL-clans relations by the following figure, every point represents a matrix and a line between two matrices represents an HL-Clan relation:

Therefore, to validate the conjecture, we must find a path between every couple of matrices. Or more simply, find a path from every matrix to the first one. Which we present in the following figures for the 6th, 8th and the 10th matrix orders (Figs. 1, 2, 3 and 4):

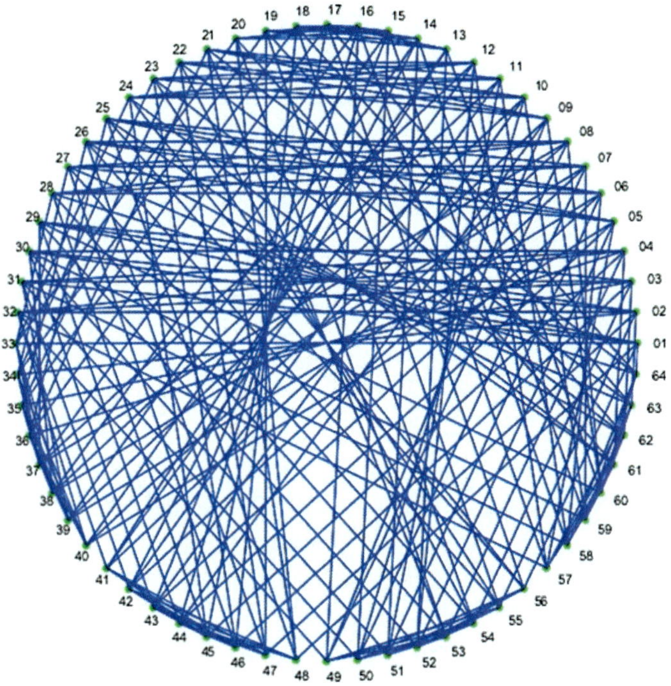

Fig. 1 The HL-clans relations

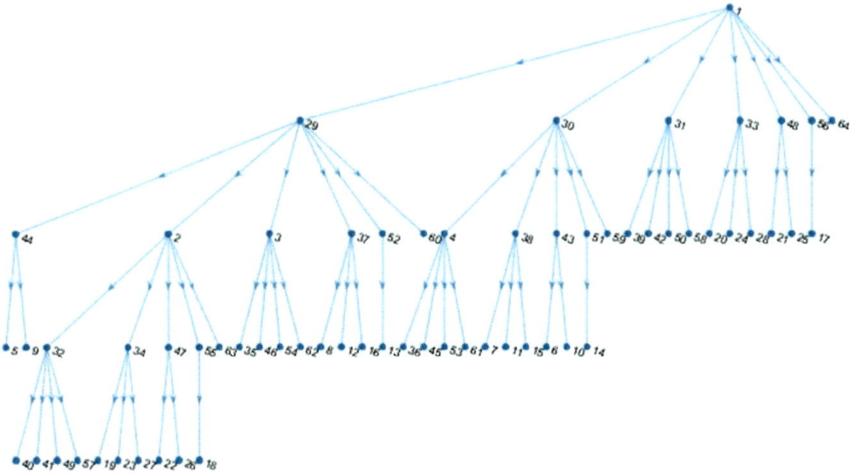

Fig. 2 The HL-clans relations paths from all matrices to the first one for the 6th order matrices

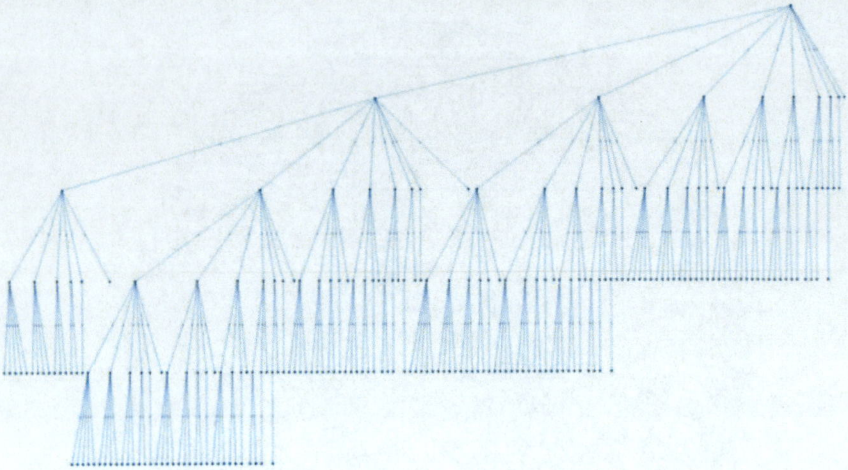

Fig. 3 The HL-clans relations paths from all matrices to the first one for the 8th order matrices

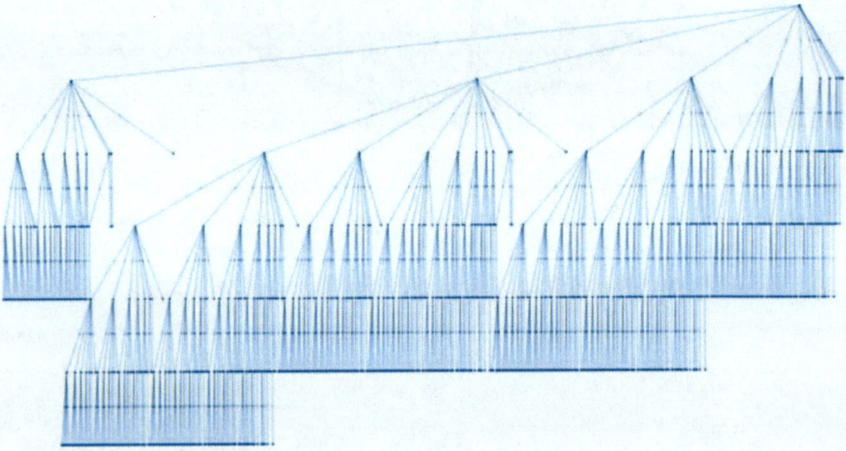

Fig. 4 The HL-clans relations paths from all matrices to the first one for the 10th order matrices

6.3 Computational Results

We illustrate in the following the results obtained from the recursive execution of the program for different matrix orders (Table 1):

Table 1 Execution results

Matrix order	Number of elementary transformations	Number of HL-clans relations	Maximum path length
6	64	448	4
8	256	2304	5
10	1024	11264	6
12	4096	53248	7
14	16384	245760	8

We can generalize the following results to a skew-symmetric matrix of order n:

1. The number of elementary matrices is 2^n;
2. The number of HL-Clans relations is $(n+1)2^n$;
3. The Maximum path length is $\frac{n}{2}+1$.

6.4 Practicals Examples

Let A and B two skew-symmetric matrices formed as follow:

$$A = \begin{pmatrix} 0 & 2 & 3 & 5 & 7 & 11 \\ -2 & 0 & 13 & 17 & 19 & 23 \\ -3 & -13 & 0 & 29 & 31 & 37 \\ -5 & -17 & -29 & 0 & 41 & 43 \\ -7 & -19 & -31 & -41 & 0 & 47 \\ -11 & -23 & -37 & -43 & 47 & 0 \end{pmatrix}, B = \begin{pmatrix} 0 & -2 & -3 & 5 & 7 & -11 \\ 2 & 0 & 13 & -17 & -19 & 23 \\ 3 & -13 & 0 & -29 & -31 & 37 \\ -5 & 17 & -29 & 0 & 41 & -43 \\ -7 & 19 & -31 & -41 & 0 & -47 \\ 11 & -23 & -37 & 43 & 47 & 0 \end{pmatrix}$$

A and B are formed to be pairwise different, we can identify the corresponding elementary transformations from the result of the SMPM program in Sect. 6.2 to the first and eighteenth matrices. Therefore from the matrices path in Fig. 2 we can deduce a valid path can going through the matrices obtained by the elementary transformations numbered by 29, 2, 55, 18 explained in the following steps (Table 2).

Table 2 HL-clan relation steps

Step	Matrix	HL-clan
1	$\begin{pmatrix} 0 & 2 & 3 & 5 & 7 & 11 \\ -2 & 0 & 13 & 17 & 19 & 23 \\ -3 & -13 & 0 & 29 & 31 & 37 \\ -5 & -17 & -29 & 0 & 41 & 43 \\ -7 & -19 & -31 & -41 & 0 & 47 \\ -11 & -23 & -37 & -43 & -47 & 0 \end{pmatrix}$	[1, 2, 3, 4, 5, 6]
2	$\begin{pmatrix} 0 & -2 & -3 & -5 & -7 & -11 \\ 2 & 0 & -13 & -17 & -19 & -23 \\ 3 & 13 & 0 & -29 & -31 & -37 \\ 5 & 17 & 29 & 0 & -41 & -43 \\ 7 & 19 & 31 & 41 & 0 & -47 \\ 11 & 23 & 37 & 43 & 47 & 0 \end{pmatrix}$	[1, 2, 3, 4, 5]
3	$\begin{pmatrix} 0 & 2 & 3 & 5 & 7 & -11 \\ -2 & 0 & 13 & 17 & 19 & -23 \\ -3 & -13 & 0 & 29 & 31 & -37 \\ -5 & -17 & -29 & 0 & 41 & -43 \\ -7 & -19 & -31 & -41 & 0 & -47 \\ 11 & 23 & 37 & 43 & 47 & 0 \end{pmatrix}$	[1, 3, 4, 5, 6]
4	$\begin{pmatrix} 0 & 2 & -3 & -5 & -7 & 11 \\ -2 & 0 & 13 & 17 & 19 & -23 \\ 3 & -13 & 0 & -29 & -31 & 37 \\ 5 & -17 & 29 & 0 & -41 & 43 \\ 7 & -19 & 31 & 41 & 0 & 47 \\ -11 & 23 & -37 & -43 & -47 & 0 \end{pmatrix}$	[1, 2, 4, 5, 6]
5	$\begin{pmatrix} 0 & -2 & -3 & 5 & 7 & -11 \\ 2 & 0 & 13 & -17 & -19 & 23 \\ 3 & -13 & 0 & -29 & -31 & 37 \\ -5 & 17 & 29 & 0 & 41 & -43 \\ -7 & 19 & 31 & -41 & 0 & -47 \\ 11 & -23 & -37 & 43 & 47 & 0 \end{pmatrix}$	

7 The SMPM's Implementation in MATLAB

```
Function result = SMPM(matDim)
% SMPM finds elementary transformations of a matrix
%order n, which means that the element-wise products
%of an arbitrary matrix A, by the elements  of the
%set returned by SMPM have equal corresponding
%minors of all orders
%       usage result = SMPM(n)
```

```
%       where n is the matrix order
clearvars;
%Initiate the 4x4 elementary transformation matrices

etm{1} = [0 1 1 1;-1 0 1 1;-1 -1 0 1;-1 -1 -1 0];
etm{2} = [0 -1 1 1;1 0 1 1;-1 -1 0 -1;-1 -1 1 0];
etm{3} = [0 1 -1 1;-1 0 1 -1;1 -1 0 1;-1 1 -1 0];
etm{4} = [0 1 1 -1;-1 0 -1 1;-1 1 0 1;1 -1 -1 0];
etm{5} = [0 -1 -1 1;1 0 1 -1;1 -1 0 -1;-1 1 1 0];
etm{6} = [0 -1 1 -1;1 0 -1 1;-1 1 0 -1;1 -1 1 0];
etm{7} = [0 1 -1 -1;-1 0 -1 -1;1 1 0 1;1 1 -1 0];
etm{8} = [0 -1 -1 -1;1 0 -1 -1;1 1 0 -1;1 1 1 0];
etm{9} = [0 -1 -1 1;1 0 -1 1;1 1 0 1;-1 -1 -1 0];
etm{10} = [0 1 1 -1;-1 0 1 -1;-1 -1 0 -1;1 1 1 0];
etm{11} = [0 -1 -1 -1;1 0 1 1;1 -1 0 1;1 -1 -1 0];
etm{12} = [0 1 1 1;-1 0 -1 -1;-1 1 0 -1;-1 1 1 0];
etm{13} = [0 -1 1 1;1 0 -1 -1;-1 1 0 1;-1 1 -1 0];
etm{14} = [0 1 -1 -1;-1 0 1 1;1 -1 0 -1;1 -1 1 0];
etm{15} = [0 1 -1 1;-1 0 -1 1;1 1 0 -1;-1 -1 1 0];
etm{16} = [0 -1 1 -1;1 0 1 -1;-1 -1 0 1; 1 1 -1 0];

%Initiate the vector of the 4th principal minors
combin4 = nchoosek(1:matDim,4);

%Prepare the initial skew-symmetric initial result
%matrix with one matrix filled with unknown values,
%for speed purpose the unknown values are marked
%by 2 instead of a symbolic value
[ii,jj] = ndgrid(1:matDim);
sqmat = zeros(matDim);
sqmat(ii>jj)= 2;
sqmat(ii==jj) = 0;
sqmat = -sqmat+tril(sqmat,-1).';
result{1,1}= sqmat;

%The 'size' function is used instead of 'height' to
%handle the case matDim = 4
[st, hg] = size(combin4);

%Looping over all principal minors
for hh=1:st
    kk=0;
    c = ~cellfun('isempty',result);
    r=sum(c(:,1));

    %Looping over all non empty matrix result
    for cc = 1:r

        %Looping over all elementary transformation matrices
        for nn=1:length(etm)
            t = result{cc,1}(combin4(hh,:), combin4(hh,:));
            %Check the matrix is totally equal to the
            %elementary transformation
            %If yes there is no need to test the other etm
```

```
                %Otherwise, the principal minor of the current
                %matrix result may have partial correspondence
                %or unknown values
                if (t == etm{nn})
                    kk = kk+1;
                    result{kk,2}= zeros(matDim,matDim);
                    result{kk ,2} = result{cc,1};
                    continue;
                else
                    v = (etm{nn} ~= t) .* t;
                    h= ismember(v, 1);
                    if (~any(h(:)))  %check that the matrix
                    %is totally null, which means all elements
                    %are either 2's or equal to pmx
                        t = etm{nn};
                        kk = kk+1;
                        result{kk,2}= zeros(matDim,matDim);
                        result{kk,2} = result{cc,1};
                        result{kk,2}(combin4(hh,:), combin4(hh,:)) = t;
                    end
                end
            end
        end
        %Switch the current matrix result with the new one filled before
        result(:,1)=[];
end

c = ~cellfun('isempty',result);
r=sum(c(:,end));
fprintf('%d Solutions generated\n',r);

prompt = 'Display solution Y/N [Y]? \n';
str = input(prompt, 's');
if   (isempty(str))    || (lower(str) == 'y')
    result{:,end}
end
```

References

1. Boussaïri, A., Chergui, B.: A transformation that preserves principal minors of skew-symmetric matrices. Electron. J. Linear Algebra **32**, 06 (2016)
2. Oeding, L.: Set-theoretic defining equations of the variety of principal minors of symmetric matrices. Algebra Number Theory **5**(1), 75–109 (2011)
3. Styan, G.P.H.: Hadamard products and multivariate statistical analysis. Linear Algebra Appl. **6**, 217–240 (1973)

On a Class of Nonlinear Elliptic Unilateral Problems Involving Only a Growth Condition on Nonlinearities

H. Sabiki, H. Moussa, and M. Rhoudaf

Abstract In this paper, we shall be concerned with the existence result of unilateral problem associated to the form $-\text{div}\Big(a(x, u, \nabla u)\Big) + g(x, u, \nabla u) = f$. The growth and the coercivity conditions on the monotone vector field a are prescribed by a N-function M as a model we give $a(x, u, \nabla u) = -\text{div}\Big(\frac{(e^{|\nabla u|}-1)}{|\nabla u|^2} \cdot \nabla u\Big)$, for $M(u) = u^p \log^q(e + t)$. We assume any restriction on M, therefore we work with Orlicz–Sobolev spaces which are not necessarily reflexive, we assume any sign condition on the nonlinearities and the obstacle is a measurable function.

1 Introduction

In the last decade, there has been an increasing interest in the study of various mathematical problems in modular spaces. These problems have many consideration in applications (see [6, 25, 27]) and have resulted in a renewal interest in Lebesgue and Sobolev spaces with variable exponent, Musialak, Orlicz space, the origins of which can be traced back to the work of Orlicz in the 1930s. In the 1950s, this study was carried on by Nakano [21] who made the first systematic study of spaces with variable exponent. Later, Polish and Czechoslovak mathematicians investigated the modular function spaces (see for example Musielak [20], Kovacik and Rakosnik [19]). The

H. Sabiki (✉)
Faculté des Sciences Kénitra, Université Ibn Tofaîl, BP 133, 14000 Kénitra, Morocco
e-mail: sabikihajar@gmail.com

H. Moussa
Laboratoire de recherche "Mathématiques Appliquées et Calcul Scientifique",
Faculté des Sciences et Techniques de Béni Mellal, Université Sultan Moulay Slimane,
Beni-Mellal, Morocco
e-mail: hichammoussa23@gmail.com

M. Rhoudaf
Laboratoire de Mathématiques et Applications Equipe: EDPs et Calcul Scientifique,
Faculté des Sciences Meknès, Université Moulay Ismail, Meknes, Morocco
e-mail: mohamedrhoudaf@gmail.com

S. Melliani and O. Castillo (eds.), *Recent Advances in Intuitionistic Fuzzy Logic Systems and Mathematics*, Studies in Fuzziness and Soft Computing 395,
https://doi.org/10.1007/978-3-030-53929-0_20

study of variational problems where the function $a(.)$ satisfies the non-polynomial growth conditions instead of having the usual p-structure arouses much interest with the development of applications to electro-rheological fluids as an important class of non-Newtonian fluids (sometimes referred to as smart fluids). The electro-rheological fluids are characterized by their ability to drastically change the mechanical properties under the influence of an external electromagnetic field. A mathematical model of electro-rheological fluids was proposed by Rajagopal and Ruzicka (we refer to [23, 25] for more details). Another important application is related to image processing [22] where this kind of the diffusion operator is used to underline the borders of the distorted image and to eliminate the noise.

Variational inequalities with gradient constraints, in particular those arising in the theory of elastoplastic torsion of bars have been studied deeply in the seventies [11, 24] and then, they have been initially developed to deal with equilibrium problems, precisely the Signorini problem [10] in that model problem, the functional involved was obtained as the first variation of the involved potential energy therefore it has a variational origin, recalled by the name of the general abstract problem. The applicability of the theory has since been expanded to include problems from economics, finance, optimization and game theory [7, 18].

The notion of unilateral problems and the notion of variational inequality in general are not equivalent if the gradient constraint is not constant. In addition, their natural relations with the associated obstacle problems become a much more delicate problem in this case.

Let Ω be a bounded open set of $I\!R^N$ $(N \geq 2)$, we consider the following nonlinear elliptic problem

$$Au + g(x, u, \nabla u) = f. \tag{1}$$

Where $Au = -div\Big(a(x, u, \nabla u)\Big)$ is a Leray–Lions operator defined on $W^{1,p}(\Omega)$, $1 < p < \infty$. Bensoussan–Boccardo–Murat [4] proved the existence of solutions for the Dirichlet problem associated to the problem (1), where g is a nonlinearity satisfies the following (natural) growth condition:

$$|g(x, s, \xi)| \leq b(|s|)(c(x) + |\xi|^p),$$

and the sign condition $g(x, s, \xi)s \geq 0$. In the case where the right-hand side f is assumed to belong to $W^{-1,p'}(\Omega)$ and g depends only on x and u see the result of Brézis and Browder in [5]. The works mentioned above employe the standard theory of monotone operators relying on the Sobolev space $W^{1,p}(\Omega)$, in order to to perform an analysis for the function $a(.)$, we use the operators satisfying non-polynomial growth conditions instead of having the usual p-structure. These require another functional setting, namely to consider monotone operators in Orlicz–Sobolev spaces. Since, in general, Orlicz–Sobolev spaces are neither reflexive nor separable.

In [17] Gossez and Mustonen solved (1) in the case where g satisfies the classical sign condition $g(x, s).s \geq 0$ and data f in $W^{-1}E_{\overline{M}}(\Omega)$.

We find also some existence results in same context for strongly nonlinear problem associated to (1) proved in [2, 3] when data f belongs either to $W^{-1}E_{\overline{M}}(\Omega)$ or $L^1(\Omega)$ with M satisfies Δ_2-condition. In the case where the Δ_2-condition are not fulfilled the above problem was studied in [1, 8, 9].

In the present paper, we deal with the existence result for the following unilateral problem but without assuming any sign condition on nonlinearities and any restriction on the N-function M of Orlicz spaces and any regularity on the obstacle ψ. More precisely, we prove the existence result for the following unilateral problem

$$(P)\begin{cases} u \in K_\psi(\Omega),\, g(x,u,\nabla u) \in L^1(\Omega),\, g(x,u,\nabla u)u \in L^1(\Omega) \\ \displaystyle\int_\Omega (a(x,u,\nabla u))\nabla(u-\varphi)\,dx + \int_\Omega g(x,u,\nabla u)(u-\varphi)\,dx \\ \qquad\qquad \leq \int_\Omega f(u-\varphi)\,dx \\ \text{for all } \varphi \in K_\psi(\Omega)\cap L^\infty(\Omega). \end{cases}$$

Where $f \in W^{-1}E_{\overline{M}}(\Omega)$ and $K_\psi = \{v \in W_0^1L_M(\Omega), u \geq \psi, a.e. \text{ in } \Omega\}$, with ψ a measurable function on Ω.

The elimination of the sign condition causes a hard difficulty, to overcame this difficulty, we use the following growth condition,

$$|g(x,s,\xi)| \leq b(|s|) + h(s)M(|\xi|), \tag{2}$$

with $h \in L^1(I\!R^+)$ and $b(|s|) \leq \overline{P}^{-1}P(|s|)$ where M and P are two N-function such that $P \ll M$ (See next section.).

Let remark in [1], the authors assume the same growth condition (2) on nonlinearities but the function b depends only on x not on u and belongs to $L^1(\Omega)$, in our case b depends on u and we assume a growth condition on b.

The model problem is to consider

$$|g(x,u,\nabla u)| \leq \overline{P}^{-1}P(|u|) + |sin(u)|e^{-u^2}M(|\nabla u|).$$

Note that this type of equations can be applied in sciences physics. Non-standard examples of $M(t)$ which occur in the mechanics of solids and fluids are $M_1(t) = tlog(1+t)$, $M_2(t) = \int_0^t s^{1-\alpha}(arcsinhs)^\alpha\,ds$ $(0 \leq \alpha \leq 1)$ and $M_3(t) = tlog(1 + log(1+t))$ (See Fuchs and Gongbao [12] and Fuchs and Seregin [13, 26].).

This chapter is organized as follows, Sect. 2 contains some preliminaries. In Sect. 3 we state and we prove our general results.

2 Preliminaries

Let $M : I\!R^+ \to I\!R^+$ be an N-function, i.e., M is continuous, convex, with $M(t) > 0$ for $t > 0$, $\frac{M(t)}{t} \to 0$ as $t \to 0$ and $\frac{M(t)}{t} \to \infty$ as $t \to \infty$. Equivalently, M admits the representation: $M(t) = \int_0^t a(s)\,ds$ where $a : I\!R^+ \to I\!R^+$ is non-decreasing, right continuous, with $a(0) = 0$, $a(t) > 0$ for $t > 0$ and $a(t) \to \infty$ as $t \to \infty$. The N-function $\overline{M}$ conjugate to M is defined by $\overline{M}(t) = \int_0^t \overline{a}(s)\,ds$, where $\overline{a} : I\!R^+ \to I\!R^+$ is given by $\overline{a}(t) = \sup\{s : a(s) \leq t\}$.

The N-function M is said to satisfy the Δ_2 condition if, for some $k > 0$,

$$M(2t) \leq k\,M(t) \quad \text{for all } t \geq 0. \tag{3}$$

It is readily seen that this will be the case if and only if for every $r > 0$ there exists a positive constant $k = k(r)$ such that for all $t > 0$

$$M(rt) \leq k\,M(t) \quad \text{for all } t \geq 0. \tag{4}$$

When (3) and (4) hold only for $t \geq t_0 > 0$, M is said to satisfy the Δ_2-condition near infinity.

Let P and M be two N-functions. $P \ll M$ means that P grows essentially less rapidly than M, i.e., for each $\varepsilon > 0$,

$$\frac{P(t)}{M(\varepsilon\, t)} \to 0 \quad \text{as } t \to \infty. \tag{5}$$

This is the case if and only if,

$$\frac{M^{-1}(t)}{P^{-1}(t)} \to 0 \quad \text{as } t \to \infty. \tag{6}$$

We will extend these N-functions into even functions on all $I\!R$. Let Ω be an open subset of $I\!R^N$. The Orlicz class $\mathcal{L}_M(\Omega)$ (resp. the Orlicz space $L_M(\Omega)$) is defined as the set of (equivalence classes of) real-valued measurable functions u on Ω such that:

$$\int_\Omega M(u(x))dx < +\infty \quad \left(\text{resp.} \int_\Omega M\left(\frac{u(x)}{\lambda}\right) dx < +\infty \text{ for some } \lambda > 0\right). \tag{7}$$

Note that $L_M(\Omega)$ is a Banach space under the norm

$$\|u\|_{M,\Omega} = \inf\left\{\lambda > 0 : \int_\Omega M\left(\frac{u(x)}{\lambda}\right) dx \leq 1\right\} \tag{8}$$

and $\mathcal{L}_M(\Omega)$ is a convex subset of $L_M(\Omega)$. The closure in $L_M(\Omega)$ of the set of bounded measurable functions with compact support in $\overline{\Omega}$ is denoted by $E_M(\Omega)$.

The equality $E_M(\Omega) = L_M(\Omega)$ holds if and only if M satisfies the Δ_2-condition, for all t or for t large according to whether Ω has infinite measure or not.

The dual of $E_M(\Omega)$ can be identified with $L_{\overline{M}}(\Omega)$ by means of the pairing $\int_\Omega u(x)v(x)dx$, and the dual norm on $L_{\overline{M}}(\Omega)$ is equivalent to $\|.\|_{\overline{M},\Omega}$. The space $L_M(\Omega)$ is reflexive if and only if M and $\overline{M}$ satisfy the Δ_2 condition, for all t or for t large, according to whether Ω has infinite measure or not.

We define the Orlicz norm $||u||_{(M)}$ by

$$||u||_{(M)} = \sup \int_\Omega u(x)v(x)\, dx \tag{9}$$

where the supremum is taken over all $v \in E_{\overline{M}(\Omega)}$ such that $||v||_{\overline{M}} \leq 1$, for which

$$||u||_M \leq ||u||_{(M)} \leq 2||u||_M \tag{10}$$

holds for all $u \in L_{M(\Omega)}$.

We now turn to the Orlicz–Sobolev space. $W^1L_M(\Omega)$ (resp. $W^1E_M(\Omega)$) is the space of all functions u such that u and its distributional derivatives up to order 1 lie in $L_M(\Omega)$ (resp. $E_M(\Omega)$). This is a Banach space under the norm

$$\|u\|_{1,M,\Omega} = \sum_{|\alpha|\leq 1} \|\nabla^\alpha u\|_{M,\Omega}. \tag{11}$$

Thus $W^1L_M(\Omega)$ and $W^1E_M(\Omega)$ can be identified with subspaces of the product of $N+1$ copies of $L_M(\Omega)$. Denoting this product by ΠL_M, we will use the weak topologies $\sigma(\Pi L_M, \Pi E_{\overline{M}})$ and $\sigma(\Pi L_M, \Pi L_{\overline{M}})$. The space $W_0^1E_M(\Omega)$ is defined as the (norm) closure of the Schwartz space $\mathscr{D}(\Omega)$ in $W^1E_M(\Omega)$ and the space $W_0^1L_M(\Omega)$ as the $\sigma(\Pi L_M, \Pi E_{\overline{M}})$ closure of $\mathscr{D}(\Omega)$ in $W^1L_M(\Omega)$. We say that u_n converges to u for the modular convergence in $W^1L_M(\Omega)$ if for some $\lambda > 0$, $\int_\Omega M\Big(\frac{\nabla^\alpha u_n - \nabla^\alpha u}{\lambda}\Big)dx \to 0$ for all $|\alpha| \leq 1$. This implies convergence for $\sigma(\Pi L_M, \Pi L_{\overline{M}})$. If M satisfies the Δ_2 condition on $I\!R^+$ (near infinity only when Ω has finite measure), then modular convergence coincides with norm convergence.

Let $W^{-1}L_{\overline{M}}(\Omega)$ (resp. $W^{-1}E_{\overline{M}}(\Omega)$) denote the space of distributions on Ω which can be written as sums of derivatives of order ≤ 1 of functions in $L_{\overline{M}}(\Omega)$ (resp. $E_{\overline{M}}(\Omega)$). It is a Banach space under the usual quotient norm.

If the open set Ω has the segment property, then the space $\mathscr{D}(\Omega)$ is dense in $W_0^1L_M(\Omega)$ for the modular convergence and for the topology $\sigma(\Pi L_M, \Pi L_{\overline{M}})$ (cf. [15]). Consequently, the action of a distribution in $W^{-1}L_{\overline{M}}(\Omega)$ on an element of $W_0^1L_M(\Omega)$ is well defined.

For $k > 0$, we define the truncation at height k, $T_k : I\!R \to I\!R$ by

$$T_k(s) = \min(k, \max(s, -k)) = \begin{cases} s & \text{if } |s| \leq k \\ \frac{ks}{|s|} & \text{if } |s| > k \end{cases} \tag{12}$$

We define for N-function M, $\mathscr{T}_0^{1,M}(\Omega)$ as a set of measurable function $u : \Omega \to I\!R$ such that $T_k(u) \in W_0^1 L_M(\Omega)$. The following abstract lemmas will be applied to the truncation operators.

Lemma 2.1 (cf. [15]) *Let $F : I\!R \to I\!R$ be uniformly lipschitzian, with $F(0) = 0$. Let M be an N-function and let $u \in W^1 L_M(\Omega)$ (resp. $W^1 E_M(\Omega)$).*

Then $F(u) \in W^1 L_M(\Omega)$ (resp. $W^1 E_M(\Omega)$). Moreover, if the set of discontinuity points D of F' is finite, then

$$\frac{\partial}{\partial x_i} F(u) = \begin{cases} F'(u)\frac{\partial u}{\partial x_i} & \text{a.e. in } \{x \in \Omega : u(x) \notin D\} \\ 0 & \text{a.e. in } \{x \in \Omega : u(x) \in D\} \end{cases}$$

Lemma 2.2 (cf. [15]) *Let $F : I\!R \to I\!R$ be uniformly lipschitzian, with $F(0) = 0$. We suppose that the set of discontinuity points of F' is finite. Let M be an N-function, then the mapping $F : W^1 L_M(\Omega) \to W^1 L_M(\Omega)$ is sequentially continuous with respect to the weak* topology $\sigma(\Pi L_M, \Pi E_{\overline{M}})$.*

Lemma 2.3 ([14]) *Let Ω be a bounded open subset of $I\!R^N$ and M is an N-function so there exists two positive constant δ and λ*

$$\int_\Omega M(\delta|v|)dx \leq \int_\Omega \lambda M(|\nabla v|)dx \text{ for all } v \in W_0^1 L_M(\Omega). \tag{13}$$

Lemma 2.4 ([16]) *Let Ω have the segment property. Then for each $v \in \mathscr{D}(\Omega)$ such that v_n converges to v for the modular convergence in $W_0^1 L_M(\Omega)$. Furthermore, if $v \in W_0^1 L_M(\Omega) \cap L^\infty(\Omega)$ then,*

$$||v_n|| \leq (N+1)||v||_{L^\infty(\Omega)}. \tag{14}$$

We give now the following lemma which concerns the Nemytskii operator type in Orlicz spaces (see [2]).

Lemma 2.5 *Let Ω be an open subset of $I\!R^N$ with finite measure.*

Let M, P and Q be N-functions such that $Q << P$, and let $f : \Omega \times I\!R \to I\!R$ be a Carathéodory function such that, for a.e. $x \in \Omega$ and all $s \in I\!R$:

$$|f(x,s)| \leq c(x) + k_1 P^{-1} M(k_2|s|),$$

where k_1, k_2 are real constants and $c(x) \in E_Q(\Omega)$. Then the Nemytskii operator N_f defined by $N_f(u)(x) = f(x, u(x))$ is strongly continuous from $\mathscr{P}(E_M(\Omega), \frac{1}{k_2}) = \{u \in L_M(\Omega) : d(u, E_M(\Omega)) < \frac{1}{k_2}\}$ into $E_Q(\Omega)$.

Below, we will use the following technical Lemmas.

Lemma 1 *Let (f_n), $f \in L^1(\Omega)$, and $\delta \in L^1(\Omega)$ such that*
(i) $f_n \geq \delta$ a.e. in Ω

(ii) $f_n \to f$ *a.e. in* Ω
(iii) $\int_\Omega f_n(x)\,dx \to \int_\Omega f(x)\,dx$
then $f_n \to f$ *strongly in* $L^1(\Omega)$.

Proof We have $|f - f_n| = 2(f - f_n)^+ - (f - f_n)$, where $(f - f_n)^+ = \sup(f - f_n, 0)$. If $(f - f_n)^+(x) \notin 0$ then $(f - f_n)^+(x) = f(x) - f_n(x) \leq f(x)$, consequently $0 \leq (f - f_n)^+ \leq f$. Since $(f - f_n)^+ \to 0$ a.e in Ω then by using Lebesgue's theorem we have $(f - f_n)^+ \to 0$ strongly in $L^1(\Omega)$. In view of (iii) we obtain $\int_\Omega |f - f_n|dx \to 0$, which is the desired result.

3 Main Results

Throughout this paper, we assume that the following assumptions hold true:

Let Ω be an open bounded subset of $I\!R^N$, $N \geq 2$, with the segment property, M be an N-function and P be an N-function such that $P << M$.

Given an obstacle function $\psi : \Omega \to \overline{I\!R}$, we consider

$$K_\psi = \{u \in W_0^1 L_M(\Omega);\ u \geq \psi\ \ a.e.\text{ in } \Omega\}, \tag{15}$$

this convex set is sequentially $\sigma(\Pi L_M, \Pi E_{\overline{M}})$ closed in $W_o^1 L_M(\Omega)$ [see [17]].

We consider the Leray–Lions operator,

$$Au = -\mathrm{div}(a(x, u, \nabla u)), \tag{16}$$

defined on $\mathcal{D}(A) \subset W_0^1 L_M(\Omega)$ into $W^{-1} L_{\overline{M}}(\Omega)$ where $a : \Omega \times I\!R \times I\!R^N \to I\!R^N$ is a Carathéodory function such that for a.e. $x \in \Omega$ and for all $\zeta, \xi \in I\!R^N (\zeta \neq \xi)$ and all $s \in I\!R$,

$$|a(x, s, \zeta)| \leq c(x) + k_1 \overline{P}^{-1} M(k_2|s|) + k_3 \overline{M}^{-1} M(k_4|\zeta|), \tag{17}$$

$$(a(x, s, \zeta) - a(x, s, \xi), \zeta - \xi) > 0, \tag{18}$$

$$a(x, s, \zeta)\zeta \geq \alpha M(|\zeta|), \tag{19}$$

with $\alpha > 0$ $k_1, k_2, k_3, k_4 \geq 0$, $c \in E_{\overline{M}}(\Omega)$.

For each $v \in K_\psi \cap L^\infty(\Omega)$ there exists a sequence $v_n \in K_\psi \cap W_0^1 E_M(\Omega) \cap L^\infty(\Omega)$ such that $v_n \to v$ for the modular convergence.

(20)

Furthermore, let $g(x, s, \xi)$ be a Carathéodory function satisfying the following assumptions:

$$|g(x, s, \xi)| \leq b(|s|) + h(s)M(|\xi|), \tag{21}$$

where $h \in L^1(I\!R^+)$ and $b(|s|) \leq \overline{P}^{-1} P(|s|)$.

Finally, we suppose that

$$K_\psi \cap L^\infty(\Omega) \neq \emptyset, \tag{22}$$

Let use give and prove the following lemma which will be needed later :

Lemma 3.1 *Under assumptions (17)–(19), and let (z_n) be a sequence in $W_0^1 L_M(\Omega)$ such that,*

$$z_n \rightharpoonup z \text{ in } W_0^1 L_M(\Omega) \text{ for } \sigma(\Pi L_M(\Omega), \Pi E_{\overline{M}}(\Omega)), \tag{23}$$

$$(a(x, z_n, \nabla z_n))_n \text{ is bounded in } (L_{\overline{M}}(\Omega))^N, \tag{24}$$

$$\int_\Omega \Big[a(x, z_n, \nabla z_n) - a(x, z_n, \nabla z \chi_s)\Big]\Big[\nabla z_n - \nabla z \chi_s\Big]\, dx \longrightarrow 0, \tag{25}$$

as n and s tend to $+\infty$, and where χ_s is the characteristic function of

$$\Omega_s = \Big\{x \in \Omega \,;\, |\nabla z| \leq s\Big\}.$$

Then,

$$\nabla z_n \to \nabla z \text{ a.e. in } \Omega. \tag{26}$$

$$\lim_{n \to \infty} \int_\Omega a(x, z_n, \nabla z_n)\nabla z_n\, dx = \int_\Omega a(x, z, \nabla z)\nabla z\, dx\,, \tag{27}$$

$$M(|\nabla z_n|) \to M(|\nabla z|) \text{ in } L^1(\Omega), \tag{28}$$

Proof Using the same argument in [1] we prove the result.

Theorem 1 *Assume that the assumptions (17)–(22) Then, there exists a measurable function u solution of the following problem:*

$$(P) \begin{cases} u \in K_\psi(\Omega), g(x, u, \nabla u) \in L^1(\Omega), g(x, u, \nabla u)u \in L^1(\Omega) \\ \displaystyle\int_\Omega a(x, u, \nabla u)\nabla(u - v)\, dx + \int_\Omega g(x, u, \nabla u)(u - v)\, dx \\ \qquad\qquad \displaystyle\leq \int_\Omega f(u - v)\, dx \\ \forall v \in K_\psi(\Omega) \cap L^\infty(\Omega). \end{cases}$$

Proof The proof is divided into 6 steps.

Step 1 Definition of an approximate problem.

Let us consider the sequence of approximate problem,

$$(P_n)\begin{cases} u_n \in K_\psi(\Omega)\cap \mathscr{D}(A) \\ \displaystyle\int_\Omega a(x,u_n,\nabla u_n)\nabla(u_n-v)\,dx + \int_\Omega g_n(x,u_n,\nabla u_n)(u_n-v)\,dx \\ \qquad\qquad \displaystyle\le \int_\Omega f(u_n-v)\,dx \quad \forall v\in K_\psi(\Omega), \end{cases} \tag{29}$$

where

$$g_n(x,s,\xi)=\frac{g(x,s,\xi)}{1+\frac{1}{n}|g(x,s,\xi)|}.$$

Note that, $g_n(x,s,\xi)$ satisfies the following conditions,

$$|g_n(x,s,\xi)|\le |g(x,s,\xi)|,\quad |g_n(x,s,\xi)|\le n.$$

By the classical result of [17], the approximate problem (P_n) has at least one solution.

Step 2 A priori estimates of the solution to the approximate problem .

Lemma 3.2 *Let u_n be a solution of the problem (P_n), then we have*

$$\int_\Omega M(|\nabla u_n|)\,dx \le C, \tag{30}$$

where C is a positive constant not depend on the n.

Proof By (20) and (22), there exists $v_0\in K_\psi\cap L^\infty(\Omega)\cap W_0^1E_M(\Omega)$.

For η small enough, let $v=u_n-\eta e^{G(u_n)}(u_n-v_0)$ where $G(s)=\displaystyle\int_0^s \frac{h(t)}{\alpha}dt$ (the function h appears in (19)), choosing v as test function in (P_n) We have,

$$\begin{aligned}\int_\Omega a(x,u_n,\nabla u_n)\nabla(e^{G(u_n)}(u_n-v_0))\,dx \\ + \int_\Omega g_n(x,u_n,\nabla u_n)e^{G(u_n)}(u_n-v_0)\,dx \\ \le \int_\Omega e^{G(u_n)}|f|(u_n-v_0)\,dx\end{aligned} \tag{31}$$

which implies that

$$
\begin{aligned}
\int_\Omega a(x,u_n,\nabla u_n)\nabla(u_n-v_0)e^{G(u_n)}\,dx \\
&+\int_\Omega a(x,u_n,\nabla u_n)\nabla u_n\frac{h(u_n)}{\alpha}e^{G(u_n)}(u_n-v_0)\,dx \\
&\le\int_\Omega |g_n(x,u_n,\nabla u_n)|e^{G(u_n)}(u_n-v_0)\,dx \\
&+\int_\Omega e^{G(u_n)}|f|(u_n-v_0)\,dx,
\end{aligned}
\tag{32}
$$

hence, by using (21),

$$
\begin{aligned}
\int_\Omega a(x,u_n,\nabla u_n)\nabla(u_n-v_0)e^{G(u_n)}\,dx \\
&+\int_\Omega a(x,u_n,\nabla u_n)\nabla u_n\frac{h(u_n)}{\alpha}e^{G(u_n)}(u_n-v_0)\,dx \\
&\le\int_\Omega h(u_n)M(|\nabla u_n|)e^{G(u_n)}(u_n-v_0)\,dx \\
&+\int_\Omega e^{G(u_n)}|f|(u_n-v_0)\,dx, \\
&+\int_\Omega \overline{P}^{-1}P(|u_n|)e^{G(u_n)}(u_n-v_0)\,dx,
\end{aligned}
\tag{33}
$$

by using (19) and the fact that $1\le e^{G(-\infty)}\le e^{G(u_n)}\le e^{G(+\infty)}\le c_0$, we get

$$
\begin{aligned}
\int_\Omega a(x,u_n,\nabla u_n)\nabla(u_n-v_0)\,dx \\
&\le c_0\Big(\int_\Omega |f||u_n|\,dx+\int_\Omega |f||v_0|\,dx,\Big) \\
&+c_0\Big(\int_\Omega \overline{P}^{-1}P(|u_n|)|u_n|\,dx+\int_\Omega \overline{P}^{-1}P(|u_n|)|v_0|\,dx\Big).
\end{aligned}
\tag{34}
$$

While $P\ll M$, we have, for all $\varepsilon>0$ there exists a constant that k_ε depending on ε such that,

$$P(t)\le M(\varepsilon t)+k_\varepsilon,\quad \forall t\ge 0, \tag{35}$$

without loss of generality, we can assume that $\varepsilon=\frac{\alpha\delta(1-c)}{9c_0\lambda+\alpha\delta(1-c)+\alpha(1-c)}<1$ where δ, λ two positive constants in Lemma 2.3 and α the constant in (19) and $0<c<1$, by convexity of M we have

$$P(t)\le \varepsilon M(t)+k_\varepsilon,\quad \forall t\ge 0. \tag{36}$$

We deduce from the above inequality (35) and Young inequality

$$c_0\int_\Omega \overline{P}^{-1}P(|u_n|)|u_n|\,dx\le 2c_0\int_\Omega M(\varepsilon|u_n|)\,dx+C_\varepsilon^1, \tag{37}$$

by Lemma 2.3 and the fact that $\frac{\varepsilon}{\delta} < 1$, we get

$$c_0 \int_\Omega \overline{P}^{-1} P(|u_n|)|u_n|\, dx \le \frac{2c_0\varepsilon\lambda}{\delta} \int_\Omega M(|\nabla u_n|)\, dx + c_\varepsilon^1. \tag{38}$$

On the other hand, f can be written as $f = f_0 - divF$ where $f_0 \in E_{\overline{M}}(\Omega)$, $F \in (E_{\overline{M}}(\Omega))^N$, using Lemma 5.7 in [15] and Young's inequality we deduce

$$\begin{aligned} \int_\Omega c_0 f_0 u_n\, dx &\le c_1 + \frac{\alpha(1-c)}{8} \int_\Omega M(|\nabla u_n|)\, dx, \\ \int_\Omega c_0 F \nabla u_n\, dx &\le c_2 + \frac{\alpha(1-c)}{8} \int_\Omega M(|\nabla u_n|)\, dx, \end{aligned} \tag{39}$$

using (38) and (39), (34) becomes

$$\begin{aligned} \int_\Omega a(x, u_n, \nabla u_n)\nabla(u_n - v_0)\, dx \le{}& \frac{\alpha(1-c)}{4} \int_\Omega M(|\nabla u_n|)\, dx \\ &+ \tfrac{3c_0\varepsilon\lambda}{\delta} \int_\Omega M(|\nabla u_n|)\, dx + c_3. \end{aligned} \tag{40}$$

Hence,

$$\begin{aligned} \int_\Omega a(x, u_n, \nabla u_n)\nabla u_n\, dx \le{}& c \int_\Omega a(x, u_n, \nabla u_n)\frac{\nabla v_0}{c}\, dx \\ &+ \tfrac{\alpha(1-c)}{4} \int_\Omega M(|\nabla u_n|)\, dx \\ &+ \tfrac{3c_0\varepsilon\lambda}{\delta} \int_\Omega M(|\nabla u_n|)\, dx + c_3. \end{aligned} \tag{41}$$

Note that c is a constant such that $0 < c < 1$. By using (18), we obtain

$$\begin{aligned} \int_\Omega a(x, u_n, \nabla u_n)\nabla u_n\, dx \le{}& c \Big\{ \int_\Omega a(x, u_n, \nabla u_n)\nabla u_n\, dx \\ &- \int_\Omega a\left(x, u_n, \frac{\nabla v_0}{c}\right)\left(\nabla u_n - \frac{\nabla v_0}{c}\right) dx \Big\} \\ &+ \tfrac{\alpha(1-c)}{4} \int_\Omega M(|\nabla u_n|)\, dx \\ &+ \tfrac{3c_0\varepsilon\lambda}{\delta} \int_\Omega M(|\nabla u_n|)\, dx + c_3, \end{aligned} \tag{42}$$

then,

$$
\begin{aligned}
(1-c)\int_\Omega a(x,u_n,\nabla u_n)\nabla u_n\,dx \le& \int_\Omega |a\left(x,u_n,\frac{\nabla v_0}{c}\right)||\left(\nabla u_n-\frac{\nabla v_0}{c}\right)|\,dx\\
&+\tfrac{\alpha(1-c)}{4}\int_\Omega M(|\nabla u_n|)\,dx\\
&+\tfrac{3c_0\varepsilon\lambda}{\delta}\int_\Omega M(|\nabla u_n|)\,dx+c_3\\
\le& \int_\Omega |a\left(x,u_n,\frac{\nabla v_0}{c}\right)||\frac{\nabla v_0}{c}|\,dx\\
&+\int_\Omega a\left(x,u_n,\frac{\nabla v_0}{c}\right)\nabla u_n\,dx\\
&+\tfrac{\alpha(1-c)}{4}\int_\Omega M(|\nabla u_n|)\,dx\\
&+\tfrac{3c_0\varepsilon\lambda}{\delta}\int_\Omega M(|\nabla u_n|)\,dx+c_3.
\end{aligned}
\tag{43}
$$

Since $\frac{\nabla v_0}{c}\in(E_M(\Omega))^N$, then by using the Young's inequality and the condition (17), we have

$$
\begin{aligned}
(1-c)\int_\Omega a(x,u_n,\nabla u_n)\nabla u_n\,dx \le& \frac{\alpha(1-c)}{4}\int_\Omega M(|\nabla u_n|)\,dx\\
&+\tfrac{\alpha(1-c)}{4}\int_\Omega M(|\nabla u_n|)\,dx\\
&+\tfrac{3c_0\varepsilon\lambda}{\delta}\int_\Omega M(|\nabla u_n|)\,dx+c_4.
\end{aligned}
\tag{44}
$$

Using again (19), we get

$$
\left(\tfrac{\alpha(1-c)}{2}-\tfrac{3c_0\varepsilon\lambda}{\delta}\right)\int_\Omega M(|\nabla u_n|)\,dx\le c_5,
\tag{45}
$$

we can easy verify that $(\frac{\alpha(1-c)}{2}-\frac{3c_0\varepsilon\lambda}{\delta})>0$ where $\varepsilon=\frac{\alpha\delta(1-c)}{9c_0\lambda+\alpha\delta(1-c)+\alpha(1-c)}$ and $0<c<1$.

Then,

$$
\int_\Omega M(|\nabla u_n|)\,dx\le C,
\tag{46}
$$

where C is a positive constant not depending on n.

It follows that the sequence $\{u_n\}$ is bounded in $W_0^1L_M(\Omega)$. Consequently, there exist a subsequence on $\{u_n\}$, still denoted by $\{u_n\}$ and a function $u\in W_0^1L_M(\Omega)$ such that :

$$
u_n\rightharpoonup u \text{ in } W_0^1L_M(\Omega) \text{ for } \sigma(\Pi L_M,\Pi E_{\overline{M}})
\tag{47}
$$

$$
u_n\to u \text{ in } E_M(\Omega) \text{ Strongly and a.e in } \Omega
\tag{48}
$$

Step 3 Boundedness of $\left(a(x,u_n,\nabla u_n)\right)_n$ in $(L_{\overline{M}}(\Omega))^N$.

Lemma 3.3 *Let u_n be a solution of the approximate problem (P_n). Then for all $k \geq 0$,*

$$\Big(a(x, u_n, \nabla u_n)\Big)_n \text{ is bounded in } (L_{\overline{M}}(\Omega))^N. \tag{49}$$

Proof Let $\varphi \in (E_M(\Omega))^N$ be arbitrary. In view of the monotonicity of a, one easily has

$$\int_\Omega a\Big(x, u_n, \nabla u_n\Big)\varphi \, dx \leq \int_\Omega a\Big(x, u_n, \nabla u_n\Big)\nabla u_n \, dx \tag{50}$$
$$+ \int_\Omega a\Big(x, u_n, \varphi\Big)\Big(\nabla u_n - \varphi\Big) \, dx,$$

and

$$-\int_\Omega a\Big(x, u_n, \nabla u_n\Big)\varphi \, dx \leq \int_\Omega a\Big(x, u_n, \nabla u_n\Big)\nabla u_n \, dx \tag{51}$$
$$-\int_\Omega a\Big(x, u_n, -\varphi\Big)\Big(\nabla u_n + \varphi\Big) \, dx,$$

since u_n is bounded in $W_0^1 L_M(\Omega)$, one easily deduce that $a(x, u_n, \nabla u_n)$ is a bounded sequence in $(L_{\overline{M}}(\Omega))^N$. Thus, up to a subsequence

$$a\Big(x, u_n, \nabla u_n\Big) \rightharpoonup \varphi_k \text{ in } (L_{\overline{M}}(\Omega))^N \text{ for } \sigma(\Pi L_{\overline{M}}, \Pi E_M) \tag{52}$$

for some $\varphi_k \in (L_{\overline{M}}(\Omega))^N$.

Step 4 Almost convergence of the gradient .

Lemma 3.4 *Let u_n be a solution of the problem (P_n), then we have the following:*
(a)

$$\lim_{j\to\infty} \lim_{n\to\infty} \int_{\{j\leq |u_n|\leq j+1\}} a(x, u_n, \nabla u_n)\nabla u_n \, dx = 0 \tag{53}$$

(b)

$$T_k(u_n) \to T_k(u) \;\; \textit{modular convergence in} \;\; W_0^1 L_M(\Omega) \;\; \textit{as} \;\; n \to \infty. \tag{54}$$

Proof of (a)
Consider the following function $v = u_n - e^{G(u_n)}T_1(u_n - T_j(u_n))^+$, for $j > 1$, where $G(u_n) = \displaystyle\int_0^{|u_n|} \frac{h(|s|)}{\alpha} \, ds$. Then, we obtain,

$$\int_\Omega a(x,u_n,\nabla u_n)\nabla u_n \frac{h(|u_n|)}{\alpha} e^{G(u_n)} T_1(u_n - T_j(u_n))^+ + \int_\Omega a(x,u_n,\nabla u_n)\nabla T_1(u_n - T_j(u_n))^+ e^{G(u_n)} + \int_\Omega g_n(x,u_n,\nabla u_n) e^{G(u_n)} T_1(u_n - T_j(u_n))^+ \le \int_\Omega f e^{G(u_n)} T_1(u_n - T_j(u_n))^+ \, dx.$$

From the growth condition (19), we have

$$\int_\Omega a(x,u_n,\nabla u_n)\nabla u_n \frac{h(|u_n|)}{\alpha} e^{G(u_n)} T_1(u_n - T_j(u_n))^+ + \int_\Omega a(x,u_n,\nabla u_n)\nabla T_1(u_n - T_j(u_n))^+ e^{G(u_n)} \le \int_\Omega b(|u_n|) e^{G(u_n)} T_1(u_n - T_j(u_n))^+ + \int_\Omega h(|u_n|) M(|\nabla u_n|) e^{G(u_n)} T_1(u_n - T_j(u_n))^+ + \int_\Omega f e^{G(u_n)} T_1(u_n - T_j(u_n))^+ \, dx.$$

Thanks to (18), we get:

$$\int_\Omega a(x,u_n,\nabla u_n)\nabla T_1(u_n - T_j(u_n))^+ e^{G(u_n)} \le \int_{\{u_n > j\}} e^{G(u_n)} \overline{P}^{-1} P(|u_n|) T_1(u_n - T_j(u_n))^+ + \int_\Omega f e^{G(u_n)} T_1(u_n - T_j(u_n))^+ \, dx. \tag{55}$$

By Young's inequality and $1 \le e^{G(u_n)} = e^{\int_0^{|u_n|} \frac{h(s)}{\alpha} ds} \le e^{\int_0^{+\infty} \frac{h(s)}{\alpha} ds} = c_0$, we obtain

$$\int_\Omega a(x,u_n,\nabla u_n)\nabla T_1(u_n - T_j(u_n))^+ \le c_0 \int_{\{u_n > j\}} (P(|u_n|) + P(T_1(u_n - T_j(u_n))^+)) + c_0 \int_\Omega f T_1(u_n - T_j(u_n))^+ \, dx. \tag{56}$$

While $P \ll M$, we get

$$\int_\Omega a(x,u_n,\nabla u_n)\nabla T_1(u_n - T_j(u_n))^+ \le c_0 \varepsilon \int_{\{u_n > j\}} (M(|u_n|) + M(T_1(u_n - T_j(u_n))^+)) + c_0 \int_\Omega f T_1(u_n - T_j(u_n))^+ \, dx. \tag{57}$$

In view of (47) and (48), we have $T_1(u_n - T_j(u_n))^+ \rightharpoonup T_1(u - T_j(u))^+$ in $W_0^1 L_M(\Omega)$ for $\sigma(\Pi L_M, \Pi E_{\overline{M}})$, in addition since $(u_n)_{n\in I\!N}$ is bounded in $L_M(\Omega)$, we deduce by Lebesgue's theorem that the right hand side of the last inequality goes to zero as n and j tend to infinity. Then, (57) becomes

$$\lim_{j\to\infty}\lim_{n\to\infty}\int_{\{j\le u_n\le j+1\}} a(x, u_n, \nabla u_n)\nabla u_n\, dx = 0, \tag{58}$$

Furthermore, consider the test function $v = u_n + e^{-G(u_n)}T_1(u_n - T_j(u_n))^-$ in (P_n), and reasoning as in the proof of (58), we deduce that

$$\lim_{j\to\infty}\lim_{n\to\infty}\int_{\{-j-1\le u_n\le -j\}} a(x, u_n, \nabla u_n)\nabla u_n\, dx = 0. \tag{59}$$

Finally, combining (58) and (59), we have

$$\lim_{j\to\infty}\lim_{n\to\infty}\int_{\{j\le |u_n|\le j+1\}} a(x, u_n, \nabla u_n)\nabla u_n\, dx = 0. \tag{60}$$

Proof of (b)

We will use the following function of one real variable, which is defined as follow:

$$h_j(s) = \begin{cases} 1 & \text{if } |s| \le j \\ 0 & \text{if } |s| \ge j+1 \\ j+1-s & \text{if } j \le s \le j+1 \\ s+j+1 & \text{if } -j-1 \le s \le -j \end{cases} \tag{61}$$

with j a nonnegative real parameter.

Let $\Omega_s = \{x \in \Omega, |\nabla T_k(u(x))| \le s\}$ and denote by χ_s the characteristic function of Ω_s, clearly, $\Omega_s \subset \Omega_{s+1}$ and meas$(\Omega \setminus \Omega_s) \to 0$ as $s \to 0$.

By Lemma 2.4 there exists a sequence $v_i \in \mathscr{D}(\Omega)$ which converges to u for the modular convergence in $W_0^1 L_M(\Omega)$.

Let use $v = u_n - \eta e^{-G(u_n)}(T_k(u_n) - T_k(v_i))^+ h_j(u_n)$ as test function in (P_n), we obtain by using (19) and (21)

$$\begin{aligned}\int_{\{T_k(u_n)-T_k(v_i)\ge 0\}} & e^{G(u_n)} a(x, u_n, \nabla u_n)\nabla(T_k(u_n) - T_k(v_i))h_j(u_n)\, dx \\ & - \int_{\{j\le u_n\le j+1\}} e^{G(u_n)} a(x, u_n, \nabla u_n)\nabla u_n (T_k(u_n) - T_k(v_i))^+\, dx \\ & \le \int_\Omega \overline{P}^{-1} P(|u_n|)(T_k(u_n) - T_k(v_i))^+ h_j(u_n) e^{G(u_n)}\, dx \\ & + \int_\Omega f(T_k(u_n) - T_k(v_i))^+ h_j(u_n) e^{G(u_n)}\, dx.\end{aligned}$$

Thanks to (60) the second integral tend to zero as n and j tend to infinity, and by Lebesgue Theorem, we deduce that the right hand side converge to zero as n and j goes to infinity.

Using the same argument as in [1], we get

$$\int_{\Omega} [a(x, T_k(u_n), \nabla T_k(u_n)) - a(x, T_k(u_n), \nabla T_k(u)\chi_s)][\nabla T_k(u_n) - \nabla T_k(u)\chi_s]\, dx \to 0$$

as $n \to +\infty$ and $s \to +\infty$, which implies by the Lemma 3.1 that

$$M(|\nabla T_k(u_n)|) \to M(|\nabla T_k(u)|) \text{ in } L^1(\Omega). \tag{62}$$

Step 5 Equi-integrability of the nonlinearities.

Thanks to (62), we obtain for a subsequence

$$\nabla u_n \longrightarrow \nabla u \text{ a.e. in } \Omega.$$

Now, we show that:

$$g_n(x, u_n, \nabla u_n) \longrightarrow g(x, u, \nabla u) \text{ strongly in } L^1(\Omega). \tag{63}$$

Let $v = u_n + e^{(-G(u_n))} \int_{u_n}^{0} h(s)\chi_{\{s<-h\}}\, ds$. Since $v \in W_0^1 L_M(\Omega)$, v is an admissible test function in (P_n). Then,

$$\begin{aligned}
&\int_{\Omega} a(x, u_n, \nabla u_n)\nabla(-e^{(-G(u_n))} \int_{u_n}^{0} h(s)\chi_{\{s<-h\}}\, ds)\, dx \\
&\qquad + \int_{\Omega} g_n(x, u_n, \nabla u_n)(-e^{(-G(u_n))} \int_{u_n}^{0} h(s)\chi_{\{s<-h\}}\, ds)\, dx \\
&\qquad \le \int_{\Omega} f(-e^{(-G(u_n))} \int_{u_n}^{0} h(s)\chi_{\{s<-h\}}\, ds)\, dx.
\end{aligned}$$

By using (21) we have

$$\begin{aligned}
&\int_{\Omega} a(x, u_n, \nabla u_n)\nabla u_n \frac{h(u_n)}{\alpha} e^{(-G(u_n))} \int_{u_n}^{0} h(s)\chi_{\{s<-h\}}\, ds\, dx \\
&\qquad + \int_{\Omega} a(x, u_n, \nabla u_n)\nabla u_n e^{(-G(u_n))} h(u_n)\chi_{\{u_n<-h\}}\, dx \\
&\qquad \le \int_{\Omega} \overline{P}^{-1} P(|u_n|) e^{(-G(u_n))} \int_{u_n}^{0} h(s)\chi_{\{s<-h\}}\, ds\, dx \\
&\qquad + \int_{\Omega} h(u_n) M(|\nabla u_n|) e^{(-G(u_n))} \int_{u_n}^{0} h(s)\chi_{\{s<-h\}}\, ds\, dx
\end{aligned}$$

$$-\int_{\Omega} f e^{(-G(u_n))} \int_{u_n}^{0} h(s)\chi_{\{s<-h\}}\, ds\, dx.$$

Using (19) and since $\int_{u_n}^{0} h(s)\chi_{\{s<-h\}}\, ds \leq \int_{-\infty}^{-h} h(s)\, ds$, we get

$$\int_{\Omega} a(x, u_n, \nabla u_n)\nabla u_n e^{(-G(u_n))} h(u_n)\chi_{\{u_n<-h\}}\, dx$$
$$\leq e^{(\frac{\|h\|_{L^1(\mathbb{R})}}{\alpha})} \int_{-\infty}^{-h} h(s)\, ds \left(\int_{\Omega} \overline{P}^{-1} P(|u_n|) + \int_{\Omega} f_0\, dx \right).$$

While $P \ll M$ and using the same argument in (38), we obtain

$$\int_{\Omega} a(x, u_n, \nabla u_n)\nabla u_n e^{(-G(u_n))} h(u_n)\chi_{\{u_n<-h\}}\, dx$$
$$\leq e^{(\frac{\|h\|_{L^1(\mathbb{R})}}{\alpha})} \int_{-\infty}^{-h} h(s)\, ds \left(\frac{\varepsilon\lambda}{\delta} \int_{\Omega} M(|\nabla u_n|)\, dx + c_1 + \int_{\Omega} f_0\, dx \right),$$

using again (19) and the fact that u_n is bounded in $W_0^1 L_M(\Omega)$, we obtain

$$\int_{\{u_n<-h\}} h(u_n) M(|\nabla u_n|)\, dx \leq c_9 \int_{-\infty}^{-h} h(s)\, ds,$$

and, since $h \in L^1(\mathbb{R}^+)$, we deduce that

$$\lim_{h\to+\infty} \sup_{n\in\mathbb{N}} \int_{\{u_n<-h\}} h(u_n) M(|\nabla u_n|)\, dx = 0. \tag{64}$$

On the other hand, let $M = e^{(\|h\|_{L^1(\Omega)})} \int_0^{+\infty} h(s)\, ds$ and $h \geq M + \|v_0\|_{L^\infty(\Omega)}$. Consider $v = u_n - e^{(-G(u_n))} \int_0^{u_n} h(s)\chi_{\{s>h\}}\, ds$. Since $v \in W_0^1 L_M(\Omega)$, v is an admissible test function in (P_n). Then, similarly to (64), we deduce that

$$\lim_{m\to+\infty} \sup_{n\in\mathbb{N}} \int_{\{u_n>h\}} h(u_n) M(|\nabla u_n|)\, dx = 0. \tag{65}$$

Moreover, we have

$$\int_{\{u_n>h\}} \overline{P}^{-1} P(|u_n|)\, dx \leq \frac{1}{h} \int_{\Omega} \overline{P}^{-1} P(|u_n|) u_n\, dx. \tag{66}$$

Since $P \ll M$, (35) and Young inequality, we get

$$\int_{\{u_n>h\}} \overline{P}^{-1} P(|u_n|)\, dx \leq \frac{2\varepsilon\delta}{h\lambda} \int_{\Omega} M(|\nabla u_n|)\, dx. \tag{67}$$

Using the fact that u_n is bounded in $W_0^1 L_M(\Omega)$, we obtain

$$\int_{\{u_n>h\}} \overline{P}^{-1} P(|u_n|)\, dx \leq \frac{2\varepsilon\delta C}{h\lambda}, \tag{68}$$

then, we conclude that

$$\lim_{h\to+\infty} \sup_{n\in\mathbb{N}} \int_{\{u_n>h\}} \overline{P}^{-1} P(|u_n|)\, dx = 0. \tag{69}$$

Combining (62), (64), (69), (65) and Vitali's Theorem, we conclude (63).

Moreover, if we take v_0 as a test function in (P_n), we get

$$\begin{aligned}\int_{\Omega} a(x,u_n,\nabla u_n)\nabla u_n\, dx + \int_{\Omega} g_n(x,u_n,\nabla u_n)u_n\, dx &\leq \int_{\Omega} a(x,u_n,\nabla u_n)\nabla v_0\, dx\\ &+\int_{\Omega} g_n(x,u_n,\nabla u_n)v_0\, dx\\ &+\int_{\Omega} f(u_n-v_0)\, dx.\end{aligned} \tag{70}$$

Hence,

$$\begin{aligned}\int_{\Omega} a(x,u_n,\nabla u_n)\nabla u_n\, dx + \int_{\Omega} g_n(x,u_n,\nabla u_n)u_n\, dx &\leq c\int_{\Omega} a(x,u_n,\nabla u_n)\frac{\nabla v_0}{c}\, dx\\ &+\int_{\Omega} g_n(x,u_n,\nabla u_n)v_0\, dx\\ &+\int_{\Omega} f(u_n-v_0)\, dx.\end{aligned} \tag{71}$$

Note that c is a constant such that $0 < c < 1$. By using (18), we obtain

$$\begin{aligned}\int_{\Omega} a(x,u_n,\nabla u_n)\nabla u_n\, dx + \int_{\Omega} g_n(x,u_n,\nabla u_n)u_n\, dx &\leq c\Big\{\int_{\Omega} a(x,u_n,\nabla u_n)\nabla u_n\, dx\\ &-\int_{\Omega} a\left(x,u_n,\frac{\nabla v_0}{c}\right)\left(\nabla u_n-\frac{\nabla v_0}{c}\right)dx\Big\}\\ &+\int_{\Omega} g_n(x,u_n,\nabla u_n)v_0\, dx+\int_{\Omega} f(u_n-v_0)\, dx,\end{aligned} \tag{72}$$

then,

$$(1-c)\int_{\Omega} a(x, u_n, \nabla u_n)\nabla u_n \, dx + \int_{\Omega} g_n(x, u_n, \nabla u_n)u_n \, dx$$
$$\leq \int_{\Omega} |a\left(x, u_n, \frac{\nabla v_0}{c}\right)||\frac{\nabla v_0}{c}| \, dx$$
$$+ \int_{\Omega} a\left(x, u_n, \frac{\nabla v_0}{c}\right)\nabla u_n \, dx$$
$$+ \int_{\Omega} g_n(x, u_n, \nabla u_n)v_0 \, dx + \int_{\Omega} f(u_n - v_0) \, dx. \tag{73}$$

Since $\frac{\nabla v_0}{c} \in (E_M(\Omega))^N$, then by using the Young's inequality and the condition (17), we have

$$\int_{\Omega} g_n(x, u_n, \nabla u_n)u_n \, dx \leq c\int_{\Omega} M(|\nabla u_n|) \, dx + \int_{\Omega} g_n(x, u_n, \nabla u_n)v_0 \, dx$$
$$+ \int_{\Omega} f(u_n - v_0) \, dx. \tag{74}$$

Which yields

$$\int_{\Omega} g_n(x, u_n, \nabla u_n)u_n \, dx \leq \beta, \tag{75}$$

where β is some positive constant, then by using Fatou's lemma, we have

$$g(x, u, \nabla u)u \in L^1(\Omega).$$

Step 6 Passing to the limit in the approximate problem
We take $v \in K_\psi \cap W_0^1E_M(\Omega) \cap L^\infty(\Omega)$, as test function in (P_n), we can write

$$\int_{\Omega} a(x, u_n, \nabla u_n)\nabla(u_n - v) \, dx + \int_{\Omega} g_n(x, u_n, \nabla u_n)(u_n - v) \, dx$$
$$\leq \int_{\Omega} f_n(u_n - v) \, dx. \tag{76}$$

which implies that

$$\int_{\Omega} a(x, u_n, \nabla u_n)\nabla(u_n - v_0) \, dx$$
$$+ \int_{\Omega} a(x, u_n, \nabla u_n)\nabla(v_0 - v) \, dx$$
$$+ \int_{\Omega} g_n(x, u_n, \nabla u_n)(u_n - v) \, dx \leq \int_{\Omega} f(u - v) \, dx. \tag{77}$$

By Fatou's lemma and the fact that

$$a(x, u_n, \nabla u_n) \rightharpoonup a(x, u, \nabla u)$$

weakly in $(L_{\overline{M}}(\Omega))^N$ for $\sigma(\Pi L_{\overline{M}}, \Pi E_M)$ on easily see that as $n \to +\infty$

$$\int_{\Omega} a(x, u, \nabla u)\nabla(u - v_0)\,dx + \int_{\Omega} a(x, u, \nabla u)\nabla(v_0 - v)\,dx + \int_{\Omega} g(x, u, \nabla u)(u - v)\,dx \le \int_{\Omega} f(u - v)\,dx. \tag{78}$$

Hence

$$\int_{\Omega} a(x, u, \nabla u)\nabla(u - v)\,dx + \int_{\Omega} g(x, u, \nabla u)(u - v)\,dx \le \int_{\Omega} f(u - v)\,dx. \tag{79}$$

thus, the proof of the Theorem 1 is now complete.

References

1. Aharoucha, L., Benkirane, A., Rhoudaf, M.: Existence results for some unilateral problems without sign condition with obstacle free in Orlicz spaces. Nonlinear Anal. **68**, 2362–2380 (2008)
2. Benkirane, A., Elmahi, A.: An existence theorem for a strongly nonlinear elliptic problems in Orlicz spaces. Nonlinear Anal. T.M.A. **36**, 11–24 (1999)
3. Benkirane, A., Elmahi, A.: A strongly nonlinear elliptic equation having natural growth terms and L^1 data. Nonlinear Anal. T.M.A. **39**, 403–411 (2000)
4. Bensoussan, A., Boccardo, L., Murat, F.: On a non linear partial differential equation having natural growth terms and unbounded solution. Ann. Inst. Henri Poincaré **5**(4), 149–169 (1988)
5. Brézis, H., Browder, F.: A property of Sobolev spaces. Commun. Partial Differ. Equ. **4**, 1077–1083 (1979)
6. Chen, Y., Levine, S., Rao, M.: Variable exponent, linear growth functionals in image restoration. SIAM J. Appl. Math. **66**, 1383–1406 (2006)
7. Daniele, P.: Variational inequalities for general evolutionary financial equilibrium. In: Giannessi, F., Maugeri, A. (eds.) Variational Analysis and Applications, pp. 279–299. Springer, New York (2005)
8. Elmahi, A., Meskine, D.: Existence of solutions for elliptic equations having natural growth terms in Orlicz spaces. Abstr. Appl. Anal. **12**, 1031–1045 (2004)
9. Elmahi, A., Meskine, D.: Non-linear elliptic problems having natural growth and L1 data in Orlicz spaces. Ann. Mat. 107–114 (2004)
10. Fichera, G.: Sul problema elastostatico di Signorini con ambigue condizioni al contorno. Rendiconti della Accademia Nazionale dei Lincei, Classe di Scienze Fisiche, Matematiche e Naturali, 8 (in Italian) **34**(2), 138–142
11. Friedman, A.: Variational Principles and Free Boundary problems. Wiley, New York (1989)
12. Fuchs, M., Gongbao, L.: Variational inequalities for energy functionals with nonstandard growth conditions. Abstr. Appl. Anal. **3**, 41–64 (1998). Zbl 0987.49019 MR 1700276
13. Fuchs, M., Seregin, G.: Variational methods for fluids for Prandtl-Eyring type and plastic materials with logarithmic hardening. Math. Methods Appl. Sci. **22**, 317–351 (1999). Zbl 0928.76087 MR 1671448
14. Gossez, J.P.: Nonlinear elliptic boundary value problems for equation with rapidly or slowly increasing coefficients. Trans. Am. Math. Soc. **190**, 217–237 (1974)
15. Gossez, J.-P.: Nonlinear elliptic boundary value problems for equations with rapidly or slowly increasing coefficients. Trans. Am. Math. Soc. **190**, 163–205 (1974)
16. Gossez, J.-P.: Some approximation properties in Orlicz-Sobolev. Studia Math. **74**, 17–24 (1982)

17. Gossez, J.-P., Mustonen, V.: Variational inequalities in Orlics-spaces. Nonlinear Anal. **11**, 379–492 (1987)
18. Harker, P.T., Pang, J.S.: Finite-dimensional variational inequality and nonlinear complementarity problems a survey of theory, algorithms and applications. Math. Program. **48**(1–3), 161–220
19. Kováčik, O., Rákosník, J.: On spaces $L^{p(x)}$ and $W^{k,p(x)}$. J. Czechoslov. Math. **41**, 592–618 (1991)
20. Musielak, J.: Orlicz Spaces and Modular Spaces. Lecture Notes in Mathematics, vol. 1034. Springer, Berlin (1983)
21. Nakano, H.: Modulared Semi-ordered Linear Spaces. Maruzen Co., Ltd., Tokyo (1950)
22. Perona, P., Malik, J.: Scale-space and edge detection using anisotropic diffusion. IEEE Trans. Pattern Anal. Mach. Intell. **12**, 629–639 (1990)
23. Rajagopal, K.R., Rúžička, M.: Mathematical modeling of electrorheological materials. Contin. Mech. Thermodyn. **13**, 59–78 (2001)
24. Rodrigues, J.F.: Obstacle Problems in Mathematical Physics. North Holland, Amsterdam (1987)
25. Růžička, M.: Electrorheological Fluids: Modeling and Mathematical Theory. Lecture Notes in Mathematics. Springer, Berlin (2000)
26. Seregin, G., Frehse, J.: Regularity of solutions to variational problems of the deformation theory of plasticity with logarithmic hardening. SFB 256. Preprint 421, Bonn 1995; Tr. St.-Petersb. Mat. Obshch. **5**, 184–222 (1998); English transl. Am. Math. Soc. Transl. Ser. 2 **193**, 127–152 (1999). Zbl 0973.74033 MR 1736908
27. Zhikov, V.: Averaging of functionals of the calculus of variations and elasticity theory. Math. USSR Izv. **29**(1), 33–66 (1987)

Social and Financial Performance in Moroccan Companies

H. Alami, A. El Hajaji, K. Hilal, and K. Mokhlis

Abstract The determination of the correlation existing between the corporate social performance (CSP) and corporate financial performance (CFP) is an important aspect in ensures capital of investors a series of information necessary to organize their behaviour in the hieht levels. Within the research, the regression econometric model was used with the aim to evaluate the relation between the CSP and CFP of firms in Morocco, using a sample of top 15 most successful Moroccan companies listed on the stock exchanges are included in the Casablanca stock exchange for 2010–2017. We measure corporate social responsibility through both an CSR index (is measured by Labour practices, decent work, Human right, Product responsibility Social profile and Social initiatives) and the financial performance of the business (is measured by return on equity and return on assets). We find a positive and significant relationship between the social performance and financial performance.

Keywords Corporate social performance · Corporate financial performance and Morocco firms

H. Alami · A. El Hajaji (✉)
LERSEM Laboratory, ENCGJ, University of Chouaïb Doukkali, El Jadida, Morocco
e-mail: a_elhajaji@yahoo.fr

H. Alami
e-mail: alamihasnaa@gmail.com

K. Hilal
LAMSC Laboratory, FST, University of Sultan Moulay Slimane, Beni-Mellal, Morocco
e-mail: hilal.khalid@yahoo.fr

K. Mokhlis
EST, University Cadi Ayyad, Safi, Marrakech, Morocco
e-mail: kh_mokhlis@yahoo.fr

S. Melliani and O. Castillo (eds.), *Recent Advances in Intuitionistic Fuzzy Logic Systems and Mathematics*, Studies in Fuzziness and Soft Computing 395,
https://doi.org/10.1007/978-3-030-53929-0_21

1 Introduction

In recent decades, companies are paying increasing attentions to corporate social responsibility (CSR) and its related performances. Different stakeholders raise pressures to companies to improve their (CSR) performances and release the related reports. Such pressures include for example, public concern, regulatory forces, consumer pressure, industry peer pressures, reputation concerns, media interest and perceived market advantage. For company itself, improving and reporting their CSR work are one kind of investments for sustainable development.

In recent years, academics in fields of several business administration have studied the relationship between corporate social performance (CSP) and corporate financial performance (CFP), The debates on the relationship between (CSR) and corporate performances are not new. There are some studies in this concern. Reference [1] provides a comprehensive conceptual model for presenting the key aspects of corporate social performances. Reference [2] applies the stakeholder theory to establish a framework for analyzing corporate social performances. If we examine the relationship between CSR and more specifically the associated financial performances, there are controversial results. Some researchers argue that there should be a positive relationship between the disclosure of CSR performances and company financial performances. The reasoning is that by reporting and improving their (CSR) work, finding mixed results on the main effect of (CSP) on (CFP) (e.g. [6, 7]). As a result, CSR performances further help the companies to achieve a good financial performance. In addition, [8] proposes an integrated model to explain the influence of (CSR) on financial performance with intellectual capital as a mediator and industry type as a moderator.

Although business and society are thought in a vicious relationship for diminishing trust among stakeholders, a few studies indicate the existence of a reciprocal nexus between corporate social performance and corporate financial performance, known as the "virtuous cycle". Despite of some empirical studies in developed countries, little research has been conducted concerning the assumption in emerging markets, where large companies have more responsibilities to create positive and sustainable shared values. Reference [9] explore the nexus between social performance and financial performance of Chinese companies for their global influential social and environmental impacts together with increasing awareness of corporate social responsibility (CSR).

Several research studies have examined corporate social responsibility (CSR) and its effects on business performance, but their results vary widely. This may stem from flawed analyses by regressing financial performance on corporate social performance, and/or perhaps from several inadequately controlled variables. Previous studies did not recognize investment in research and development as a critical variable because there is considerable empirical evidence to indicate that it has a strong positive impact on profitability. This misidentification creates biased estimates of the financial impact of CSR. Between 2002 and 2004, [10] examine 1000 Taiwanese cases in which firms include their R&D expenditures as one of their business strategies for sustainable development; they also identify their charitable expenditures as contributions to CSR. Based on theoretical assertions and empirical evidence in the literature, [10] identify a positive relationship between CSR and financial performance. When the model is properly specified, it found that while CSR does not have much positive impact on short-term financial performance, it does offer a remarkable long-term fiscal advantage. The central question we seek to answer in this paper is what degree influences the social performance on the finance performance.

This paper seeks to show the strong bond between corporate financial performance in the presence of social performance. It introduces a new approach by theorizing about and analysing the suggested linkage and presents many cases and examples to demonstrate what it might imply and how it might be applied to or be implemented in the real world.

In this paper has aims. First, it aims to show the strong linkage between corporate financial performance, corporate financial and social performance. In other words, this paper strives to determine the reasons for the correlation existing between financial performance and social performance. It shows how social performance might play an important role in profit changes. Even though we acknowledge that social performance may not be the only reason for these profit changes, it is arguably one of the most important drivers with influential effects in the financial world.

Second, this paper aims to show how to measure the effect or the degree of the effect of social performance. To this end, we benefit from the emerging concept of treasury loss, which pertains to the difference between accounting profits and taxable profits. We also discuss its legal implementations and implications, as well as its significance and relevance to the subject matter.

Third, this paper strives to document real-life situations, cases (for example Moroccan companies) or other evidence related to how social performance might influence the implementation of corporate financial reporting of profit volumes/sizes, which are the leading drivers of and widely accepted proxies for corporate financial performance.

2 Literature Review

2.1 Corporate Social Responsibility (CSR)

The concept of social responsibility, which reaches back to the beginning of the 20th century, gave birth to a management dedicated to the study of relations between the company and its environment. The notion of CSR is added to many other concepts that refer almost to the same notion (ethics, durable development, corporate citizenship ...). Its success with business leaders reflects the desire to better reconcile the economic, social and environmental. Our society is today looking for new alternatives in the face of alarming reports of the deterioration of the planet and growing disparity in the distribution of wealth. CSR seeks to propose new economic and social models that improve the living conditions of humanity while respecting the limits of ecosystem capacities.

According to [4], unlike other works, Wood's model does not distinguish between the interrelationship of society and business and its insistence. The company has expectations about corporate behavior and promoted by externalities. Wood proposes to integrate new quantifiable and measurable dimensions (societal impacts, societal programs and societal policies) of social performance into Carroll's model. The goal is to simplify the model and make it more operational. Like the notion of CSR, the concept of social performance has been the subject of a rich and abundant academic literature.

According to [5], "The academic work on social responsibility will first focus on a double questioning of the limits of the concept by seeking on the one hand to assess the boundaries, and on the other hand to understand the fundamentals. They will then focus on characterizing the management of this responsibility by companies from a more managerial and less philosophical point of view, through the notion of social sensitivity that reflects the company's ability to prevent and manage societal problems in its business environment [3]. In this perspective, the notion of societal performance is at the same time a synthesis of previous work aimed at reconciling the two previous perspectives and as an extension of this reflection by putting at the heart of the concerns the problem of measuring the actions concretely implemented. Implemented by the companies".

Social and environmental performance is therefore defined as "an organizational configuration of principles of social responsibility, processes of social sensitivity and program, policies and observable outcomes that are related to the social relations of the enterprise" (see [25]).

Corporate social responsibility (CSR) has become one of the most important aspects on the global agenda, both at company and other public or private institutions level, where the empowerment of technology, globalization, CSR has gained greater good importance as an idea, as a corporate strategy and as a practical programme in corporations [11–13]. A display of concepts has emerged around the phenomenon: stakeholder theory that comes from ethics [14, 15]; shareholder value [16]; CSR [1, 17]; corporate citizenship related to the political concept of citizen [18]; corporate social performance related to sociology [19]; and corporate codes of conduct [20].

In recent years, there has been a significant change in the way companies operate, because managers/administrators are now more aware that companies are in constant interaction with society and social preferences have changed. In the field one of the issues that has gained greater relevance in business administration is CSR, a multidimensional construct that includes a commitment to society, customers, government and the relationship between CSR and Earnings Management [21].

In this context, the question of how and to what extent the PS affects FP is becoming essential and decisive for the future of companies and for the future of the very concept of CSR.

2.2 The Link Between PSE and PFE: The Social Impact Hypothesis

According to the Chicogo School, the sole social responsibility of the company is to increase its profits. Therefore, any investment must generate a return on investment or added value that is measurable. This also involves devaluing the return on investment of socially responsible practices. However, the prevailing paradigm suggests that CSR represents a cost, and that companies that increase their social performance incur costs and lose their competitive advantage, which reduces their profitability and competitiveness. According to this theory, CSR is no longer profitable for the company, CSR incurs an additional financial burden for organizations. This lowers economic performance leading to a decline in competitiveness and profitability.

The second school, which can be defined as "social", establishes that a good PES is a prerequisite for the legitimacy of the firm in its environment; and therefore PSE and PFE can only be positively correlated in the long run (Freeman 1984): this is the stakeholder theory of the firm. This assumption of social impact therefore implies a "lead and lags" relationship between PS and FP insofar as it is the PES that develops first (in terms of image and reputation—good or bad) and then has an impact on the PFE (positive or negative).

According to this school, the entrepreneur has an economic, social, human and environmental responsibility towards all stakeholders of the company (employees, shareholders, consumers and suppliers, local authorities, etc.). Companies are therefore responsible for the externalities they generate. They must practice responsible management and improve their social and environmental performance. Proponents of this theory establish that a good PES is a pre-requisite for the legitimacy of the firm in its environment; and so CSR must serve the interests of all stakeholders, not just shareholders. Therefore, the PSE/PFE relationship is of obvious academic interest, but it also has important managerial issues. When the company faces significant social pressures and engages in CSR programs, it must justify to its shareholders the relevance and timeliness of these actions in terms of economic impacts especially. The goal for the management of the firm is to answer the following question: Can the company's social and environmental commitment contribute to improving the EFP?

2.3 Financial Performance

Corporate financial performance and its measurement have always been important for businesses and their stakeholders. Corporate profits, one way or another, have been widely recognized as financial highlights that signal corporations' financial performances. The higher a business's profits, the more successful it is considered to be. However, a business's financial success relies on many things. Regardless, profits always rank first or at least top of list of many other highlights. This focus on profits is especially true in LEs in general and in public companies in particular. The financial information released by public companies whose stocks are traded in stock and capital markets is of immeasurable significance for their shareholders, investors and creditors, among other stakeholders. In addition, the full disclosure of corporate financial performance information is first and foremost published as a component of the financial information released to the public.

Any of many different mathematical measures to evaluate how well a company is using its resources to make a profit. Common examples of financial performance include operating income, earnings before interest and taxes, and net asset value. It is important to note that no one measure of financial performance should be taken on its own. Rather, a thorough assessment of a company's performance should take into account many different measures.

As an extension of corporate financial reporting practices and implementations, we observe that firms often differ in terms of what they report. Irrespective of their significance, these differences require some customizations in the reported incomes/revenues, expenses/costs and, in turn, profits. Indeed, even though firms report their financial performance activities in fairly convenient ways, they might still have problems with their reported bottom lines. The root cause of these problems lies rather in the definitions of accounting and taxable profits. Accounting profits are reported corporate profit figures that follow corporate reporting customs, regulations, arrangements and, ultimately, practices. However, taxable profits are the corporate profit figures that happen to be created, as opposed to those that follow tax laws. Once divergences or inconsistencies between tax regulations and financial reporting regulations occur, the differences between taxable and accounting practices will be inevitable.

In [22] adopts a new vision to develop a better understanding of accounting and taxable profits, which are both related to corporate financial performance. This vision shows that the goal incongruence issue, crossing the interests of the principal with those of the agent, might cause significant variation in the level of corporate profits and thus introduce a dichotomy: accounting profits versus taxable profits. We therefore consider any practice that involves the improper handling of corporate financial performance. Handling a given corporation's financial earnings indicators therefore relates but is not limited to issues such as financial recognition, financial reporting, accruals, earnings quality, mandatory disclosure, voluntary disclosure, tax shelters, earnings repatriations (relocations or shifts), earnings strips and any other (legally) unacceptable practices.

In [23] comprehensively quantifies the public finance performance of local governments in China during the course of fiscal decentralization reform. In [24] examine the effect of a pilot project that introduced Performance-Based Financing from 2011 to 2013 in Burkina Faso.

Measures of financial performance take a variety of forms. They differ from each other on several dimensions, and many issues concern the choice of which particular financial measure to employ. For example, measures may be absolute (e.g., sales, profit), return-based (e.g., profit/sales, profit/capital, profit/equity), internal (e.g., profit/sales), external (e.g., market value of the firm), a level for a single period (e.g., one year), a mean or a growth rate over several years or a variability (e.g., standard deviation) about a mean or a trend. We construe financial performance broadly, and in our empirical study we also introduce several index for measures finance performance, for example (e.g., Gross Profit Margin, Net Profit, Net Profit Margin, Aging Accounts Receivable and Current Ratio).

3 The Empirical Model

We conducted the analysis using a sample of listed Morocco firms. We selected the top 15 most successful Moroccan companies listed on the stock exchanges are included in the Casablanca stock exchange for 2010–2017. Attijariwafa Bank, BMCE Bank, BMCI, Centrale Danone, Cosumar, Itissalat Al-Maghrib, LafargeHolcim Maroc, Lesieur Cristal, Lydec, Managem, OCP Group, Oulmes, SMI, Taqa Morocco and Wafa Assurance. Considering the historical monthly average CSP[1] and CFP[2] index over the period 10/2010 to 10/2017 (see https://fr.investing.com) and let r_t is the continuously compounded performance between monthly $t-1$ ant t.

$$r_t = 100 \times \log\left(\frac{P_t}{P_{t-1}}\right),$$

where P_t and P_{t-1} denote the value in monthly t denote the value in monthly $t-1$. We have 84 values observations and 83 performance observations in total. Based on these information, the finance-social performance was determined, and also the performance for the 83 monthly period analyzed, The graphical representation of performance is illustrated as Figs. 1 and 2. We used the regression econometric model meant to evaluate the relation between the social performance and finance performance for 15 firms morocco.

Besides the statistics appointed for the development of the CSP and CFP index during the considered period based on these informations for seven years period analyzed.

In the first level the data processed because of the help of Microsoft Excel were imported in a new application created in EVIEWS analytical analysis were used or the two data series considered and the evolution of the two indicators performance was graphically represented. The results are displayed below:

[1]The companies concerned were covering the following six areas: Respect for human rights, valuing human capital, environmental protection, business ethics, the effectiveness and independence of governance, engagement to business development territories.

[2]The companies concerned were evaluated on the following five criteria: Gross Profit Margin, Net Profit, Net Profit Margin, Aging Accounts Receivable and Current Ratio.

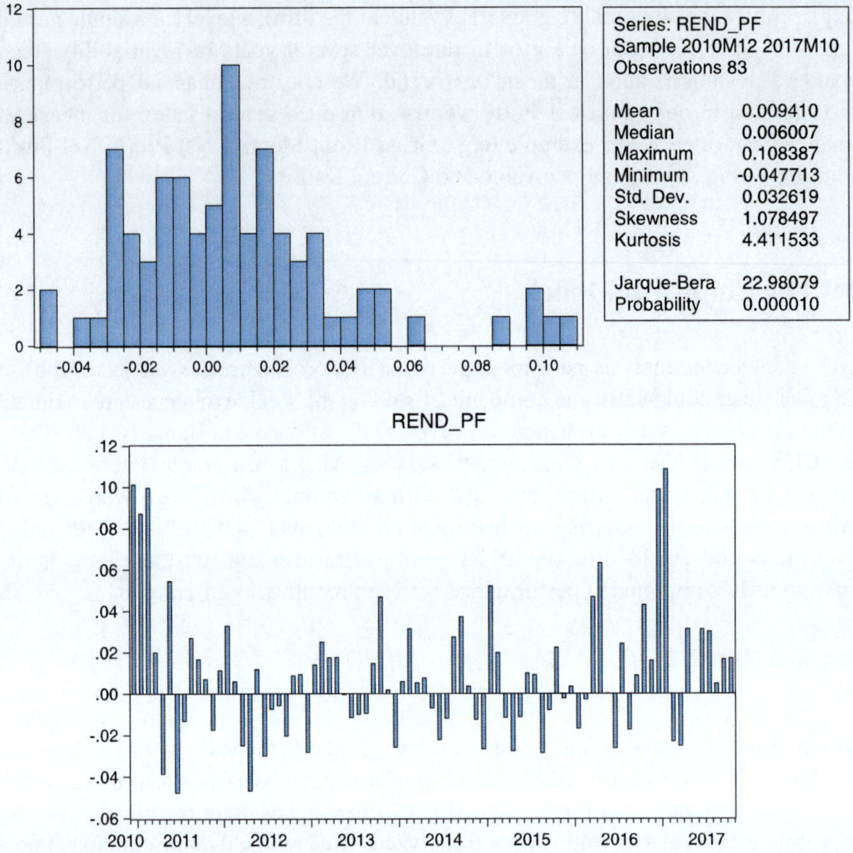

Fig. 1 Statistical tests for finance performance

Tables provide the descriptive statistics of social-finance performance. The mean value is close to zero while the standard deviation is much larger. The Jarque–Bera statistic rejects the null hypothesis of Gaussian distribution at 1% significance level implying the fat-tail distribution, also evidenced by the positive Skewness and high values of Kurtosis. The statistics of Augmented Dickey–Fuller (ADF) and Phillips–Perron unit root tests based on the lowest AIC value reject the null hypothesis of a unit root at 1% significance level indicating that the return series are stationary. Out of the graphs above it can be noticed that the repartition of the two data series is a very similar one which means that there is a pronounced dependence between the two sizes. This idea can be sustained also by the argument of the graphic representation of the evolution of the two analyzed indicators (Fig. 3).

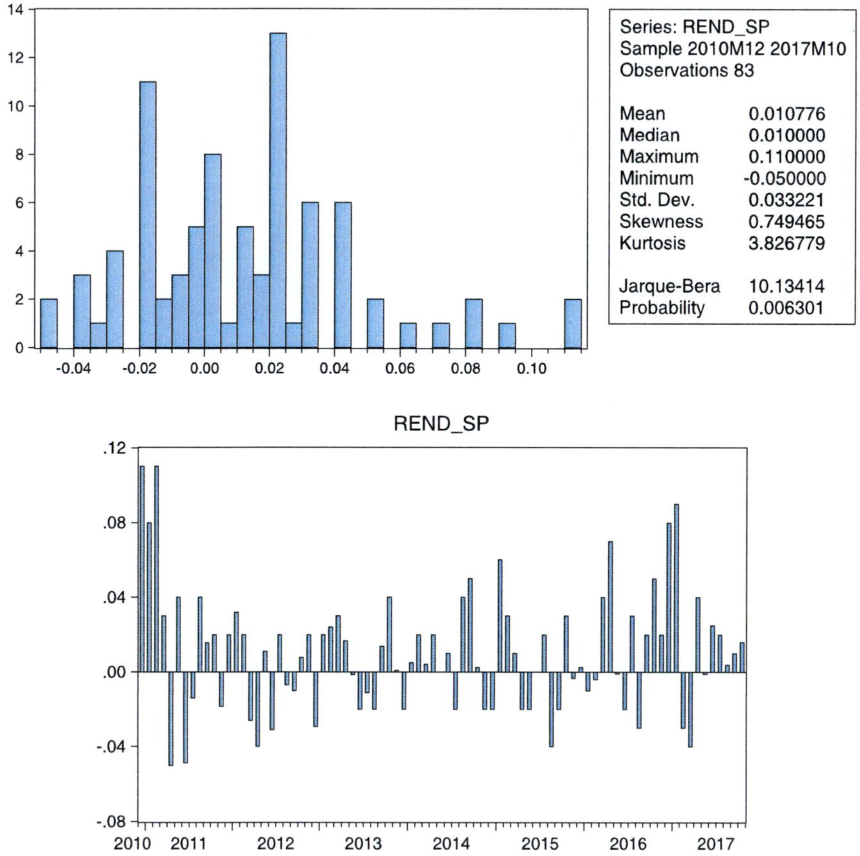

Fig. 2 Statistical tests for social performance

From the above graph it is obvious that the evolution of the two data series is more than similar which leads to the conclusion that there is a pronounced dependence between the two analysed indicators. In order to establish the type of the econometric model to be utilized, to define the existing relation between the finance-social performance. A graph representation of the data series has been achieved, by drawing up the relative regression line (Fig. 4).

Based on the observations made, the linear simple regression model was defined as:

$$REND\_FP = \alpha + \beta . REND\_SP + \epsilon .$$

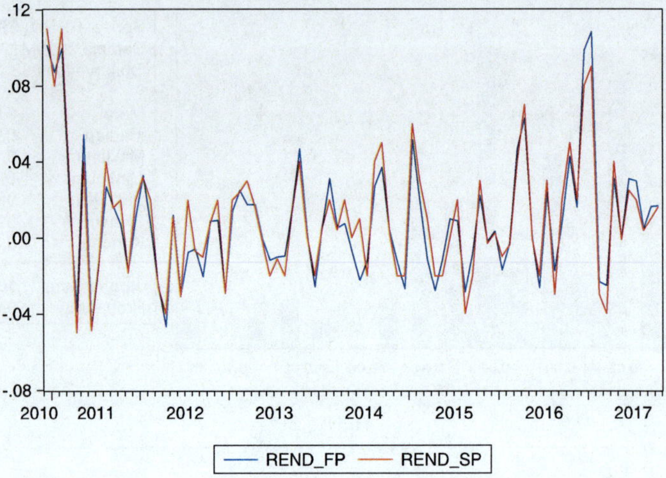

Fig. 3 Evolution of the finance-social performance. *Source* Author's Computation Using E-Views 9.0. the resulting data from a questionnaire with a scale of 1 to 10

Fig. 4 Relative regression line

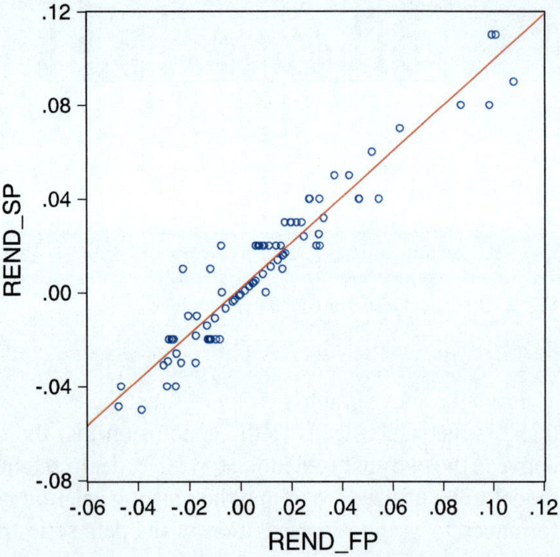

In Eviews, by using for estimation the least squares method, the parameters of the model identified above were estimated. Subsequently, the model was tested for validity by using specific tests:

Dependent Variable: REND_FP Method: Least Squares Date: 10/31/17 Time: 22:05 Sample: 2010M12 2017M10 Included observations: 83				
Variable	Coefficient	Std. Error	t-Statistic	Prob.
C	-0.000719	0.001095	-0.656764	0.5132
REND_SP	0.939989	0.031527	29.81490	0.0000
R-squared	0.916489	Mean dependent var		0.009410
Adjusted R-squared	0.915458	S.D. dependent var		0.032619
S.E. of regression	0.009484	Akaike info criterion		-6.454555
Sum squared resid	0.007286	Schwarz criterion		-6.396270
Log likelihood	269.8640	Hannan-Quinn criter.		-6.431140
F-statistic	888.9282	Durbin-Watson stat		1.778127
Prob(F-statistic)	0.000000			

Estimation Command:
=========================
LS REND_FP C REND_SP

Estimation Equation:
=========================
REND_FP = C(1) + C(2)*REND_SP

Substituted Coefficients:
=========================
REND_FP = -0.000719202177723 + 0.939988723008*REND_SP

According to the results of the previous stage, the regression model that describes the linear dependence between the evolution of the social-finance performance can be written as:

$$REND\_FP = -0.000719202177723 + 0.939988723008 \times REND\_SP + \epsilon.$$

The value recorded by Prob (0.0000) indicates the fact that the variable is significant from statistical point of view. Thus, we can conclude that the influence which the general evolution of the finance performance has upon the considered social performance is a significant one. The connection between the two variables is a direct one which shows that for a decrease of 1% of the social performance yield, the finance performance yield records an increase of 93.99%.

The value being recorded by R-squared shows that, at a ratio of 91.64%, the finance performance yield is explained by the social performance yield the difference up to 100% representing the influence of other factors not included in the present model.

Regarding the validity of this econometric model, it is confirmed by the tests performed by default in Eviews. It is appropriate to observe the values of R^2, F-statistic (888.9282) or Prob(F-statistic) (0, 00) tests. All this fully confirm the strength of the model and also the possibility to be used in future analyses, to estimate the evolution of the finance performance depending on the evolution of the social performance. The model justifies, based on official data recorded on the 38 years of historical monthly performance, the idea, outlined in economic literature, that such econometric model can be successfully used to define and systematize, under a mathematical form, the relationships existing between various indicators specific to the performance. Also, it can be stated that the values of finance performance can be estimated starting from the values of social performance.

4 Discussion and Conclusion

In this paper we have presented a new frequency approach based on the econometric analysis by studying the monthly data for social-finance performance from October 2010 to October 2017. By analyzing the correlation between the social performance and finance performance, we have observed that there exist a high degree of interaction between them during the study period and that the value of finance performance can be estimated starting from the value of social performance. Our study justifies and confirms the possibility that the presented econometric approach can be used in future analysis, to estimate the evolution of the finance performance depending on the evolution of the social performance and define and systematize, under a mathematical form, the relationships existing between various indicators specific to the social-finance performance.

References

1. Carroll, A.B.: A three-dimensional model of corporate performance. Acad. Manag. Rev. **4**(4), 497–505 (1979)
2. Clarkson, M.B.E.: A stakeholder framework for analyzing and evaluating corporate social performance. Acad. Manag. Rev. **20**, 92–117 (1995)
3. Kviz, F.J.: Basic Statistics for Nurses, by Rebecca Grant Knapp, p. 308. Wiley, New York (1978)
4. Haigh, C., Jones, N.: Techno-research and cyber-ethics: challenges for ethics committees. Res. Ethics **3**, 80–83 (2007)
5. Gond, J.-P., Igalens, J.: Le Management de la RSE. Pearson, Paris (2011)
6. Chen, L., Feldmann, A., Ou, T.: The relationship between disclosures of corporate social performance and financial performance: evidences from GRI reports in manufacturing industry. J. Product. Econ., Int. (2015). https://doi.org/10.1016/j.ijpe.2015.04.004
7. Browera, J., Kashmiri, S., Mahajan, V.: Signaling virtue: does firm corporate social performance trajectory moderate the social performance-financial performance relationship? J. Busin. Res. **81**, 86–95 (2017)
8. Chin-Shien, L., Ruei-Yuan, C., Van Thac, D.: An integrated model to explain how corporate social responsibility affects corporate financial performance. J. Sustainab. **7**, 8292–8311 (2015)
9. Xiong, B., Lu, W., Skitmore, M., Chau, K.W., Ye, M.: Virtuous nexus between corporate social performance and financial performance: a study of construction enterprises in China. J. Cleaner Product. 1–11 (2016)
10. Lin, C.-H., Yang, H.-L., Liou, D.-Y.: The impact of corporate social responsibility on financial performance: Evidence from business in Taiwan. Technol. Soc. **31**, 56–63 (2009)
11. GÓmez, L.M., Vargas-Preciado, L., Crowther, D.: Corporate social responsibility practices and programs as a key strategic element in organizational performance. In: Corporate Social Responsibility and Corporate Governance, pp. 223–247 (2017)
12. Augusto, L.: Corporate social responsibility in Portuguese companies: online communication practices. In: Corporate Social Responsibility and Corporate Governance, pp. 157–183 (2017)
13. Carrillo-Durán, M.-V., Tato-Jimánez, J.-L.: Current trends and future challenges of communication policies and practices regarding the management of corporate social responsibility (CSR): the case of Latin American SMEs. In: Corporate Social Responsibility and Corporate Governance, pp. 41–63 (2017)

14. Reed, M.S., Vella, S., Challies, E., de Vente, J., Frewer, L., Hohenwallner-Ries, D., Huber, T., Neumann, R.K., Oughton, E.A., Sidoli del Ceno, J., van Delden, H.: A theory of participation: what makes stakeholder and public engagement in environmental management work? Soc. Ecol. Restor. (2017). https://doi.org/10.1111/rec.12541
15. Phillips, R., Freeman, R.E., Wicks, A.C.: What stakeholder theory is not. Busin. Ethics Quart. **13**(4), 479–502 (2003)
16. Cook, A., Glas, C.: Leadership change and shareholder value: how markets react to the appointments of women. Human Resour. Manag. 0090–4848 (2011)
17. Izquierdo, R.J.S., Vicedo, J.C.: Corporate social responsibility of financial organizations in the social economy: a case study on savings banks. Serv. Busin. 1862–8516 (2012)
18. Veleva, V.R.: Managing corporate citizenship: a new tool for companies. Corp. Soc. Responsib. Environ. Manag. 1535–3958 (2010)
19. Harrington, B.: Organizational Performance and Corporate Social Capital: A Contingency Model, pp. 83–106. Elsevier, Amsterdam (2001)
20. Erwin, P.M.: Corporate codes of conduct: the effects of code content and quality on ethical performance. J. Busin. Ethics 0167–4544 (2011)
21. Hong, Y., Andersen, M.L.: The relationship between corporate social responsibility and earnings management: an exploratory study. J. Busin. Ethics 0167–4544 (2011)
22. Kaymaz, Ö, Kaymaz, Ö., Sayar, A.R.Z.: Corporate Financial Reporting and Performance: A New Approach. Palgrave Macmillan, UK (2015). ISBN: 978-1-349-70332-6,978-1-137-51533-9
23. Zhang, S., Zhang, J., Chen, S.: Assessment of local public finance performance in China when undesirable fiscal phenomena are taken into account. China World Econ. **20**(6), 82–101 (2012)
24. Steenlanda, M., Robyn, P.J., Compaore, P., Kabore, M., Tapsoba, B., Zongo, A., Haidarad, O.D., Finka, G.: Performance-based financing to increase utilization of maternal health services: Evidence from Burkina Faso. JSSM - Popul. Health **3**, 179–184 (2017)
25. Wood, D.J.: Corporate social performance revisited. Acad. Manag. J. **16**, 691–718 (1991)